NANOTECHNOLOGY IN ELECTRONICS, PHOTONICS, BIOSENSORS AND ENERGY SYSTEMS

SELECTED TOPICS IN ELECTRONICS AND SYSTEMS

Editor-in-Chief: **M. S. Shur**

ISSN: 1793-1274

*Published**

**The complete list of the published volumes in the series can be found at*
https://www.worldscientific.com/series/stes

Selected Topics in Electronics and Systems – Vol. 67

NANOTECHNOLOGY IN ELECTRONICS, PHOTONICS, BIOSENSORS AND ENERGY SYSTEMS

Editors

F. Jain
University of Connecticut, USA

C. Broadbridge
Southern Connecticut State University, USA

M. Gherasimova
University of Bridgeport, USA

H. Tang
Yale University, USA

World Scientific

NEW JERSEY · LONDON · SINGAPORE · BEIJING · SHANGHAI · HONG KONG · TAIPEI · CHENNAI · TOKYO

Published by

World Scientific Publishing Co. Pte. Ltd.

5 Toh Tuck Link, Singapore 596224

USA office: 27 Warren Street, Suite 401-402, Hackensack, NJ 07601

UK office: 57 Shelton Street, Covent Garden, London WC2H 9HE

British Library Cataloguing-in-Publication Data
A catalogue record for this book is available from the British Library.

Selected Topics in Electronics and Systems — Vol. 67
NANOTECHNOLOGY IN ELECTRONICS, PHOTONICS, BIOSENSORS AND ENERGY SYSTEMS

ISBN 978-981-128-375-8 (hardcover)
ISBN 978-981-128-376-5 (ebook for institutions)
ISBN 978-981-128-377-2 (ebook for individuals)

For any available supplementary material, please visit
https://www.worldscientific.com/worldscibooks/10.1142/13607#t=suppl

Preface

This special issue on *Nanotechnology in Electronics, Photonics, Biosensors, and Energy Systems* is comprised of research papers selected from the 31st annual symposium of the Connecticut Microelectronics and Optoelectronics Consortium (CMOC), held virtually on March 1, 2023 and hosted by Information Technology Staff, University of Connecticut (Storrs Campus).

Organized by a team of seven academic institutions and over thirteen companies across the United States, this symposium attracted authors from both academia and industry with topics representative of CMOC's dynamic and relevant mission.

The oral and poster papers presented span from use of graphene saturable absorbers in fiber ring laser systems, materials for supercapacitors, nanoelectronic and nanophotonic devices, electronic nose sensor array, bio-nano-systems, artificial intelligence/machine learning, and emerging technologies, to applications in each of these fields.

Systems implementing additively manufactured RF devices for communication, packaging, remote sensing, compact multi-bit FETs and memories are also included in this special issue on high performance materials for implementing high-speed electronic systems.

Plasmonic nanostructures with electrical connections have potential applications as new electro-optic devices. Quantum dot based devices are discussed with regard to optical logic gates, mid-infrared photodetectors, gain and index tailored external cavity high power lasers.

In the area of material synthesis, nanostructured filters for plastic particle filtration are illustrated. Additive manufacturing techniques such as inkjet printing to sustainably accelerate the massive deployment of 5G/mm-Wave systems including multiple-input, multiple-output (MIMO) tile-based phased array are presented.

Quantum dot random access nonvolatile memories, Gate all around (GAA) quantum dot channel (QDC), and spatial wavefunction switched (SWS) FETs for high-speed multi-bit logic and compute in memory applications are additional topics included.

In summary, the papers presented in this special issue broadly illustrate relevant aspects of high performance materials and emerging nanodevices for implementing high-speed electronic systems. We would like to take this opportunity to express our thanks to the authors, participants, and reviewers for their contributions and active participation, net-working, and knowledge sharing on a variety of research areas.

Guest Editors:
F. Jain (*University of Connecticut*)
C. Broadbridge (*Southern Connecticut State University*)
M. Gherasimova (*University of Bridgeport*)
H. Tang (*Yale University*)

v

Contents

Heating Effects on Nanofabricated Plasmonic Dimers with Interconnects

Rahul Raman[*], John Grasso[†] and Brian G. Willis[‡]

*Chemical and Biomolecular Engineering, University of Connecticut,
191 Auditorium Road, Storrs, CT 06269, USA*
**Rahul.raman@uconn.edu*
†John.grasso@uconn.edu
‡brian.willis@uconn.edu

Plasmonic nanostructures with electrical connections have potential applications as new electro-optic devices due to their strong light–matter interactions. Plasmonic dimers with nanogaps between adjacent nanostructures are especially good at enhancing local electromagnetic (EM) fields at resonance for improved performance. In this study, we use optical extinction measurements and high-resolution electron microscopy imaging to investigate the thermal stability of electrically interconnected plasmonic dimers and their optical and morphological properties. Experimental measurements and finite difference time domain (FDTD) simulations are combined to characterize temperature effects on the plasmonic properties of large arrays of Au nanostructures on glass substrates. Experiments show continuous blue shifts of extinction peaks for heating up to 210°C. Microscopy measurements reveal these peak shifts are due to morphological changes that shrink nanorods and increase nanogap distances. Simulations of the nanostructures before and after heating find good agreement with experiments. Results show that plasmonic properties are maintained after thermal processing, but peak shifts need to be considered for device design.

Keywords: Plasmonic dimers; nanofabrication; thermal annealing; localized surface plasmon resonance.

1. Introduction

Nanostructures made of materials such as Cu, Ag and Au may exhibit localized surface plasmon resonances (LSPR) that enhance interactions of light with matter [1, 2]. The spectral position of the resonances can be tuned by the size and shape of nanostructures so that they may act as tiny antenna to collect and concentrate electromagnetic (EM) radiation from the ultraviolet to the infrared. At resonance, EM fields are strongly enhanced around plasmonic nanostructures, and these fields may enable new types of electro-optic devices that convert radiation into electrical signals for applications as sensors or for solar energy harvesting. For example, LSPR-generated EM fields may stimulate production of hot carriers that traverse Schottky barriers to create photocurrents for light sensors [3]. EM field enhancements are known to be especially strong at sharp tips and in nanogaps formed

---

[‡]Corresponding author.

between closely spaced particles. When nanogaps are formed between sharp tips, EM field enhancements can exceed 10^3 [4]. Achieving such strong field enhancements may be critical for creating efficient devices.

The vast majority of studies related to plasmonics have investigated nanoparticles suspended in solution or randomly dispersed on solid supports. However, when ordered structures are created by nanofabrication techniques, it is possible to add electrical interconnections to plasmonic nanorods for electro-optic functions [5]. Nanofabrication, thin film deposition, and reactive processes provide opportunities for engineering plasmonic nanostructures to tune their optical properties and create nanogaps for enhanced EM fields. For example, atomic layer deposition (ALD) can be used to tune nanogap sizes and integrate plasmonic nanostructures with other materials [6].

In general, nanofabrication and thin film engineering to create functional devices will require thermal processes. In some cases, thermal processing may exceed temperatures of 200°C. These thermal treatments may affect plasmonic properties and need to be factored into device design. Previous research on ALD-processed plasmonic nanostructures found significant blue shifts after thermal processing that were unexpected based on uniform and conformal growth [6]. In this study, we investigate optical and morphological changes of nanostructures after thermal treatments to determine how temperature affects plasmonic properties. As a model system, we investigate nanostructures made of Au on clear glass substrates. Optical extinction curves are measured before and after thermal treatments at temperatures up to 210°C to determine how heating affects plasmonic properties. We also investigate structural changes using high-resolution electron microscopy.

2. Experiment

Nanostructures were fabricated as large arrays of interconnected nanorod dimers using electron beam lithography with a F125 electron beam writer (Elionix, Japan), and using poly(methyl methacrylate) (PMMA) photoresist (Kayak, USA). The design uses a 550 nm square unit cell repeated over an array 200×200 um^2 square. Substrates were 75 mm diameter fused silica wafers (GM Glass, USA). E-spacer was used for charge dissipation (Showa Denko, Japan). After development with methyl isobutyl ketone and isopropyl alcohol (MIBK/IPA), samples were rinsed with IPA, dried with N_2 gas and processed with a 75 W O_2 plasma in a barrel etcher before metal deposition. A thin film stack of 4 ± 1 nm Ti and 45 ± 5 nm Au was deposited in a high vacuum electron beam evaporator (Denton Vacuum, USA). Samples were further processed by liftoff using Remover PG (Kayak) to reveal the nanostructure arrays. A second layer of photolithographic processing was used to add electrical connections and guide marks for optical beam alignment. The second layer used a Shipley resist s1805 (Shipley, USA), and an MLA-150 lithography tool (Heidelberg, Germany). Metal deposition was done in the same electron beam evaporation tool using Ti/Au layers of 10/200 nm, respectively. Liftoff used the same procedures as the first level. After liftoff, samples were rinsed with IPA and dried with N_2 gas before optical measurements.

Samples were heated in air on a hot plate at atmospheric pressure with a range of temperature setpoints for 20 minutes. A thermocouple was used to calibrate sample temperatures by attaching it to a blank substrate with similar properties as the samples with nanofabricated devices. Samples were cooled in air for 10 minutes before being mounted to an ellipsometer for optical measurements.

Optical extinction measurements were taken with a JA Woollam M2000V spectroscopic ellipsometer operating in transmission mode. Reference transmission spectra were taken adjacent to feature arrays, and extinction was calculated as $(1-T/T_R)$, where T_R is the reference transmission of clear glass. The optical beam is approximately 200 um in diameter and collects data from more than 10^5 nanostructures in each array. Secondary electron microscopy (SEM) images of samples before and after heating were taken with a high-resolution Verios SEM (ThermoFisher Scientific). A thin (few nm) layer of AuPd was sputtered onto samples for charge dissipation. SEM images of before-heating samples were taken from an adjacent die, whereas post-SEM images are from the same die used for optical data before and after heating. SEM images were analyzed using PROSEM software (GeniSys, Germany) to calculate average nanostructure sizes and interparticle distances. The reported data are averages of 20 different feature measurements.

Simulations of optical extinction spectra used finite difference time domain (FDTD) methods as implemented in Lumerical commercial software (Ansys, USA). Geometric structures were extracted from high-resolution images recorded with the Verios SEM. Representative structures were extracted from images and used in the FDTD simulations. Two-dimensional (2D) images were converted to 3D shapes by extruding the image outlines along the z-direction a distance corresponding to the thin film thickness. This procedure gives a flat profile for the tops of simulated features. The model nanostructures were placed on a SiO_2 substrate and transmission spectra were calculated. Periodic boundary conditions were applied to the horizontal boundaries of the 3D simulation cell, and a plane wave polarized along the length direction of the nanorods was directed towards the nanostructures in the vertical direction. The same unit cell size of 550 nm square was used for experiments and simulations. Optical constants were taken from the Lumerical library using data from Palik for Ti and SiO_2, and Johnson and Christy for Au [7, 8].

3. Results and Discussion

Figure 1 shows a schematic of the design for interconnected plasmonic dimers, which consist of nanorod dimers with interconnect lines running through the centers of the nanorods. The interconnect lines are 45–50 nm in width, which is close to the resolution limit of the nanolithography methods used. The center positions of the interconnect connections are chosen based on prior work that determined center contacts provide the least perturbation to nanorod plasmonic resonances [9, 10]. The configuration of nanorod dimer pairs with small nanogaps promotes strongly enhanced EM fields in the regions between the tips of the nanorods [11]. Electrical interconnects allow for future applications of voltages and measurements of photocurrents while exciting plasmon resonances with light.

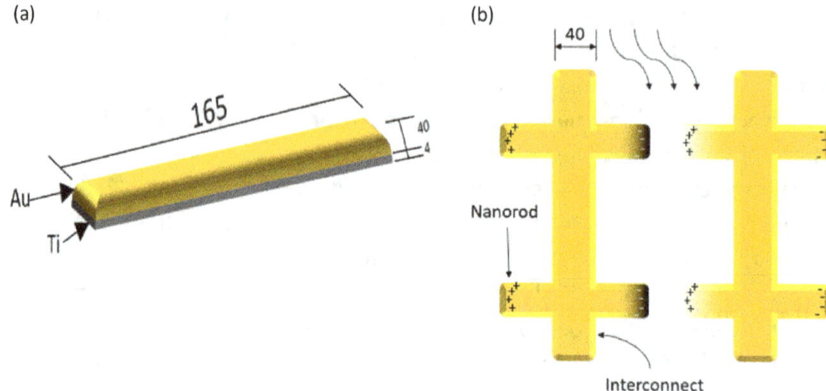

Fig. 1. Nanorods are paired to form plasmonic dimers with interconnects. Dimensions are given in nm. (a) Side view of nanorods. (b) Top view of nanorods with interconnects.

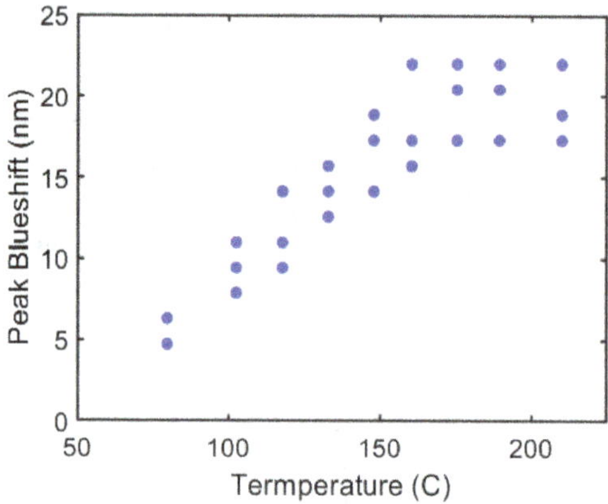

Fig. 2. Blue shifts of peak wavelengths relative to spectra taken before heating are shown vs. temperature.

Optical extinction spectra were investigated using a series of thermal treatments up to 210°C with incremental heating starting from 80°C. Heating causes blue shifts of the extinction peaks that increase with temperature, but no significant changes of extinction magnitudes were observed. A plot of blue shift vs. temperature is shown in Fig. 2. The data are scattered, but the trend shows a linear increase of blue shifts with temperatures that exceed 20 nm at the maximum temperature of 210°C. The maximum temperature of the experiments was limited by the hot plate specifications, but the effect does not saturate, and higher temperatures may lead to even larger blue shifts. The shifts of the extinction spectra indicate some changes in the optical or morphological properties of the nano-structures. High-resolution electron microscopy and FDTD simulations were used to further investigate the blue shifts.

High-resolution electron microscopy was used to investigate morphological changes of the nanostructures and to extract nanostructure shapes for FDTD modeling. Figure 3 shows

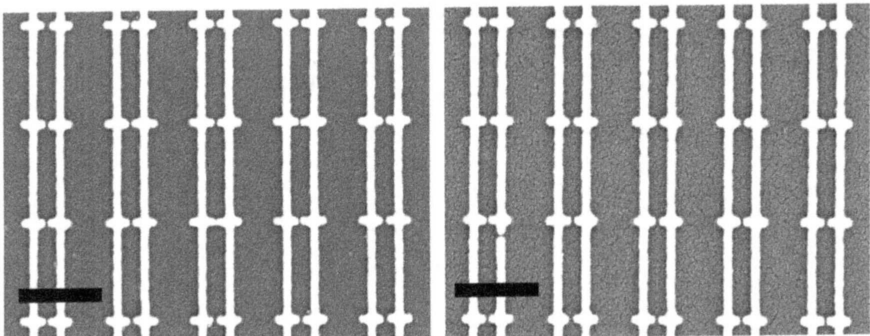

Fig. 3. SEM images from different stages of heating. (Left) Before heating. (Right) After heating to 210°C. Scale bar is 500 nm.

example SEM images before and after heating for a sample heated to the maximum temperature of 210°C. The microscopy studies reveal significant heat-induced changes of the nanostructures, including shrinking along length directions and increases in width directions. Shrinking of dimer pairs in length also causes increases of nanogap distances between nanorods. The increased interparticle distances due to nanorod shrinking are a significant factor in the blue shifts. Previous studies of nanoparticle dimers have shown that resonance peak wavelengths are highly sensitive to interparticle distances with an exponential dependence for very small nanogaps [12].

Measurements of nanostructures before and after heating show significant morphological changes. Before heating, average nanorod lengths and widths were 161.5 ± 5.5 nm and 50.5 ± 0.5 nm, respectively. Interparticle separation was 16 ± 5 nm. After heating to the maximum temperature of 210°C, the average length shrinks to 154.5 ± 7 nm, and the width increases to 55.5 ± 1 nm. The change of length is −7 nm, which is slightly larger than the width increase of +5 nm. The shrinking and broadening of the nanostructures cause nanogap distances to increase by 5.5 nm to 21.5 ± 4 nm. The magnitudes of the nanogap increases are roughly consistent with the magnitudes for length contractions. The morphological changes are consistent with thermodynamic driving forces to minimize surface area and round sharp corners toward more spheroidal shapes. Although temperatures are far below the melting point of Au (1064°C), the small sizes and large surface-to-volume ratios of the nanorods may promote rapid surface diffusion and mass transport driven by surface energy reduction. Other studies of Au nanorods have also reported the tendency for Au nanorods to spheroidize at temperatures higher than 80°C [13].

FDTD simulations were investigated to connect morphological features with optical properties. Optical extinction data taken before and after heating to the highest temperature of 210°C are shown in Fig. 4, along with FDTD simulations. The simulations use representative geometric structures extracted from the experimental SEM images in Fig. 3. Experimental spectra for before-heating measurements have an extinction peak at 831 nm, while FDTD simulations peak at 840 nm. The thickness modeled is Ti/Au 4/50 nm, which is within the expected range for experimental thickness. The peak locations are very close,

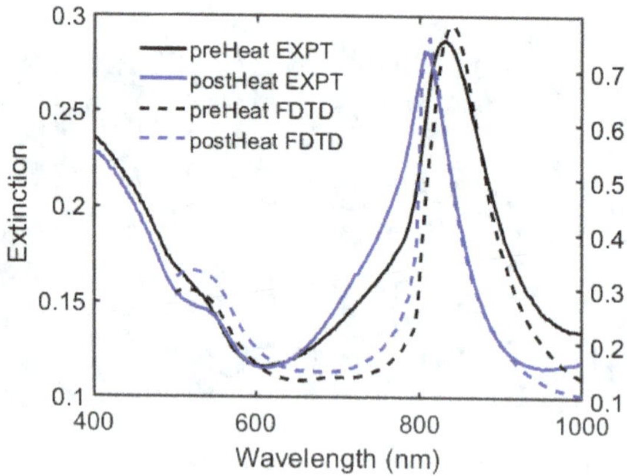

Fig. 4. Experimental (solid) and FDTD (dashed) extinction spectra before and after heating to 210°C. The left axis is for experiments and the right is for FDTD simulations.

but could be closer by using a thicker Au layer in the simulations, which would blue shift the resonance a small amount. A slightly larger nanogap would also blue shift the resonance. Deviations of experiments from a flat top surface could also account for some of the difference.

Full-widths at half-maximum (FWHM) were extracted from experiment and simulation curves by fitting the peak maxima and high wavelength leading edges to Lorentzian line shapes. This partial peak fitting was used to avoid the influence of diffraction effects that distort the blue side of the extinction peaks. The diffraction effect occurs near 800 nm due to the coincidence of the SiO_2 index of refraction (1.45) and the periodicity of the lattice at 550 nm. Using this approach, before-heating experiments have FWHM of 100, while FDTD simulations are 90 nm. These values are larger than what is obtained for a full peak fit due to the diffraction effects. The larger peak width for experiments is expected due to inhomogeneous broadening from variations of the nanorod sizes across the large arrays. Electron-beam lithography proximity error may account for ± 5 nm peak wavelength variation across the 200×200 um^2 array. A Ti adhesion layer used in experiments was included in the FDTD model. Simulations show that Ti layer thickness strongly affects the FDTD curve FWHM, and a thicker layer would broaden the peaks and bring simulations closer to experiments. The FWHM of the before-heating experiments and FDTD simulations are remarkably close considering that the experiments are sampling $> 10^5$ structures, while simulations are only modeling a single representative structure extracted from SEM images. Considering the large number of nanostructures sampled by the optical bream, the overall agreement between experiment and theory is very good.

After heating, nanorods shrink in length and expand in width, while nanogaps increase. These changes are expected to blue shift the resonances, and experiments bear this out. After heating, experimental spectra blue shift by 23 nm to 808 nm. FDTD simulations using structures extracted from post-heating SEM images in Fig. 3 show a similar blue shift of 28 nm to 812 nm. The leading edges of the two plots overlap significantly, and the (partial fit) FWHM are similar, with experiment at 81 and FDTD simulation at 86 nm. The smaller FWHM for experiment is an artefact of the peak fitting range, and a full fit to the FDTD curve gives FWHM close to 52 nm, but the peak shape is distorted by the diffraction effect. The diffraction effect is not as apparent in the experiments due to the irregular nature of the scatters and the finite size of the lattice. However, both pre- and post-heating experimental curves show a shoulder on the blue side of the peak that may be due to a partial influence of the diffraction effect. This effect may also account for the broad tail between 700 and 800 nm in the experimental spectra. Interestingly, the experimental post-heating spectra have a narrower FWHM than the before-heating spectra, which is predicted by the simulations. The narrowing of the experimental peak further indicates that a diffraction effect may be influencing the spectra. The close match of experiments and simulations is evidence that the optical properties (refractive index) of the Au nanorods are not changed by heating in air. This is unlikely to be the case for Cu or Ag materials where extensive oxidation may occur without an inert atmosphere. The good agreement also suggests that the Ti adhesion layer at the SiO_2 interface is reasonably well modeled as a metallic layer.

Besides the main LSPR peak, experiments and simulations show a second feature near 530 nm that is assigned to the transversal resonance of the interconnect lines that run through the centers of the nanorods. This second peak is not seen for plasmonic dimers without interconnects. The experimental peaks are partly obscured by a rising background below 500 nm, but the features are clearer in spectra from different samples. The peak locations are similar for experiments and simulations. Baseline extinction levels away from the peaks are also similar for experiments and simulations, with an extinction level near 0.1. In contrast, extinction peak maxima are more than a factor of two different experiments. The discrepancy of extinction maxima is partly due to inhomogeneous broadening from size variation across the large arrays. Another factor is the difference between the optical properties of nanostructures vs. the bulk optical constants used for simulations. Both experiments and simulations show a slight reduction of maximum extinction levels after heating but the difference is small.

We also investigated the effects of the Ti adhesion layer on the optical properties of plasmonic nanostructures using FDTD simulations. Experimentally, a thin layer of Ti is necessary to promote adhesion between Au nanostructures and SiO_2 (quartz) substrates. The Ti layer is known to dampen plasmon resonances, but direct Au/SiO_2 interfaces are unlikely to survive nanofabrication processes such as liftoff, where aggressive chemicals and ultrasonic agitation are used. Figure 5 shows a comparison of FDTD simulations with and without a 4 nm Ti layer between Au and SiO_2. Peak wavelengths are similar for post-heating extinction curves, but pre-heating data without Ti are slightly shifted to the blue at 836 nm compared to 840 nm with Ti. As expected, FWHM are larger for simulations with Ti. For partial peak fitting in the region 700-1000 nm, FWHM calculated before heating increases from 55 to 80 nm when Ti is added. For post-heating simulations, the increase is from 33 to 52. These values are dependent on the peak fitting region, but qualitatively, they show that Ti layers significantly broaden extinction curves. Including a Ti layer also decreases peak heights. For pre-heating plots, the peak is 16% smaller than for pure Au.

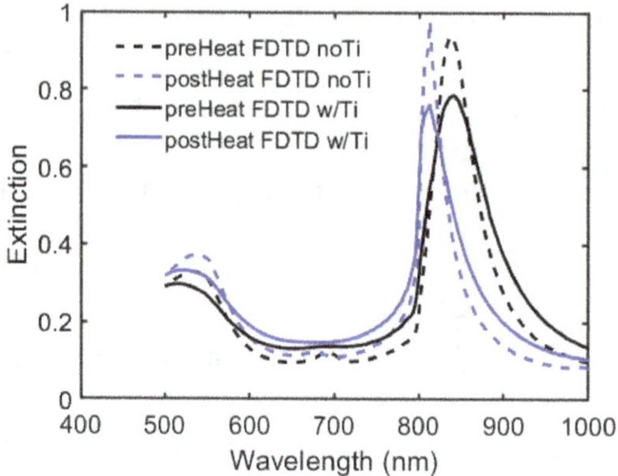

Fig. 5. Comparison of FDTD simulations before heating (black) and after heating (blue) with and without a 4 nm Ti layer.

After heating, simulations without Ti predict an increased extinction level, whereas peak extinction decreases when Ti is considered. Minimizing Ti layer thickness without losing device yield may enhance optical properties and device performance.

4. Conclusion

Heating nanofabricated Au plasmonic dimers with interconnects shows changes of optical extinction spectra that blue shift with increasing temperatures. High-resolution electron microscopy imaging shows that heating causes shape changes that reduce lengths and increase widths of nanorods, while increasing the distances between nanostructures. FDTD simulations using nanostructures extracted from images with and without heating show good agreement with experiments, which indicates that the morphological changes are the cause of the blue shifts. Peak widths are broader for experiments due to inhomogeneous broadening, but the differences are small when considering the large number of nano-structures sampled in the optical beam. The arrays provide strong plasmonic resonances that can be improved further by optimizing uniformity and reducing the thickness of the Ti adhesion layer. Heating causes blue shifts, but peak intensity and width are not significantly affected. Thermal processing induces blue shifts that need to be accounted for in device design, but these studies show that plasmonic resonances are not degraded.

Acknowledgments

The authors acknowledge the National Science Foundation (NSF) (Grant No. 2150158) and the Office of Naval Research (Grant No. N00014-22-1-2567). This work was per-formed in part at the Harvard University Center for Nanoscale Systems (CNS); a member of the National Nanotechnology Coordinated Infrastructure Network (NNCI), which is supported by the National Science Foundation under NSF award no. ECCS-2025158. Electron microscopy was performed at the UConn/Thermo Fisher Scientific Center for Advanced Microscopy and Materials Analysis (CAMMA).

References

1. E. Ozbay, Science **311** (5758), 189-193 (2006).
2. S. A. Maier, *Plasmonics: Fundamentals and Applications* (Springer, New York, 2007).
3. M. W. Knight, H. Sobhani, P. Nordlander and N. J. Halas, Science **332** (6030), 702-704 (2011).
4. D. R. Ward, F. Huser, F. Pauly, J. C. Cuevas and D. Natelson, Nature Nanotechnology **5** (10), 732-736 (2010).
5. P. Pertsch, R. Kullock, V. Gabriel, L. Zurak, M. Emmerling and B. Hecht, Nano Letters **22** (17), 6982-6987 (2022).
6. C. Zhang, T. Gao, D. Sheets, J. N. Hancock, J. Tresback and B. Willis, Journal of Vacuum Science & Technology B **39** (5), 053203 (2021).
7. E. D. Palik and G. Ghosh, *Handbook of Optical Constants of Solids*, ed. Edward D. Palik. (Academic Press, Orlando, 1985).
   P. B. Johnson and R. W. Christy, Physical Review B **6** (12), 4370-4379 (1972).
8. J. C. Prangsma, J. Kern, A. G. Knapp, S. Grossmann, M. Emmerling, M. Kamp and B. Hecht, Nano Letters **12** (8), 3915-3919 (2012).
9. D. T. Zimmerman, B. D. Borst, C. J. Carrick, J. M. Lent, R. A. Wambold, G. J. Weisel and B. G. Willis, Journal of Applied Physics **123**, 063101 (2018).
10. P. Nordlander, C. Oubre, E. Prodan, K. Li and M. I. Stockman, Nano Letters **4** (5), 899-903 (2004).
11. C. Tabor, R. Murali, M. Mahmoud and M. A. El-Sayed, The Journal of Physical Chemistry A **113** (10), 1946-1953 (2009).
12. S. S. E. Collins, M. Cittadini, C. Pecharromán, A. Martucci and P. Mulvaney, ACS Nano **9** (8), 7846-7856 (2015).

Utilizing Machine Learning for Rapid Discrimination and Quantification of Volatile Organic Compounds in an Electronic Nose Sensor Array

John Grasso[*,‡], Jing Zhao[†,§] and Brian G. Willis[*,¶]

[*]*Chemical and Biomolecular Engineering, University of Connecticut,
191 Auditorium Road, Storrs, CT 06269, USA*
[†]*Department of Chemistry, University of Connecticut,
55 N. Eagleville Road, Storrs, CT 06269, USA*
[‡]*John.grasso@uconn.edu*
[§]*Jing.zhao@uconn.edu*
[¶]*brian.willis@uconn.edu*

Volatile organic compounds (VOCs) are ubiquitous in the surroundings, originating from both industrial and natural sources. VOCs directly impact the quality of both indoor and outdoor air and play a significant role in processes such as fruit ripening and the body's metabolism. VOC monitoring has seen significant growth recently, with an emphasis on developing low-cost, portable sensors capable of both vapor discrimination and concentration measurements. VOC sensing remains challenging, mainly because these compounds are nonreactive, appear in low concentrations and share similar chemical structures that results in poor sensor selectivity. Therefore, individual gas sensors struggle to selectively detect target VOCs in the presence of interferences. Electronic noses overcome these limitations by employing machine learning for pattern recognition from arrays of gas sensors. Here, an electronic nose fabricated with four types of functionalized gold nanoparticles demonstrates rapid detection and quantification of eight types of VOCs at four concentration levels. A robust two-step machine learning pipeline is implemented for classification followed by regression analysis for concentration prediction. Random Forest and support vector machine classifiers show excellent results of 100% accuracy for VOC discrimination, independent of measured concentration levels. Each Random Forest regression analysis exhibits high R^2 and low RMSE with an average of 0.999 and 0.002, respectively. These results demonstrate the ability of gold nanoparticle gas sensor arrays for rapid detection and quantification.

Keywords: Electronic nose; chemiresistor; VOC; sensor array; nanoparticles.

1. Introduction

Development of chemical sensors targeting volatile organic compounds (VOCs) has gained increased attention in recent years, driven by useful applications in health management, environmental monitoring, public safety, agriculture and food production [1]. Potential for VOC monitoring has been demonstrated in several areas, such as the early detection of particular cancers via non-invasive breath analysis, the use of VOCs as markers for the detection of explosive compounds, and the development of reliable amine gas sensors for

---

[¶]Corresponding author.

food quality monitoring [2–4]. Despite success in several sectors, poor selectivity and sensitivity exhibited by individual gas sensors hinder their widespread application. Complex analytical techniques, such as mass spectroscopy and gas chromatography, provide comprehensive analysis, detection and quantification; but they are limited in applications mainly on account of bulky equipment, high operating costs, and time-consuming analysis [1]. Thus, there is a need for low-cost, portable gas sensors that can both distinguish and quantify VOC compounds for use in several applications.

The pursuit of portable VOC sensors is ongoing, with significant effort focusing on improving sensitivity, selectivity, response kinetics, and stability [1]. Highly sensitive gas sensors based on metal oxides are commercially available; however, discrimination between analytes remains challenging [5]. Electronic nose ('e-nose') sensor arrays overcome limitations of selectivity by utilizing machine learning algorithms to detect patterns and discriminate vapors [6]. E-nose systems exploit the overlapping cross-selectivity of individual gas sensors to form odor fingerprints. Both vapor identification and concentration predication can be realized from e-nose sensors. Capman *et al.* developed a graphene-based variable capacitor sensor array functionalized with 36 chemical receptors capable of distinguishing five VOC analytes at four concentration levels using supervised machine learning classification [7]. Data classes were separated into analyte-concentration pairs, thus a single classifier performs both discrimination and concentration prediction. This approach restricts concentration output to discrete levels fixed during training and requires large data sets in each class to minimize under-fitting of the classification model. Alternatively, deep learning methods, such as convolutional neural networks and extreme learning machines, have demonstrated simultaneous VOC classification and concentration estimation [8, 9]. Wang *et al.* discriminated 12 VOCs at 10 concentration levels ranging from 10 to 100 ppm using eight commercially available metal oxide-based gas sensors. Additionally, a pipeline combining classification and regression was integrated into an electronic nose of three sensors with both high classification accuracy and near-unity R^2-score of regression [6]. These e-noses are effective but limited in application due to high power consumption, high operating temperatures, limited chemical selectivity of metal oxide-based sensors, and the complex integration of specific deep learning algorithms into generalized electronic nose sensor arrays.

Room temperature chemiresistive gas sensors show great potential and are attracting significant interest. Organo-functionalized gold nanoparticles (AuNPs) possess favorable sensing properties relying on vapor sorption and swelling of AuNP networks [5]. The characteristic structure of organic-AuNPs allows for flexible configurability through incorporation of a wide variety of organic ligands carrying different functional groups [5]. Tunability of ligand polarity and other chemical properties results in adjustable selectivity toward target analytes that enhances overall performance of e-nose sensors.

In this study, simultaneous identification and concentration prediction of eight VOCs is successfully demonstrated with an electronic nose chemiresistor array fabricated with four types of organic-functionalized AuNPs as sensing elements. A large number of devices with fast, reversible responses are used as input for a machine learning pipeline that performs both vapor classification and concentration prediction. The approach can readily be implemented into any electronic nose sensor array.

2. Experimental Methods

2.1. *Sensor Fabrication*

Chemiresistor arrays were fabricated on four-inch silicon wafers using standard photo-lithography and liftoff processes following a procedure reported earlier [2]. Arrays of 140 circular interdigitated electrodes, 256 μm in diameter with an electrode-to-electrode separation distance of 2 μm, were defined for each sensor chip. A 10 nm titanium adhesion layer and 200 nm gold electrode layer were deposited on the silicon surface via electron-beam evaporation. The silicon wafer was diced into individual sensor chips 13 × 13 mm^2 in size. Before depositing AuNPs, sensor chips were rinsed with acetone/isopropyl alcohol, dried in nitrogen, and subsequently treated with UV-ozone to reduce surface contamination. AuNPs were then deposited onto the electrodes using drop casting from a micro-pipette.

Organo-functionalized AuNPs were prepared via colloidal synthesis followed by ligand exchange reactions or purchased from Nanopartz. Four ligands with varied chemical structures were investigated including: 4-(dimethylamino) pyridine (DMAP), 3-mercapto-propionic acid (MPA), 4-aminothiophenol (ATP), and 12-mercaptododecanoic acid NHS ester (MDN). Figure 1 illustrates a simplified representation of functionalized AuNPs and displays the chemical structure of the studied ligands. Diverse functional groups and alkyl chain lengths provide increased cross-reactivity within the sensor arrays by varying affinity for target molecules. Each sensor array is divided into four quadrants, one AuNP solution is drop-cast in each quadrant producing a sensor chip functionalized with up to 35 sensing elements of each AuNP chemiresistor. Figure 1 shows an SEM image of as-deposited AuNPs bridging microelectrodes to produce a functioning chemiresistor.

2.2. *Sensor Experiment*

Sensor arrays were mounted on ceramic PGA 144 chip carriers (Spectrum Semiconductor Materials, USA) whose leads were wire bonded (Model 747630E, WestBond, USA) to electrode contact pads. The PGA chip carriers were inserted into a breakout board con-nected to a high-speed switch matrix (3706A, Keithley, USA) for DC resistance mea-surements. A custom flow cell fabricated from aluminum with an internal volume of 0.25 mL sealed the sensing chamber via an o-ring at the chip carrier surface. Figure 2 displays a flow schematic of the sensing apparatus.

The outlet of the sensing chamber is connected to a vacuum pump (Model MP 301Z, Welch) and flow is choked through a 250 μm orifice (Lenox Laser, USA) restricting the rate to approximately 500 ml/min. Upstream of the sensor chamber, a 3-way solenoid valve connects two Tedlar bags (SVC Inc., USA); one bag contains nitrogen (UHP300, Airgas, USA) for baseline measurements while the other bag contains prepared VOC samples. VOC vapors are delivered to the sensor for 15 seconds followed by a nitrogen

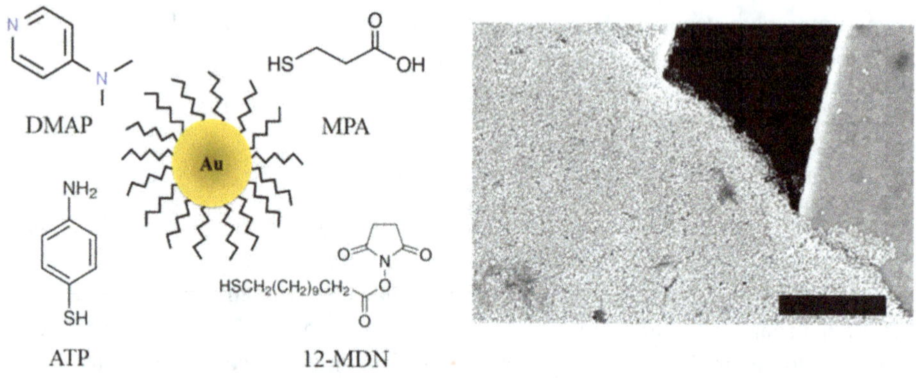

Fig. 1. (Left) Simplified illustration of organo-functionalized gold nanoparticles. Chemical structures of the four studied ligands; (right) SEM image of 12-MDN-AuNP chemiresistor sensor. Scale bar is 1 μm.

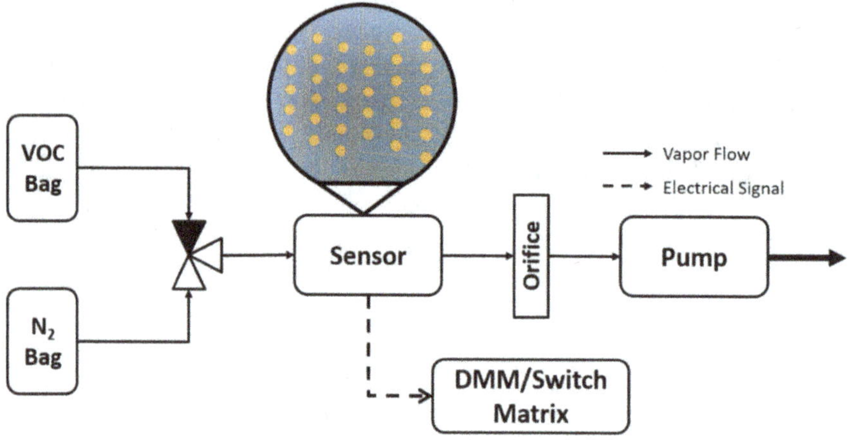

Fig. 2. Flow schematic of custom sensing apparatus. Flow is delivered to sensors from a downstream vacuum pump operating in choked flow regime. Upstream of sensing chamber, a three-way valve (normally open to N_2) switches between two Tedlar bags containing nitrogen and VOC samples. Sensors are electrically connected to a DMM/switch matrix for DC resistance measurements. Inset: Optical image of sensor electrodes before AuNP deposition.

purge of 15 seconds repeated 12 times for a series of vapor pulses. Real-time, near simultaneous resistance measurements for all sensor elements are recorded at a sampling rate of 2.3 seconds. Savitzky–Golay filtering is used for baseline normalization. Sensor responses are extracted as change in resistance during vapor sorption, normalized by the baseline resistance ($\Delta R/R_0$).

Eight VOC compounds at four dilute concentrations (p/p_0 = 0.01, 0.05, 0.1, and 0.2) were studied: vapor concentration is expressed as a ratio of partial pressure to saturated vapor pressure at room temperature. Investigated VOCs include: acetone, acetonitrile, butyraldehyde, chloroform, ethanol, methylene chloride, propionaldehyde, and toluene.

Saturated vapor standards were prepared by injecting 10 mL of liquid analyte through a septum into 5 L Tedlar bags and inflating to 80% capacity with nitrogen and equilibrating overnight. Glass syringes were used to extract saturated vapor for preparation of dilute samples. Saturated vapor was injected into a separate Tedlar bag filled with a known volume of nitrogen to prepare dilutions. Samples were tested from lowest to highest concentration.

2.3. *Machine Learning Methods*

The data set contains 384 observations for each device, made up of 48 observations for each of eight vapors. Observations include 12 repeats for each of four concentrations. Following feature extraction in MatLab, devices that did not respond to all target analytes were removed, reducing the number of working devices to 89, and the total number of data points to 34,176. Inactive sensor elements occur due to randomness of AuNP deposition from solvent evaporation during drop casting. Prior to data partitioning, z-score standardization is performed on all included features to maintain a consistent scale for sensor responses. Data are split into training and testing data sets at a ratio of 3:1. Training data are used with five-fold stratified cross-validation for hyperparameter tuning of methods. Testing data are withheld during training and inputted to final trained model to evaluate its performance to unknown data.

A two-step data analysis pipeline is implemented for simultaneous classification and concentration prediction of target VOCs in Python. First, vapor identification is performed through several machine learning classification algorithms, including: random forests (RF), k-nearest neighbor (KNN), linear discriminant analysis (LDA), and support vector machines (SVM). Each analyte is equally represented in the data set with a 48×89 matrix of sensor response values per vapor. Following classification, separate random forest regression models are built for each vapor class and trained. Test data concentration prediction is evaluated for each model and reported metrics include coefficient of determination (R^2) and root mean square error (RMSE).

3. Results

3.1. *Chemiresistor Sensor Responses*

Figure 3 shows real-time sensor responses for one selected chemiresistor sensor to four VOCs at a concentration of $p/p_0 = 0.2$. This sensor demonstrates the consistency, rapid responsivity, and recovery of AuNP sensors. Repeated exposure to VOCs did not affect the response magnitude or its tendency to return to the original baseline. The average sensor response for this sensor is $1.11 \pm 0.00\%$, $1.75 \pm 0.05\%$, $2.05 \pm 0.03\%$, and $3.25 \pm 0.05\%$ toward acetonitrile, acetone, butyraldehyde, and chloroform at $p/p_0 = 0.2$, respectively, as depicted in Fig. 3. Standard deviation is calculated from 12 repeated measurements of each VOC. DMAP-AuNPs exhibit the strongest responses to chloroform, followed by butyraldehyde, acetone, and acetonitrile.

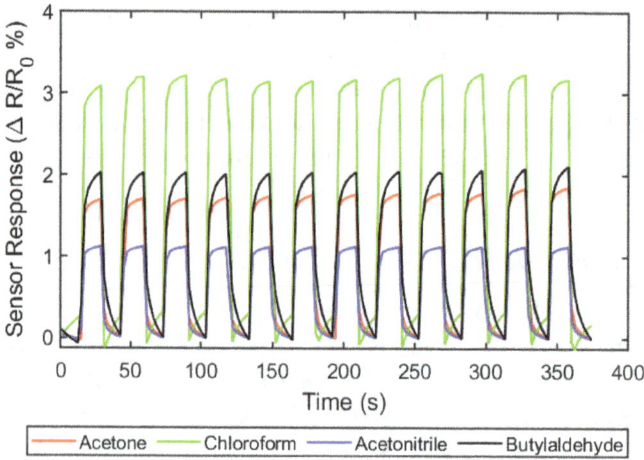

Fig. 3. Sensor response for individual AuNP-DMAP chemiresistor sensor to four VOCs at $p/p_0 = 0.2$; acetone (red), chloroform (green), acetonitrile (blue), and butyraldehyde (black).

Ligand diversity across the sensor array enables analyte patterns to be resolved. Figure 4 visualizes the pattern of sensor responses to acetone averaged over all devices for each AuNP functionalization, at each concentration level. General selectivity of each AuNP type toward acetone is shown; DMAP exhibits the lowest response, followed by MPA, ATP, and the highest response from 12-MDN. Additionally, sensor responses are proportional to vapor concentration, independent of AuNP functionalization. This property is exploited during classification where vapor identification is determined for all vapor observations, independent of concentration.

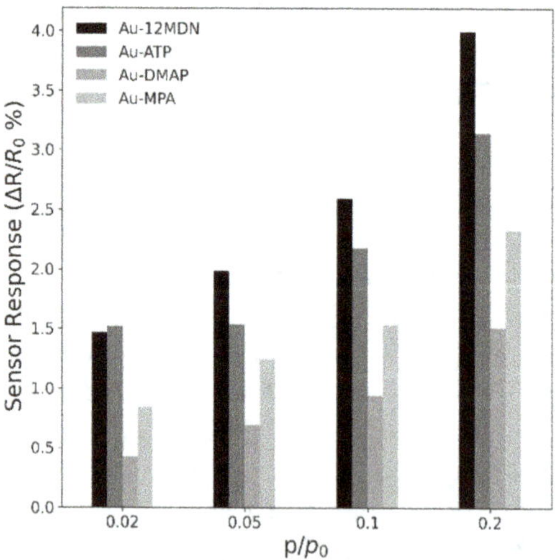

Fig. 4. Average acetone sensor response for four types of functionalized gold nanoparticles at varied vapor concentrations. Number of devices for each AuNP type: 12-MDN, 26; ATP, 20; DMAP, 20; MPA, 23.

3.2. *Vapor Classification and Concentration Prediction*

Vapor type defines the data classes used during classification where all concentration levels are lumped into the same class. Compared to the approach by Capman *et al.* in which analyte-concentration classes are used, lumping concentration and vapor type into a single class increases the training data size from 9 to 36 observations per class. This increases model robustness and reduces the possibility of under fitting during training. Validation accuracy evaluates model performance and is reported as the average classification accuracy of repeated five-fold cross-validations for the best classifier found during hyperparameter turning on the training data. Testing accuracy reports the ability of the trained model to predict vapor type when unknown data from the testing data set are input to the model. Table 1 summarizes the results of several supervised machine learning methods.

Table 1. Summary of classification accuracy for validation and testing data.

Method	Validation Accuracy (%)	Testing Accuracy (%)
LDA	97.4 ± 2.4	95.8
SVM	99.7 ± 0.8	100
KNN	98.1 ± 2.0	97.9
RF	100.0 ± 0.0	100

All classifiers reported validation accuracies greater than 97%; with RF achieving 100% validation. Both SVM and RF reported testing accuracies of 100%, indicating both high accuracy and precision toward vapor discrimination, independent of analyte concentration. Classification performance is unassociated with analyte concentration. Two VOC species, butyraldehyde and propionaldehyde, share similar chemical end groups but differ in alkyl chain length. These species are accurately distinguished despite chemical similarities, owing to the large number of gas sensors and varying organic ligand structure of the AuNP chemiresistors.

Eight regression models were constructed from the training data for each VOC vapor with 9 observations used for each concentration level. Excellent agreement between actual and predicted concentration is obtained on the testing data for each vapor type. An average R^2 of 0.999 ± 0.001 and RMSE of 0.002 ± 0.001 was obtained for all regression models. Figure 5 illustrates the combined regression parity plot. Predictions are tightly confined near the parity line with most deviations occurring at the concentration level of 0.05 p/p_0.

Simultaneous training of classification and regression models for vapor identification and concentration prediction has demonstrated excellent performance when unknown data are analyzed. High classification accuracy and low regression error establishes this machine learning framework as a viable pipeline to produce gas sensor arrays capable of both vapor classification and quantification.

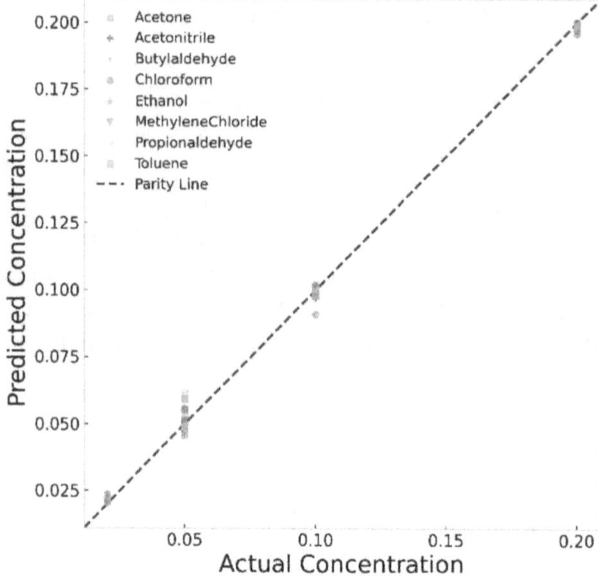

Fig. 5. Random forest regressor parity plot for testing data. Each VOC was tested at all concentrations.

4. Conclusions

This work has shown the successful classification and concentration prediction of VOCs with a 140-device gold nanoparticle-based chemiresistor gas sensor array. Four types of organic-functionalized AuNPs were studied to improve selectivity and pattern recognition via supervised machine learning. The framework for a two-step machine learning pipeline for simultaneous discrimination and concentration prediction is presented and verified with an experiment using eight VOCs at four concentration levels. The results show 100% classification accuracy for both random forest and support vector machine methods. Varying concentration levels do not degrade the ability of the classifier to discriminate vapors. This supports a hypothesis that the analyte-concentration-dependent sensor response does not significantly affect the unique odor-prints produced by the sensor array. Regression analysis shows excellent agreement between actual and predicted concentrations with near-unity R^2 and an average RMSE of 0.002 ± 0.001. These results suggest that a small number of concentration levels can be used to train an e-nose sensor to dynamically predict concentration, unrestricted to discrete prediction levels. Further work is needed to expand dynamic range and evaluate long term stability, but results are encouraging.

Acknowledgments

The authors acknowledge the National Science Foundation (NSF) (Grant No. 2150158) and the Office of Naval Research (Grant No. N00014-22-1-2567). This work was performed in part at the Harvard University Center for Nanoscale Systems (CNS); a member of the National Nanotechnology Coordinated Infrastructure Network (NNCI), which is supported by the National Science Foundation under NSF Award no. ECCS-2005158. Electron microscopy was performed at the UConn/Thermo Fisher Scientific Center for advanced Microscopy and Materials Analysis (CAMMA).

References

1. M. Khatib and H. Haick, "Sensors for Volatile Organic Compounds," *ACS Nano*, vol. 16, pp. 7080-7115, 2022.
2. T. Gao, C. Zhang, Y. Wang, J. A. Diaz, J. Zhao and B. G. Willis, "Machine Learning Assisted Nanoparticle-Based Chemiresistor Array for Explosive Detection," *IEEE Sensors*, vol. 20, no. 23, pp. 14016-14023, 2020.
3. R. S. Andre, L. A. Mercante, M. H. Facure, R. C. Sanfelice, L. Fugikawa-Santos, T. M. Swager and D. S. Correa, "Recent Progress in Amine Gas Sensors for Food Quality Monitoring: Novel Architectures for Sensing Materials and Systems," *ACS Sensors*, vol. 7, pp. 2104-2131, 2022.
4. Y. Jian, N. Zhang, T. Liu, Y. Zhu, D. Wang, H. Dong, L. Guo, D. Qu, X. Jiang, T. Du, Y. Zheng, M. Yuan, X. Fu, J. Liu, W. Dou, F. Niu, R. Ning, G. Zhang, J. Fan, H. Haick and W. Wu, "Artificially Intelligent Olfaction for Fast and Noninvasive Diagnosis of Bladder Cancer from Urine," *ACS Sensors*, vol. 7, pp. 1720-1731, 2022.
5. H. Schlicke, S. C. Bittinger, H. Noei and T. Vossmeyer, "Gold Nanoparticle-Based Chemiresistors: Recognition of Volatile Organic Compounds Using Tunable Response Kinetics," *ACS Applied Nano Materials*, vol. 4, pp. 10399-10408, 2021.
6. T. Wang, H. Zhang, Y. Wu, X. Chen, X. Chen, M. Zeng, J. Yang, Y. Su, N. Hu and Z. Yang, "Classification and Concentration Prediction of VOCs With High Accuracy Based on an Electronic Nose Using an ELM-ELM Integrated Algorithm," *IEEE Sensors*, vol. 22, no. 14, pp. 14458-14469, 2022.
7. N. S. Capman, X. V. Zhen, J. T. Nelson, V. R. S. K. Chaganti, R. C. Finc, M. J. Lyden, T. L. Williams, M. Freking, G. J. Sherwood, P. Buhlmann, C. J. Hogan and S. J. Koester, "Machine Learning-Based Rapid Detection of Volatile Organic Compounds in a Graphene Electronic Nose," *ACS Nano*, vol. 16, pp. 19567-19583, 2022.
8. T. Wang, H. Zhang, Y. Wu, W. Jiang, X. Chen, M. Zeng, J. Yang, Y. Su, N. Hu and Z. Yang, "Target Discrimination, Concentration Prediction, and Status Judgment of Electronic Nose System Based on Large-Scale Measurement and Multi-Task Deep Learning," *Sensors & Actuators*, vol. 351, 2022.
9. W. Tang, Z. Chen, Z. Song, C. Wang, Z. Wan, C. L. J. Chan, Z. Chen, W. Ye and Z. Fan, "Microheater Integrated Nanotube Array Gas Sensor for Parts-Per-Trillion Level Gas Detection and Single Sensor-Based Gas Discrimination," *ACS Nano*, vol. 16, pp. 10968-10978, 202.

Modeling of Enhancement Mode HEMT with Π-Gate Optimization for High Power Applications

Md. Maruf Hossain*, Md. Maruf Hossain Shuvo, Twisha Titirsha and Syed Kamrul Islam

*Department of Electrical Engineering and Computer Science,
University of Missouri, Columbia, MO 65211, USA
mh5md@umsystem.edu

This paper presents technology computer-aided design (TCAD) modeling of an enhancement-mode aluminum gallium nitride (AlGaN)/gallium nitride (GaN) high electron mobility transistor (HEMT) with extensive π-gate optimization for high-power and radio frequency (RF) applications. Effects of the gate voltages on threshold (V_{th}), transconductance (g_m), breakdown voltage (V_{BR}), cutoff frequency (f_T), maximum frequency of oscillation (f_{max}) and minimum noise figure (NF$_{min}$) are systematically investigated with different gate structures (π–Shaped p-GaN MISHEMT, π–Shaped p-GaN HEMT, π–Gate HEMT). A comparative study demonstrates that π–Gate with additional p-GaN and insulating layer makes the device effectively operate in the enhancement mode having a threshold voltage (V_{th}) = 1.72 V with a breakdown voltage (V_{BR}) = 341 V, exhibiting better gate control with maximum transconductance ($g_{m.max}$) of 0.321 S/mm. In addition, the proposed device architecture with an optimized gate structure maintains a balance between a positive device threshold and a high breakdown voltage and achieves a better noise immunity with the minimum noise figure of 0.64 dB while operating at 10 GHz with a cutoff frequency (f_T) of 33.4 GHz, and a maximum stable operating frequency (f_{max}) of 82.3 GHz. Moreover, the device achieved an outstanding V_{th}, $g_{m.max}$, V_{BR}, f_T, f_{max} and NF$_{min}$ making it suitable for high-power, high-speed electronics, and low-noise amplifiers.

Keywords: High electron mobility transistor; π-gate optimization; two-dimensional electron gas (2DEG), AlGaN/GaN.

1. Introduction

The inherent polarization properties of gallium nitride (GaN) result in the emergence of a two-dimensional electron gas (2DEG) channel and ensure high current densities [1, 2]. Effectively exploiting the excellent material characteristics of GaN, the high electron mobility transistors (HEMTs) have gained prominence for potential applications for high-power and high-frequency operations. Moreover, higher electron mobilities and saturation velocities allow GaN devices to operate in the radio frequency (RF) range and withstand high-power operations in microwave communications and low noise amplifier [3, 4]. The conventional Schottky gate aluminum gallium nitride (AlGaN)/GaN HEMTs exhibit high gate leakage current and low breakdown voltage with the shortened source-to-drain distance [5]. Extensive research has been conducted over the last few decades on various

---

*Corresponding author.

types of GaN-based HEMT device architectures to obtain enhancement mode (E-mode) operation for high power applications requiring sufficiently high breakdown voltage and low leakage current [6–8]. Three frequently used techniques to develop E-mode HEMTs are fluorine ion implantation [9], recessed gate etching [10], and inclusion of a Schottky p-type GaN (p-GaN) in gate [11, 12]. P-GaN HEMTs were one of the three approaches that worked well for E-mode operation because of their steady characteristics [13]. However, the limitations of Schottky p-GaN HEMT include low threshold voltage, high leakage current in gate during ON-state operations and poor noise performance. In contrast, the adoption of a metal-insulator-semiconductor HEMT (MISHEMT) has gained popularity as a method of reducing gate leakage currents [14–16]. MISHEMT is assembled by adding an insulating oxide layer between the metal gate and the semiconductor layer of a HEMT. As the triggering of the 2DEG is contingent upon proper gate arrangements, optimization of gate architecture is an important area to explore in p-GaN MISHEMTs [17, 18]. The π-gate architecture has demonstrated several advantages over T-gate and rectangular gate structures in HEMT devices [19]. For instance, the trapping of electrons within π–structure limits the kinetic energy of electrons thereby minimizing the hot-electron effect [20]. In this work, a modified π-gate structure is proposed for p-GaN MISHEMT.

The modified π–gate architecture is obtained by splitting up the T–gate architecture into pillar forms [21]. The insertion of a p-GaN cap layer underneath the gate ensures normally OFF operation. In addition, a 10 nm Al_2O_3 insulation layer is included between the gate metal and the p-GaN cap layer which demonstrates a reduction of the gate leakage current [22–24]. Furthermore, the insulation layer in the proposed model increases the oxide capacitance and boosts the channel charge density, transconductance, and drain current. The major contributions of this work are summarized in the following:

- A novel MISHEMT semiconductor device is proposed by modification and optimization of the π-shaped p-GaN MISHEMT.
- Direct-current (DC) and small-signal analyses of the proposed device are performed with technology computer-aided design (TCAD) tools achieving promising results.
- A comparative study with the T-gate HEMT has confirmed the effectiveness of the proposed device with superior noise immunity for RF and high-power applications.

2. Proposed Device Architecture and Simulation Setup

A cross-sectional view of π-shaped p-GaN MISHEMT, modeled in this work, is presented in Fig. 1(a). The two variants of π-shaped p-GaN HEMT and π-gate HEMT are shown in Figs. 1(b) and 1(c), respectively.

Figure 1(a) illustrates the 10 nm-thin insulation layer of Al_2O_3 in between the gate and the p-GaN layer. The other device parameters, such as device dimensions, material properties, layer thickness, and gate pillar footprints, are kept the same to examine the effect of different gate structures on device performance. Dimensions of every layer of the proposed device and the gate architectures of two existing devices are illustrated in Fig. 1.

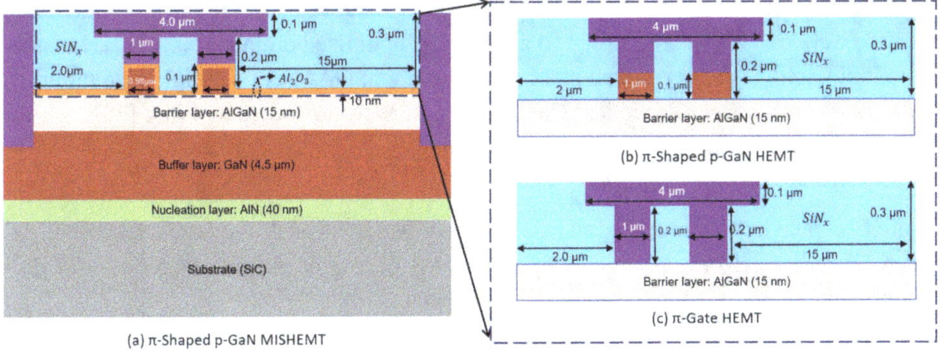

Fig. 1. Device architecture of (a) proposed π-shaped p-GaN MISHEMT, (b) π-shaped p-GaN HEMT, and (c) conventional π-Gate HEMT.

The proposed device is mounted on a silicon carbide (SiC) substrate to achieve a high breakdown voltage, better thermal stability, and high-frequency operations. To reduce the lattice mismatch, a 20 nm nucleation layer of aluminum nitride (AlN) is introduced between the buffer layer and the substrate. Typically, a nucleation layer improves the crystal quality and minimizes the electrical and the thermal influences of the substrate. The thickness of the undoped GaN buffer layer is 4.5 μm. Between the 4.5 μm undoped GaN buffer layer and the lightly n-type doped 15 nm AlGaN barrier, a 2DEG channel is formed, which ensures substantial carrier confinement at the hetero-interface of the quantum well. The drain-to-source distance is 20 μm, which remains to be the same in all four device architectures explored. The length and the thickness of each p-GaN cap layer underneath the gate legs are 90 nm and 80 nm, respectively. The Al2O3 insulating layer has a thickness of 10 nm. The extended flat plate above the gate legs is 0.5 μm extending 0.1 μm on each side. The gate structure of the π-shaped p-GaN HEMT is represented in Fig. 1(b), where the π-gate is directly deposited over the p-GaN cap layer [20]. In comparison to π-shaped p-GaN MISHEMT architecture presented in Fig. 1(a), there is no Al2O3 insulating layer between the gate and the p-GaN layer in π-shaped p-GaN HEMT. Figure 1(c) represents the gate architecture of a conventional π-gate HEMT without the p-GaN cap layer and the insulating layer underneath the gate [25]. The simulation employs a variety of physics-based models, including FLDMOB for electric field-dependent carrier mobility, CONMOB for concentration-dependent carrier mobility, and Lombardi CVT. In addition, the Shockley–Read–Hall (SRH) recombination model for capturing trap-assisted recombination, and Fermi–Dirac statistics for particle distribution, were employed to match the simulation to the experimental behavior [26, 27]. For modeling of lower field electron mobility, the Albrecht model is used in GaN [28]. Furthermore, to replicate the sheet carrier densities in 2DEG, charges resulting from different polarizations (e.g., piezoelectric, and spontaneous) of AlGaN and GaN layers and at the AlGaN/GaN interface have been calibrated. The Schottky gate work function is adjusted to improve pinch-off voltage and achieve better performance. To match the electron concentration of the simulated device to that of the experimental data, the charges confined in bulk GaN layer and at AlGaN/GaN interface were carefully considered. To achieve the optimized drain current characteristics two mobility models are exploited in TCAD simulation: GANSAT for high-field mobility and nitride-specific field-dependent Caughey–Thomas for low-field mobility [29].

2.1. *Device energy-band diagram*

Figure 2 depicts the conduction and the valence bands of an E-mode GaN MISHEMT under the gate electrode. A band-bending process occurs at the interface due to a discontinuity in the bandgap between AlGaN and GaN.

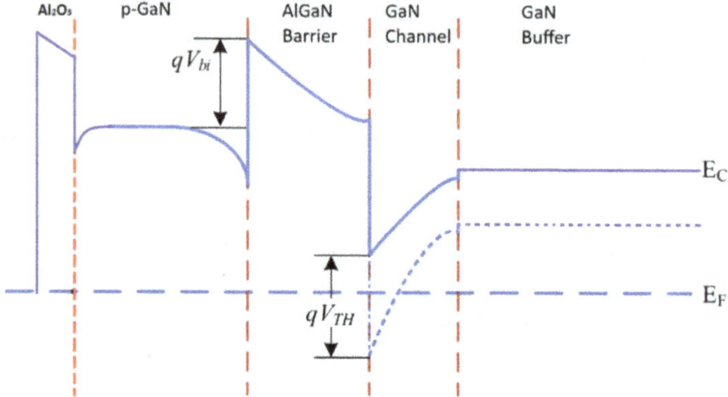

Fig. 2. Energy band diagram of proposed π-shaped p-GaN MISHEMT at zero gate bias.

3. Device Physics

3.1. *2DEG charge density*

The barrier layer thickness is a significant factor to be considered because it influences the 2DEG charge density of the HEMT in the absence of the gate bias. As the thickness of the barrier layer (AlGaN) increases, the 2DEG density/sheet-charge density (N_s) increases. Function of 2DEG density (N_s) can be expressed as in Eq. (1):

$$N_{(S)}(x)=\frac{\sigma(x)}{e}-\left(\frac{\varepsilon_0\varepsilon_r(x)}{t_{AlGaN}q^2}\right)\left[q_{\varphi_b}(x)+E_F(x)-\Delta E_C(x)\right], \tag{1}$$

where ΔE_C is the conduction band energy difference between AlGaN and GaN, q_{φ_b} is the gate contact Schottky-barrier height, E_F is the Fermi energy level, ε_r is the dielectric constant of AlGaN, t_{AlGaN} is barrier layer (AlGaN) width, q is the charge of an electron, and σ(x) is sheet charge at the AlGaN/GaN interface. The electron mobility of the device changes with the thickness of the barrier layer (t_{AlGaN}). Initially, the electron mobility increases with barrier thickness which starts to decrease after barrier thickness exceeds a certain level. Therefore, proper selection of the barrier thickness plays a significant role in attaining high sheet-charge capacity and electron mobility.

When gate voltage (V_G) is applied, the 2DEG carrier concentration (n_s) can be calculated using Eq. (2):

$$n_s = \frac{\varepsilon}{q t_{AlGaN}}\left(V_G - V_{th} - \frac{E_F}{q}\right), \tag{2}$$

where V_G is the applied gate voltage, V_{th} is the threshold voltage, ε is the dielectric constant of gate oxide and p-GaN cap layer, and E_F is the Fermi energy level.

3.2. *Threshold voltage*

The threshold voltage of the device also depends on the capacitance underneath the gate legs and is given by Eq. (3):

$$V_{th} = \emptyset_b - \frac{\Delta E_C}{q} - \frac{q t_{AlGaN}^2}{\varepsilon} N_D + \frac{q t_{AlGaN}}{\varepsilon}\sigma + V_{bi}, \tag{3}$$

where N_D is the doping concentration of barrier layer (AlGaN), and V_{bi} is the built-in potential of the p-GaN/AlGaN layer. It is evident from both equations that the materials with higher dielectric constant improve the threshold voltage of the device. The built-in potential barrier is the main factor that contributes to the positive threshold voltage that can be interpreted from Fig. 3. The higher the value of the built-in potential, the higher the threshold voltage of the HEMT. A higher value of threshold voltage can be achieved by increasing the doping concentration of the p-GaN layer. Built-in potential V_{bi} can be calculated using Eq. (4):

$$V_{bi} = \frac{E_g(\text{GaN})}{q} - \frac{E_F - \Delta E_C}{q} - \frac{q n_{sp}^2}{2\varepsilon(N_A + N_D)}, \tag{4}$$

where n_{sp} is the 2DEG sheet carrier density and N_A is the acceptor concentration at p-GaN cap layer.

3.3. *Transconductance and drain current model of HEMT*

The device transconductance can be calculated [30] as

$$g_m = \frac{\partial I_{DS}}{\partial V_G} = \frac{2.q^2.W.V_{DS}.\mu_o.\lambda.L_G}{C_G\left(\frac{\lambda.L_G.q}{C_G(V_G-V_{th})^2} + \frac{k(L_{DS}-L_G)}{N_s}\right)^2 (V_G-V_{th})^3}. \tag{5}$$

Device transconductance depends on the gate capacitance. Total gate capacitance (C_G) can be calculated using the three series capacitances including the oxide capacitance (C_{ox}), p-GaN cap layer capacitance (C_{GaN}), and AlGaN barrier layer capacitance (C_{AlGaN}) by using (6):

$$\frac{1}{C_G} = \frac{1}{C_{ox}} + \frac{1}{C_{GaN}} + \frac{1}{C_{AlGaN}} = \frac{t_{ox}}{\varepsilon_o \varepsilon_{ox}} + \frac{t_{GaN}}{\varepsilon_o \varepsilon_{GaN}} + \frac{t_{AlGaN}}{\varepsilon_o \varepsilon_{AlGaN}} \tag{6}$$

Functions are derived based on the total gate capacitance to determine the effect of the p-GaN cap layer and the dielectric constant of gate oxide on the drain current and the transconductance of the device. The drain current can be calculated using the following equations [30]:

$$I_{ds} = q.W.(\mu.E).N_s(x)$$

$$= q.W.\mu_o . \frac{V_{ds}.R_{ds}.N_s(x)}{L_{GD}(R_{GD}+R_{GS}+R_G+2R_C)}$$

$$= q.W.\mu_o . \frac{V_{ds}.\frac{L_{GD}}{N_S\mu_o}.N_s(x)}{L_{GD}\left(\frac{L_G}{N_G\mu_G}+k\frac{L_{DS}-L_G}{N_S\mu_o}\right)}$$

$$= \frac{q.W.\mu_o.V_{ds}}{\frac{L_G}{N_G}\frac{\mu_o}{\mu_G}+\frac{k(L_{DS}-L_G)}{N_S}}$$

$$= \frac{q.W.\mu_o.V_{ds}}{\frac{\lambda.L_G.q}{C_G(V_G-V_{th})}+\frac{k(L_{DS}-L_G)}{N_S}} \tag{7}$$

Here W is the gate width, q is the electron charge, N_s is the electron concentration in the channel, μ_o is the electron mobility through the channel and k is contact resistance coefficient.

4.   TCAD Modeling and Simulation Results

DC analysis of the proposed devices is carried out in terms of threshold voltage (V_{th}), breakdown voltage (V_{BR}), maximum value of the transconductance ($g_{m,max}$), gate-induced gate leakage current (I_{GIDL}) and the output-conductance (g_d). For small signal analysis, the unity current gain cut-off frequency (f_T), the maximum frequency of oscillation (f_{max}) and the minimum noise figure (NF_{min}) are observed. Three different gate structures are compared to find the overall better-performing and more reliable enhancement-mode device.

4.1.   *DC analysis*

Figure 3 depicts the I_{ds} versus V_{gs} characteristics of all devices. The conventional π-gate HEMT has depletion-type threshold voltages of −0.7 V. After adding a p-GaN cap layer underneath the gate, the device operates in an enhancement mode by forming a p-n heterojunction diode comprising of the p-GaN gate and the n-type AlGaN barrier layer, resulting in a built-in potential. This potential barrier inside the heterojunction diode aids in the removal of any polarization charge in the channel, allowing normally OFF operation. The threshold voltage increases further if a thin Al_2O_3 oxide layer is added between the metal gate and the p-GaN cap layer. The threshold voltage of the proposed π-shaped p-GaN MISHEMT and the π-shaped p-GaN HEMT are 1.74 V and 1.31 V, respectively.

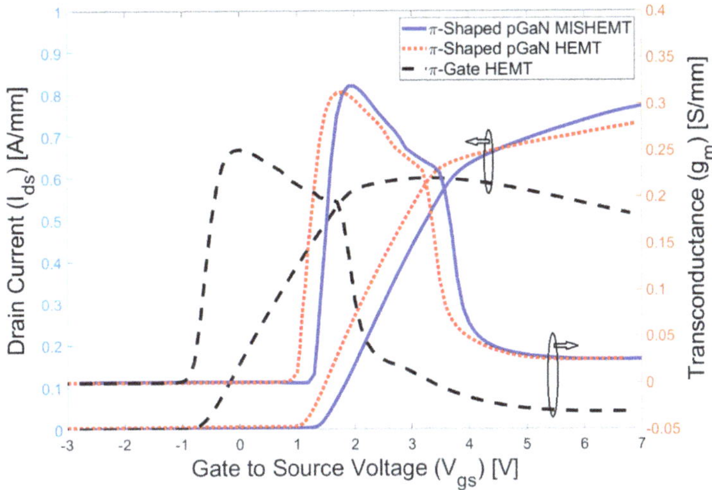

Fig. 3. Transfer characteristics of the proposed π-shaped p-GaN MISHEMT, π-shaped p-GaN HEMT and π-gate HEMT with V_{gs} swipe from -3.0 to 7.0 V at fixed gate bias V_{ds} = 12 V.

Keeping all the devices in similar biasing conditions, transconductance is measured. Figure 3 also shows the changing behavior of the transconductance depending on the gate voltage. Transconductance with a higher value ensures better control of the gate voltage over the drain current and faster switching speed. The proposed π-shaped p-GaN MISHEMT shows a higher value of transconductance of 0.321 S/mm, while at the same bias condition, π-shaped p-GaN HEMT, and π-gate show transconductances of 0.305 S/mm and 0.231 S/mm, respectively.

The characteristic curve in Fig. 4 is obtained by keeping the gate bias (V_{gs}) constant at 3 V and measuring the drain current (I_{ds}) by swiping the drain voltage (V_{ds}) from 0 to 12 V. As the conventional π-gate has the lowest threshold voltage, its drain current curve shows a higher drain saturation current at a given gate voltage compared to the π-shaped p-GaN HEMT and π-shaped p-GaN MISHEMT. The maximum drain current for simple π-gate HEMTs at 3 V gate voltage is 0.86 A/mm. On the other hand, both E-mode devices, π-shaped p-GaN HEMT and π-shaped p-GaN MISHEMT, show 0.72 (A/mm) and 0.71 (A/mm), respectively.

Figure 4 represents the device output conductance, which is inversely related to the device on-resistance. The output conductance with a higher value ensures better control of the drain voltage over the drain current and lower on-resistance. Although all three devices have different peak drain current density levels, their output conductances (g_d) are comparable to one another.

At a closed-gate condition with a threshold bias of $V_g = -10$ V for all devices, Fig. 5 plots the sub-threshold drain leakage versus the drain voltage. At the breakdown voltage, the subthreshold drain leakage current increases linearly with the drain voltage. Adding a p-type doped GaN layer improves the breakdown voltage to 330 V. A further improvement

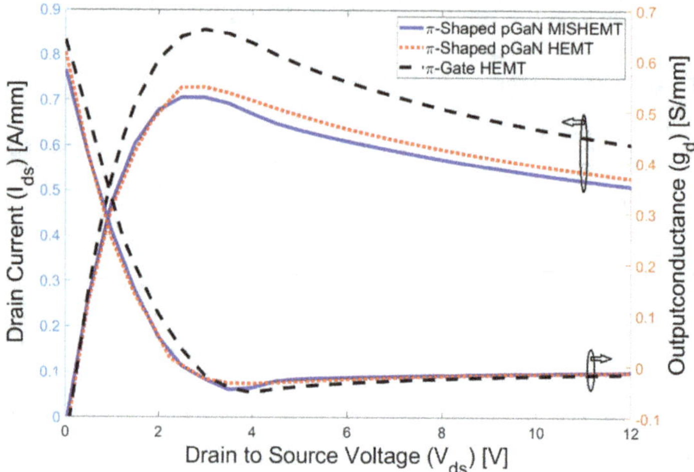

Fig. 4. Device drain-current characteristics with V_{ds} swipe from 0 to 12 V at fixed gate bias $V_{\mathrm{gs}} = 3.0$ V.

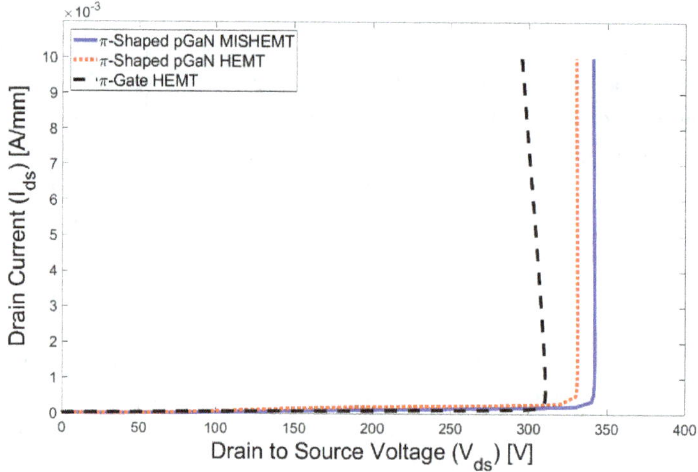

Fig. 5. Device off-state breakdown behavior with V_{ds} swipe from 0 to 500 V at gate biased $V_{\mathrm{ds}} = -10$ V.

occurs after adding a thin Al_2O_3 insulating layer underneath the π-gate legs, which improves the breakdown voltage to 341 V by improving the gate Schottky barrier.

From reverse and forward current measurements, the drain current (I_{ds}) is plotted as a function of the gate-to-source voltage (V_{gs}) as shown in Fig. 6. These curves present the gate-induced drain leakage current (I_{GIDL}) at the off-state gate bias. The drain leakage current of a π-shaped p-GaN MISHEMT is several orders of magnitude lower than that of the π-shaped p-GaN MISHEMT or conventional π-gate HEMT. Table 1 summarizes all the DC performance parameters discussed above for three different device structures.

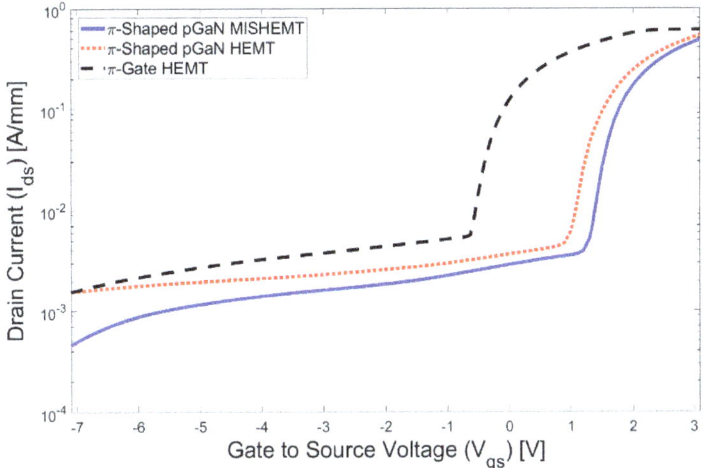

Fig. 6. Gate induced drain leakage behavior of the devices.

Table 1. DC Performance Parameters of Three Different Device Structures

Devices	V_{th} (V)	$g_{\text{m,max}}$ (S/mm)	I_{GIDL} (A)	V_{BR} (V)	$I_{\text{ds(max)}}$ (A)
π-gate HEMT	−0.7	0.231	2.6×10^{-3}	295	0.856
π-gate p-GaN HEMT	1.31	0.305	7.21×10^{-3}	330	0.727
π-gate p-GaN MISHEMT	**1.74**	**0.321**	$\mathbf{8.3 \times 10^{-3}}$	**341**	**0.705**

4.2. *Small signal analysis*

For small signal analysis, cut-off frequency (f_t), maximum operating frequency (f_{max}) and minimum noise figure of merit (NF_{min}) were studied in this work. The π-gate structure modifies the electric field profiles which effectively mitigates the hot carrier effects in HEMT and limits the thermal degradation which improves the device reliability.

As the π-gate structure is obtained by modification of T-gate, it is more of a MOS structure than a metal-semiconductor contact. Therefore, the ability to control the gate is reduced to some extent leading to a decrease in 2DEG confinement which in turn reduces the peak transconductance (g_m). In addition, the fringing fields produced in between the two legs of the π-gate increase the parasitic capacitance, which is an intentional trade-off for achieving an overall balanced device in terms of both DC and RF performance. Adding p-GaN and an Al_2O_3 insulating layer underneath the legs increases the parasitic capacitance further, which results in a lower f_t and f_{max}.

Device cut-off frequency is the operating frequency where device current gain becomes unity. This means that above the cut-off frequency HEMTs cannot be used as an amplifier. Equations (8) and (9) represent the expression of current gain and cut-off frequency [31]:

$$A_i = \frac{g_m}{j\omega\left(c_{gs}+c_{gd}+c_p\right)}, \tag{8}$$

$$f_T = \frac{g_m}{2\pi\left(c_{gs}+c_{gd}+c_p\right)}. \tag{9}$$

where C_{gs} is the gate to source capacitance, C_{gd} is the capacitance between the gate and the drain, C_p is the parasitic capacitance. From the expression of cut-off frequency, it can be interpreted that summation of three capacitance is inversely related to the f_T.

One of the key challenges in achieving high-gain millimeter-wave power amplification is increasing the frequency of oscillation (f_{max}). Although the proposed device structure has better transconductance and positive threshold voltage due to the addition of a p-GaN cap layer and an Al_2O_3 insulation layer, this modification increases the device gate capacitance and parasitic capacitance, resulting in a decrease in device f_{max}. At the maximum frequency of oscillation f_{max}, the unilateral power gain of the device becomes unity [31] which can be expressed as in Eq. (10):

$$f_{max} = \frac{F_t}{2\sqrt{(R_i+R_s+R_g)/R_{gd}+2\pi f_T R_g C_{gd}}}, \tag{10}$$

where C_{gd} is the gate to drain capacitance, R_i, R_s, R_g and R_{gd} represent the gate-charging, source, gate, and output resistance, respectively.

Figure 7 shows the change in the current gain of the mentioned devices in terms of operating frequency. Table 2 represents the extracted values of cut-off frequency from Fig. 7 for three different gate structures. The cutoff frequency for π-shaped p-GaN MISHEMT is 33.4 GHz. For π-shaped p-GaN HEMT and π-gate HEMT, these values

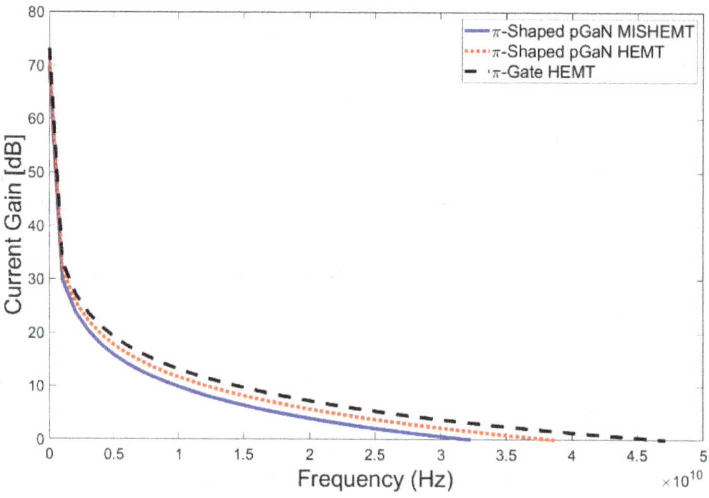

Fig. 7. Device current gain as a function of the operating frequency.

Table 2. RF Performance Parameters of Different Devices

Devices	f_T (GHz)	f_{max} (GHz)	F_{min} (dB)		
			5 GHz	7 GHz	10 GHz
π-gate HEMT	48.1	96.3	0.47	0.65	0.94
π-gate p-GaN HEMT	39.2	91.4	0.40	0.55	0.78
π-gate p-GaN MISHEMT	**33.4**	**82.3**	**0.32**	**0.45**	**0.64**

become 39.2 GHz and 48.1 GHz, respectively. Conventional π-gate HEMT shows a higher cut-off frequency as it has less gate capacitance because there is no p-GaN cap layer or Al_2O_3 insulating layer between the gate metal and the AlGaN barrier layer.

Figure 8 represents the power gain of different devices as a function of operating frequency. The maximum frequencies for the proposed π-shaped p-GaN MISHEMT is 82.3 GHz. The value f_{max} for simple π-gate architecture is 96.3 GHz, and that of a π-shaped p-GaN HEMT is 91.4 GHz. As the addition of the p-GaN cap layer and the oxide layer between the gate and barrier layers increases the gate capacitance and the parasitic capacitances, a decrease in the cut-off frequency occurs resulting in a lower f_{max} compared to the other two device structures.

Figure 9 represent the NF_{min} values for different device structures at different operating frequencies. At 10 GHz operating frequency, the minimum noise figure for a simple π-structure and for a π-shaped p-GaN HEMT obtained as −0.94 dB and 0.78 dB, respectively. At the same operating condition, the minimum noise figure for the π-shaped p-GaN MISHEMT becomes 0.64 dB. The proposed device architecture has degraded NF_{min} due to the split in fingering fields and additional oxide layer underneath the gate, contributing to the increased parasitic capacitance in the π-gate.

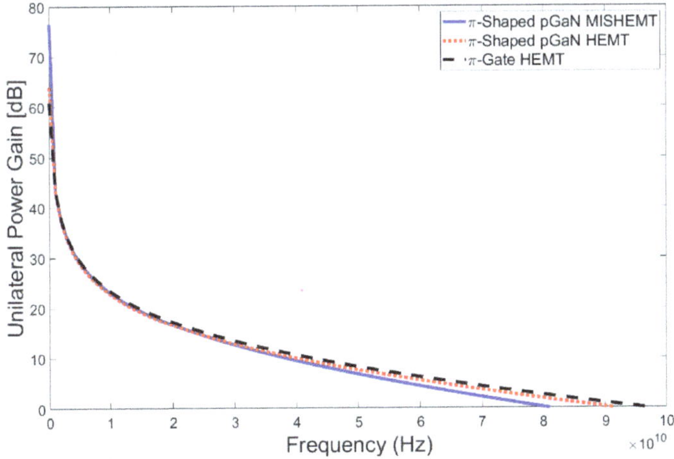

Fig. 8. Device unilateral power gain as a function of the operating frequency.

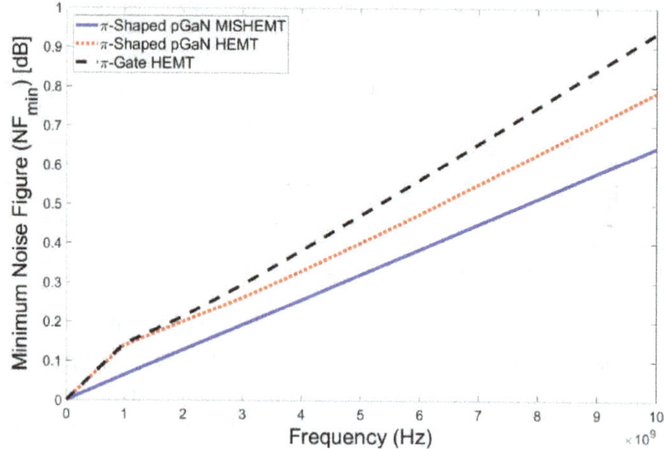

Fig. 9. Minimum noise figure-of merit of the devices at different operating frequency.

5. Conclusion

This work represents a π-gate p-GaN MISHEMT for reliable enhancement mode operation with improved gate control ability and stable small signal performance. TCAD-based intensive analysis was carried out to compare the performance of the proposed device architecture with conventional π-gate counterparts. The transformation of the gate electrode to a π-shaped structure with a p-type doped cap layer and an insulating oxide layer beneath the gate legs improves the threshold voltage to 1.72 V, the breakdown voltage to 341 V, and the transconductance to 0.321 S/mm. These properties outperform those of Schottky-gated p-GaN-gated HEMT. The RF performance of the device is characterized by the cut-off frequency, the maximum oscillation frequency, and the minimum noise figure, which are reported to be 33.4 GHz, 82.3 GHz, and 0.64 dB, respectively. With these remarkable properties, the proposed structure of π-shaped p-GaN MISHEMT demonstrates its potential for utilization in RF circuits, particularly in high-frequency integrated power amplifiers.

References

1. P. Gangwani, S. Pandey, S. Haldar, M. Gupta, and R. S. Gupta, "Polarization dependent analysis of AlGaN/GaN HEMT for high power applications," *Solid State Electronics*, vol. 51, no. 1, pp. 130–135, Jan. 2007, doi: 10.1016/j.sse.2006.11.002.

2. J. P. Ibbetson, P. T. Fini, K. D. Ness, S. P. DenBaars, J. S. Speck, and U. K. Mishra, "Polarization effects, surface states, and the source of electrons in AlGaN/GaN heterostructure field effect transistors," *Applied Physics Letters*, vol. 77, no. 2, pp. 250–252, Jul. 2000, doi: 10.1063/1.126940.

3. A. S. A. Fletcher and D. Nirmal, "A survey of Gallium Nitride HEMT for RF and high power applications," *Superlattices and Microstructures*, vol. 109, pp. 519–537, Sep. 2017, doi: 10.1016/j.spmi.2017.05.042.

4.　K. J. Chen *et al.*, "GaN-on-Si power technology: devices and applications," *IEEE Transactions on Electron Devices*, vol. 64, no. 3, pp. 779–795, Mar. 2017, doi: 10.1109/TED.2017.2657579.

5.　A. Mimouni, T. Fernández, J. Rodriguez-Tellez, A. Tazon, H. Baudrand, and M. Boussuis, "Gate leakage current in GaN HEMT's: a degradation modeling approach," *Electrical and Electronic Engineering*, vol. 2, no. 6, pp. 397–402, Jan. 2013, doi: 10.5923/j.eee.20120206.09.

6.　T. P. C. and R. J. G. K. Matocha, "High-voltage normally off GaN MOSFETs on sapphire substrates," *IEEE Transactions on Electron Devices*, vol. 52, no. 1, pp. 6–10, Jan. 2005, doi: 10.1109/TED.2004.841355.

7.　E. Bahat-Treidel, O. Hilt, F. Brunner, J. Wurfl, and Gü. Trankle, "Punchthrough-voltage enhancement of AlGaN/GaN HEMTs using AlGaN double-heterojunction confinement," *IEEE Transactions on Electron Devices*, vol. 55, no. 12, pp. 3354–3359, Dec. 2008, doi: 10.1109/TED.2008.2006891.

8.　Yong Cai, Yugang Zhou, K. J. Chen, and K. M. Lau, "High-performance enhancement-mode AlGaN/GaN HEMTs using fluoride-based plasma treatment," *IEEE Electron Device Letters*, vol. 26, no. 7, pp. 435–437, Jul. 2005, doi: 10.1109/LED.2005.851122.

9.　Y. Cai, Y. Zhou, K. M. Lau, and K. J. Chen, "Control of threshold voltage of AlGaN/GaN HEMTs by fluoride-based plasma treatment: from depletion mode to enhancement mode," *IEEE Transactions on Electron Devices*, vol. 53, no. 9, pp. 2207–2215, Sep. 2006, doi: 10.1109/TED.2006.881054.

10.　W. B. Lanford, T. Tanaka, Y. Otoki, and I. Adesida, "Recessed-gate enhancement-mode GaN HEMT with high threshold voltage," *Electronics Letters*, vol. 41, no. 7, p. 449, 2005, doi: 10.1049/el:20050161.

11.　L.-Y. Su, F. Lee, and J. J. Huang, "Enhancement-mode GaN-based high-electron mobility transistors on the Si substrate with a p-type GaN cap layer," *IEEE Transactions on Electron Devices*, vol. 61, no. 2, pp. 460–465, Feb. 2014, doi: 10.1109/TED.2013.2294337.

12.　S. Shamsir, F. Garcia, and S. K. Islam, "Modeling of enhancement-mode GaN-GIT for high-power and high-temperature application," *IEEE Transactions on Electron Devices*, vol. 67, no. 2, pp. 588–594, Feb. 2020, doi: 10.1109/TED.2019.2961908.

13.　R. Hao *et al.*, "Normally-off *p*-GaN/AlGaN/GaN high electron mobility transistors using hydrogen plasma treatment," *Applied Physics Letters*, vol. 109, no. 15, p. 152106, Oct. 2016, doi: 10.1063/1.4964518.

14.　C.-C. Hsu, P.-C. Shen, Y.-N. Zhong, and Y.-M. Hsin, "AlGaN/GaN MIS-HEMTs with a p-GaN cap layer," *MRS Advances*, vol. 3, no. 3, pp. 143–146, Jan. 2018, doi: 10.1557/adv.2017.626.

15.　G. Kurt *et al.*, "Normally-off AlGaN/GaN MIS-HEMT with low gate leakage current using a hydrofluoric acid pre-treatment," *Solid State Electronics*, vol. 158, pp. 22–27, Aug. 2019, doi: 10.1016/j.sse.2019.05.008.

16.　Y. C. Lin *et al.*, "Optimization of gate insulator material for GaN MIS-HEMT," in *2016 28th International Symposium on Power Semiconductor Devices and ICs (ISPSD)*, IEEE, Jun. 2016, pp. 115–118. doi: 10.1109/ISPSD.2016.7520791.

17.　C.-J. Yu, C.-W. Hsu, M.-C. Wu, W.-C. Hsu, C.-Y. Chuang, and J.-Z. Liu, "Improved DC and RF performance of novel MIS *p*-GaN-Gated HEMTs by gate-all-around structure," *IEEE Electron Device Letters*, vol. 41, no. 5, pp. 673–676, May 2020, doi: 10.1109/LED.2020.2980584.

18.　T. Pu *et al.*, "Normally-off AlGaN/GaN heterojunction metal-insulator-semiconductor field-effect transistors with gate-first process," *IEEE Electron Device Letters*, vol. 40, no. 2, pp. 185–188, Feb. 2019, doi: 10.1109/LED.2018.2889291.

19. K. Sehra, V. Kumari, M. Gupta, M. Mishra, D. S. Rawal, and M. Saxena, "Optimization of π – gate AlGaN/AlN/GaN HEMTs for low noise and high gain applications," *Silicon*, vol. 14, no. 2, pp. 393–404, Jan. 2022, doi: 10.1007/s12633-020-00805-7.

20. K. Sehra, V. Kumari, M. Gupta, M. Mishra, D. S. Rawal, and M. Saxena, "A Π-shaped p-GaN HEMT for reliable enhancement mode operation," *Microelectronics Reliability*, vol. 133, p. 114544, Jun. 2022, doi: 10.1016/j.microrel.2022.114544.

21. A. D. Latorre-Rey, J. D. Albrecht, and M. Saraniti, "A Π-shaped gate design for reducing hot-electron generation in GaN HEMTs," *IEEE Transactions on Electron Devices*, vol. 65, no. 10, pp. 4263–4270, Oct. 2018, doi: 10.1109/TED.2018.2863746.

22. Y.-Z. Yue, Y. Hao, and J.-C. Zhang, "AlGaN/GaN MOS-HEMT with stack gate HfO_2/Al_2O_3 structure grown by atomic layer deposition," in *Compound Semiconductor Integrated Circuits Symposium*, IEEE, Oct. 2008, pp. 1–4. doi: 10.1109/CSICS.2008.59.

23. H.-C. Chiu *et al.*, "Normally-Off p-GaN gated AlGaN/GaN MIS-HEMTs with ALD-grown Al_2O_3/AlN composite gate insulator," *Membranes*, vol. 11, no. 10, p. 727, Sep. 2021, doi: 10.3390/membranes11100727.

24. Zhiwei Bi, Yue Hao, Hongxia Liu, Linjie Liu, and Qian Feng, "Characteristics analysis of gate dielectrics in AlGaN/GaN MIS-HEMT," in *International Conference of Electron Devices and Solid-State Circuits (EDSSC)*, IEEE, Dec. 2009, pp. 419–422. doi: 10.1109/EDSSC.2009.5394226.

25. K. Sehra, V. Kumari, V. Nath, M. Gupta, D. S. Rawal, and M. Saxena, "Comparison of linearity and intermodulation distortion metrics for T - and Pi - gate HEMT," in *International Conference on Electrical, Electronics and Computer Engineering (UPCON)*, IEEE, Nov. 2019, pp. 1–6. doi: 10.1109/UPCON47278.2019.8980221.

26. R. N. Hall, "Electron-hole recombination in germanium," *Physical Review*, vol. 87, no. 2, pp. 387–387, Jul. 1952, doi: 10.1103/PhysRev.87.387.

27. W. Shockley and W. T. Read, "Statistics of the recombinations of holes and electrons," *Physical Review*, vol. 87, no. 5, pp. 835–842, Sep. 1952, doi: 10.1103/PhysRev.87.835.

28. J. D. Albrecht, R. P. Wang, P. P. Ruden, M. Farahmand, and K. F. Brennan, "Electron transport characteristics of GaN for high temperature device modeling," *Journal of Applied Physics*, vol. 83, no. 9, pp. 4777–4781, May 1998, doi: 10.1063/1.367269.

29. M. Farahmand *et al.*, "Monte Carlo simulation of electron transport in the III-nitride wurtzite phase materials system: binaries and ternaries," *IEEE Transactions on Electron Devices*, vol. 48, no. 3, pp. 535–542, Mar. 2001, doi: 10.1109/16.906448.

30. Y. Zhao *et al.*, "Temperature-dependent characteristics for the *p*-type CuO gate HEMT and high-k HfO_2 MIS-HEMT on the Si substrates," *AIP Advances*, vol. 11, no. 10, p. 105204, Oct. 2021, doi: 10.1063/5.0064695.

31. K. Jena, R. Swain, and T. R. Lenka, "Effect of thin gate dielectrics on DC, radio frequency and linearity characteristics of lattice-matched AlInN/AlN/GaN metal–oxide–semiconductor high electron mobility transistor," *IET Circuits, Devices & Systems*, vol. 10, no. 5, pp. 423–432, Sep. 2016, doi: 10.1049/iet-cds.2015.0332.

Encryption Using Optical Pseudo-Random Binary Sequence Based on Optical Logic Gate

Shunyao Fan[*], Ashiq Rahman and Niloy K. Dutta

*Department of Physics, University of Connecticut,
Storrs, Connecticut 06269, USA*
Shunyao.fan@uconn.edu

In this paper, we propose a scheme for high-speed all-optical Pseudo-Random Binary Sequence (PRBS) generator and use it for generating keystream for encryption. This PRBS generator design is based on Linear Feedback Shift Registers (LFSR) and optical XOR and AND gates. The optical logical gates are based on quantum dot-semiconductor optical amplifier Mach-Zehnder interferometer (QD-SOA-MZI). With two photon absorption (TPA) in quantum dot-semiconductor optical amplifier (QD-SOA), this kind of optical logic gates performs well when processing data in an ultra-fast timescale and therefore able to function as high speed PRBS generator. Result shows that it's possible for this scheme to realize all-optical encryption and decryption at high process rate up to 320 Gb/s. We simulated different ways of generating keystream with schemes such as cascaded generator, parallel generator and alternating step generator. These generators use more than one LFSR. Result shows that the schemes we use can function as stable and complex keystream generators.

Keywords: Optical logic gates; quantum dot-semiconductor optical amplifier; pseudo-random binary sequence.

1. Introduction

High-speed communication systems require higher processing rate, and therefore all-optical data processing is expected to play a significant role. Hence it is important that we replicate some useful tools in optical networks with all-optical logic gates. Pseudo-Random Binary Sequence (PRBS) is a sequence of binary digits (1's and 0's) that is generated with a deterministic algorithm yet difficult to predict and exhibits statistical behavior similar to a truly random sequence. Therefore, PRBS is widely used in electronics, including simulation of noise in signal transmission, data encryption/decryption, and in bit error rate testers (BERTs) [1]. PRBS generators can be achieved by Linear Feedback Shift Registers (LFSR) and logic gates [2, 3]. High-speed all-optical logic gates have been demonstrated. Hence, based on all-optical logic gates, we may build a PRBS generator that has a high process rate.

---

[*]Corresponding author.

In this paper, we provide a model to simulate encryption and decryption using PRBS generator that is based on LFSR and quantum dot-semiconductor optical amplifier Mach–Zehnder interferometer (QD-SOA-MZI). We also simulate different designs for encryption key-steam generator based on PRBS.

2.   Optical Logic Gate Based on QDSOA-MZI

The principle of logic XOR operation utilizing cross-phase modulation (XPM) process in SOAs has been previously discussed and analyzed [3]. The logic XOR operation is achieved utilizing the cross-gain modulation (XGM) and cross phase modulation (XPM) processes in QD-SOAs. As shown in Fig. 1, Data streams A and B are carried by two optical pulse streams at wavelength λ_1 are separately injected into the two arms of the MZI through port A and B. There is a clock stream at wavelength λ_2 injected into port C and evenly split into the two arms of the MZI. Due to XGM and XPM, the phases and amplitudes of two clock streams are modulated as they travel with data streams in the QD-SOAs. When they recombine at port D, their interference will produce different results based on initial conditions.

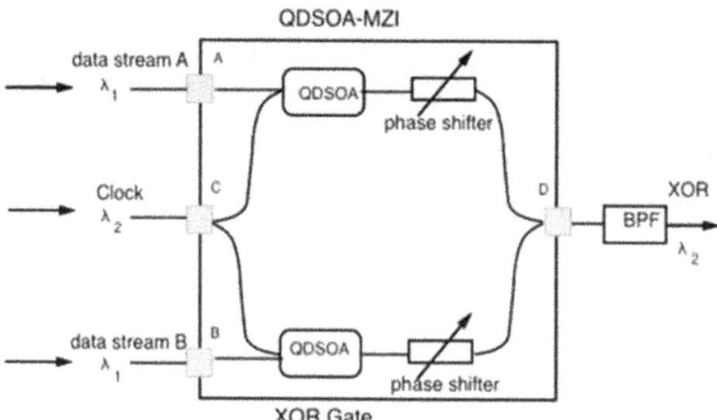

Fig. 1. Schematic diagram of optical gates based on QDSOA-MZI. The truth table of XOR is also shown. The band pass filter is an optical filter which transmits the light at wavelength λ_2 and blocks light at wavelength λ_1.

The Truth Table of XOR

Data B	Data A	Output XOR
0	0	0
1	0	1
0	1	1
1	1	0

To achieve XOR operation, we initially set the MZI unbalanced with a phase difference π between the two branches. Thus, when input data A and B are the same, the two clock streams are experiencing the same gain and phase shift in QD-SOAs and when they recombine at port D, considering the initial phase difference π, they will undergo destructive interference and the output result is "0". Correspondingly, when input data A and B are not the same, the gain and phase modulation of the two clock streams are different and their interference will output "1" at wavelength λ_2. The band pass filter is an optical filter which transmits the light at wavelength λ_2 and blocks light at wavelength λ_1.

We can also use this same scheme to realize the logic AND operation. To achieve this, we input data stream A into port A, and inject a delayed version of data stream A into port B. MZI is initially set so that there is no output if there is no differential phase change in the two arms for an input signal propagating through both arms. Signal B (which also carry the result of the AND operation) is injected at port C of the SOA-MZI. When there is a signal (A) input "1", in one of the arms there is a phase shift induced on to the control signal and a delayed phase shift appears at the other arm. Thus signal A (if it is 1) produces a phase gate for signal B which travels through both arms and interfere at the output. Thus if B=1 and A=1, the output is "1". If A=0 there is no phase gate and hence the output is "0" for both B=0 and B=1, and if A=1 B=0, the output is "0". This is the logic function AND.

Then input data stream B into port C and we will get a pattern of A AND B out from port D. When data stream A input "1", in one of the arms there is a phase shift induced on the control signal and a delayed phase shift appears at the other arm. Thus, for signal A is "1", produces a phase gate for signal B which travels through both arms and interfere at the port D. With data stream A and data stream B both input "1", the output is "1". If data stream A input "0", there will be no phase gate, and therefore no output. If data stream B input "0" the output will also be "0". Thus, this scheme functions as an AND gate.

3. Devices and Rate Equations

The device we choose here to construct the all-optical logic gate is the InGaAs/InGaAsP/InP QD-SOA, in which InAs quantum dots are embedded in InGaAsP layers. The gain of this type of device around 1.55 μm is typically ~15 dB and the noise figure is low at ~7 dB [4].

Figure 2 illustrates the optical gain, the TPA process and carrier transitions between the wetting layer (WL), the QD excited state (ES) and the QD ground state (GS). The device gain is determined by the carrier density of the QD ground state. The TPA generates carriers in the bulk region. These carriers then relax to the WL, and eventually are captured into QDs on ultrafast timescale [5]. Generally, carriers in the barriers are free to move in 3D and are captured very rapidly by the 2D wetting layer at a relaxation timescale of ~70 fs [5]. Ju et al. have shown that the carrier dynamics in the bulk region due to TPA can also refill the WL and QDs on ultrafast timescales and thus significantly reduce pattern effects for optical signal-processing operating at Tbit/s [5]. They introduced a three coupled rate equations model including the barrier region, as well as the WL and QDs. In our model, we ignore the barrier dynamics and assume that carriers are injected directly from the contacts into the WL [5]. Since the only recipient of the pump current is the WL, and the QD excited state serves as a carrier reservoir for the ground state with ultra-fast carrier

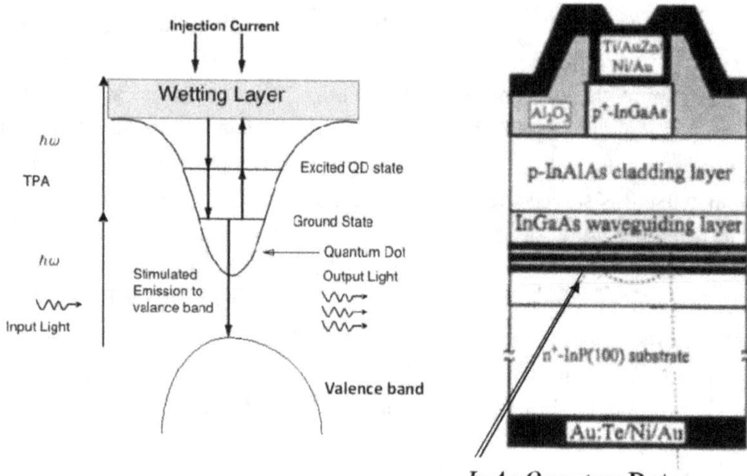

InAs Quantum Dots

Fig. 2. Schematic of two-photon absorption (TPA) and a typical device structure. The InGaAs waveguide layer which has a bandgap intermediate between the cladding layers (InAlAs) and the quantum dots (InAs) and the layer (InGaAs) surrounding the quantum dots serve as the wetting layer. Similar layer structure is feasible using the InP/InGasAsP/InGaAs material system with InGaAs quantum dots.

relaxation to the latter, the device gain dynamics is affected by their carrier densities and transition rates.

The change in carrier densities of the three energy levels including the TPA process are described by the following coupled rate equations Eq. (1). See Refs. [6–8] for more details.

$$\frac{dw}{dt} = \frac{1}{eVN_{wm}} - \frac{w}{\tau_{wr}} - \frac{w}{\tau_{w\text{-}e}}(1-h) + \frac{N_{esm}}{N_{wm}}\frac{h}{\tau_{e\text{-}w}}(1-w) + \frac{\kappa}{2\hbar\omega N_{wm}}[\frac{S(t)}{A}]^2;$$

$$\frac{dh}{dt} = -\frac{h}{\tau_{esr}} - \frac{N_{wm}}{N_{esm}}\frac{w}{\tau_{w\text{-}e}}(1-h) - \frac{h}{\tau_{e\text{-}w}}(1-w) + \frac{N_{gsm}}{N_{esm}}\frac{f}{\tau_{g\text{-}e}}(1-h) - \frac{h}{\tau_{e\text{-}g}}(1-f); \tag{1}$$

$$\frac{df}{dt} = -\frac{f}{\tau_{gsr}} - \frac{f}{\tau_{g\text{-}e}}(1-h) + \frac{N_{esm}}{N_{gsm}}\frac{h}{\tau_{e\text{-}g}}(1-f) - \frac{\Gamma_d}{A_d}a(2f-1)\frac{1}{N_{gsm}}\frac{S(t)}{\hbar\omega}.$$

where w, h and f represent the occupation probability of the wetting layer, the QD excited state and ground state, respectively; N_{wm}, N_{esm} and N_{gsm} are the maximum densities of carriers in each state; the spontaneous radiation lifetime of each state is denoted by τ_{ar} ("a" being "w", "es" or "gs"); $\tau_{a\text{-}b}$ denotes the relaxation time between any state "a" and state "b"; Γ_d is the active layer confinement factor; I is the injected current; a is the differential gain; V is the volume of the active layer; A_d is the effective cross-sectional area of the active layer; k is the TPA coefficient; $\hbar$ is the reduced Plank constant; A is the modal area and S(t) is the total input light power. The TPA generated carriers are considered by the last term in first equation.

The gain of QD-SOA including nonlinear process such as carrier heating (CH) and spectral hole burning (SHB) effects is expressed as Eq. (2). See Refs. [9] and [10] for more details:

$$g(t) = \frac{a(N - N_t)}{1 + (\varepsilon_{CH} + \varepsilon_{SHB})S(t)},$$
(2)

where N and N_t are the GS carrier density, the transparency GS carrier density respectively; ε denotes the gain suppression factor. The refractive index of the active region is affected by the injected light and the change of temperature due to carrier heating. As a result, it will cause a phase change to any probe wave injected into the QD-SOA. This is shown in the following equation: (See Ref. [11] for more details)

$$\phi(t) = -\frac{1}{2}[\alpha G_L(t) + \alpha_{CH}\Delta G_{CH}(t)],$$
(3)

where $G_L(t)$ is the linear gain factor of the device given by $g(t)L$, L being the effective length of the active layer; α is the linewidth enhancement factor of the device corresponding to band-to-band transition and α_{CH} is the linewidth enhancement factor of the device related to carrier heating process [12].

The output of MZI from the combination of two data streams can be expressed as

$$P_{out} = \frac{P_{cb}(t)}{4}[G_1(t) + G_2(t) + 2\sqrt{G_1(t)G_2(t)}\cos(\phi_1(t) - \phi_2(t) + \phi_0)],$$
(4)

where P_{cb} is the light power of the input clock signal; $G_1(t)$ and $G_2(t)$ are the calculated total linear gain factors. The parameters we use for simulation is shown in what follows.

Table 1. The parameters used in the model.

Parameter	Description	Value
τwr	Lifetime for WL recombination	0.2 ns[a]
τesr	Lifetime for ES recombination	0.2 ns[a]
τgsr	Lifetime for GS recombination	0.1 ns[a]
b	TPA coefficient	70 cm/GW[b]
Gd	Confinement factor for active QD region	0.1
a	Gain differential	$8.6*10{-}15$ cm[2][c]
$\tau w\text{-}e$	WL to ES relaxation lifetime	1 ps[d]
$\tau g\text{-}e$	GS to ES relaxation lifetime	10 ps[d]
a	Linewidth enhancement factor for gain dynamics	4
aCH	Linewidth enhancement factor for CH	1.2
L	Length of active region	1.0 mm
eCH	Gain suppression factor for CH	$0.3*10{-}17$ cm[3][e]
eSHB	Gain suppression factor for SHB	$7.5*10{-}17$ cm[3][e]

4. PRBS Model

Due to the compact and stable structure of the SOA-MZI based optical logic gate [see Ref. 3], it can be used to build the PRBS generator. The PRBS generator used here is based on a linear feedback shift register (LFSR). As shown in Fig. 3, the LFSR has m data storing units (delay lines in optics), these storing units will store one binary data bit for one clock period. The system is synchronized with one clock. For each period, we let the bits in nth and the mth unit go through a XOR process, the output will be reshaped with an AND gate and band pass filter then feed back to the first unit of LFSR. The output PRBS signal can be tapped from the end of the LFSR [see Refs. 3, 13–16].

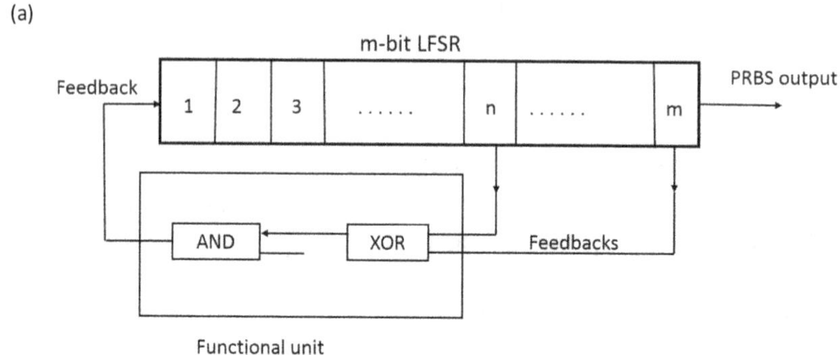

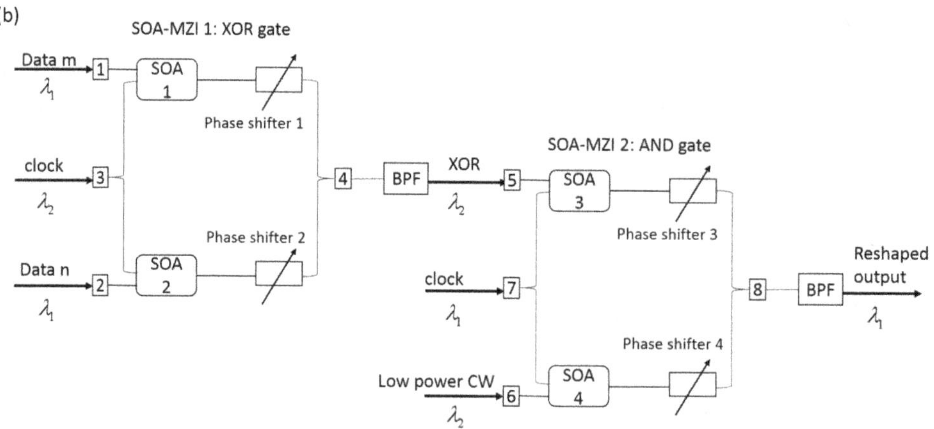

Fig. 3. Schematic diagram of PRBS generator. (a) Block diagram of a LFSR; (b) functional unit, two SOA-MZIs operating as XOR and AND gates. BPF: bandpass filter.

For a LFSR with m units, the PRBS repetition bit period is $T=2^m-1$. We can use PRBS-m indicate the size of the sequence. Most used PRBS has a LFSR with 7 or more units. For the simulations in this paper, we choose PRBS-7 and it has a repetition period of 127 bits. Figure 4 shows a Simulation result of PRBS based on QDSOA-MZI and LFSR.

(a)   (b)

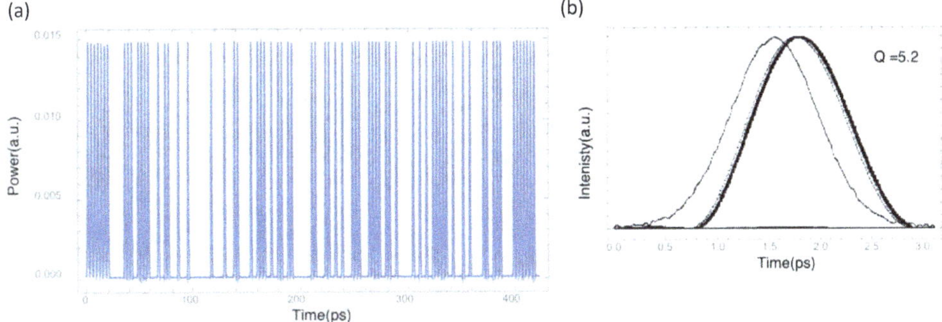

Fig. 4. (a) Simulation result of PRBS, operating at 320 Gb/s. The initial input of the LFSR is seven "1" s, and it's using 4th and 7th unit. (b) Eye-diagram of the PRBS sequence at 1.0 ps pulse.

The principle reason for fast gain and phase recovery for QD-SOA is the fast well to excited state of dot transition time. The calculated Q-factor as a function of this parameter is shown in Fig. 5. Q2 approximately equals the ratio of signal to noise (in power).

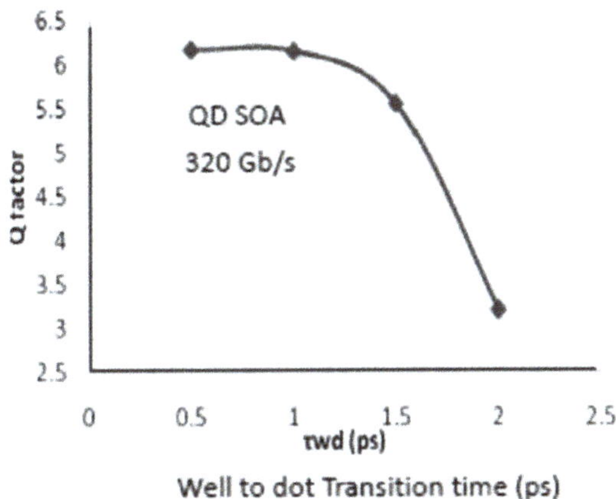

Fig. 5. Calculated Q-factor for PRBS operation at 320 Gb/s as a function of well to dot transition time. Note that Q ~ 6 is feasible (Q = 6 corresponds to an error rate of 10-9 in a transmission system).

5. Encryption/Decryption with Keystream

Using XOR algorithm, we can encrypt a data by apply XOR operation to the data with a keystream and decrypt it by apply XOR operation to the encrypted data with the same keystream. The processes of encryption and decryption are shown in Fig. 6. A simulated result is shown in Fig. 7.

For this encryption/decryption method, the regeneration of a reliable long period keystream with statistical behavior similar to a truly random sequence is essential. There for PRBS are suitable for the quest. Yet, PRBS based on one LFSR is a linear system and the generated key-stream is easily predictable. Large period, large linear complexity

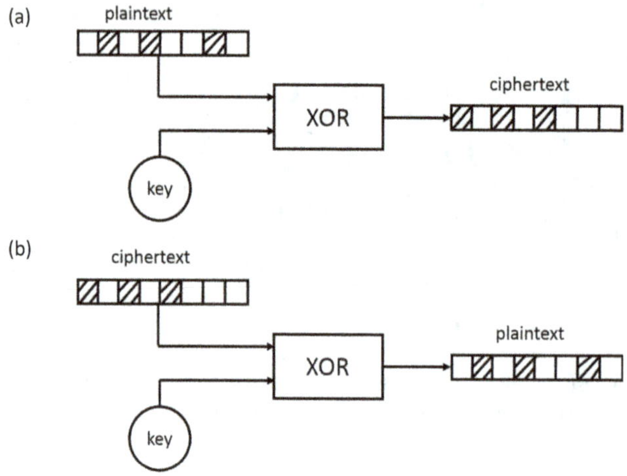

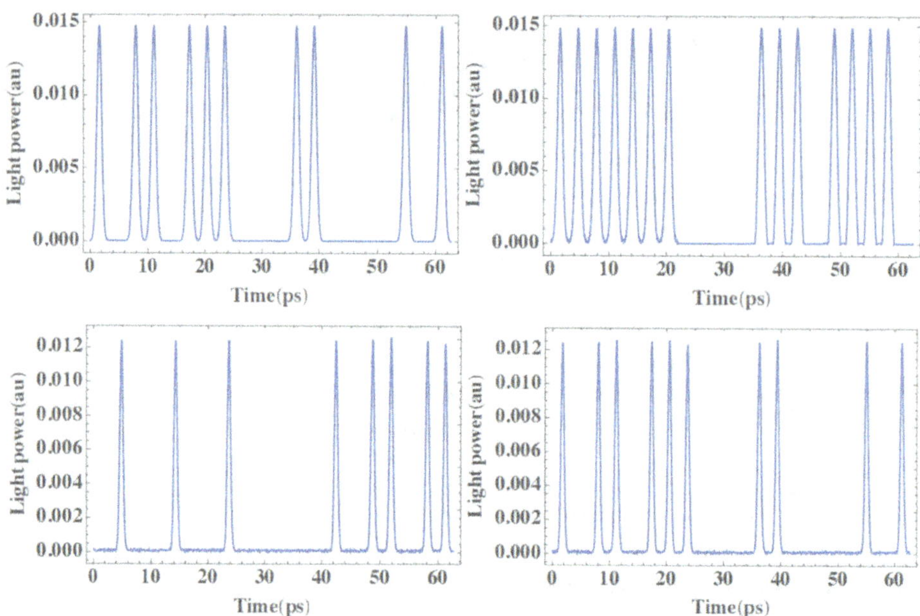

Fig. 6. (a) Schematic diagram of encryption process. (b) Schematic diagram of decryption process. The key streams are the same.

Fig. 7. Input Data (Top-Left). Key (Top-Right). Encrypted Data (Bottom-Left). Decrypted Data (Bottom-Right). All these results are for 320 Gb/s data rate.

and good statistical properties are three necessary conditions for a key-stream generator to be considered secure in cryptography [17]. One way to make better encryption is to combine several LFSRs or using the output of one (or more) LFSRs to control the clock of one (or more) other LFSRs. This method provide a better key-stream generators which effectively increase the complexity of the key-stream. In Fig. 8, one design that can be used is cascaded designed generator. We use the output of previous PRBS (with single LSFR) as the input of next PRBS. This design use 3 LSFRs, therefore can generate longer period (larger m) keys, while provide us with more alternatives.

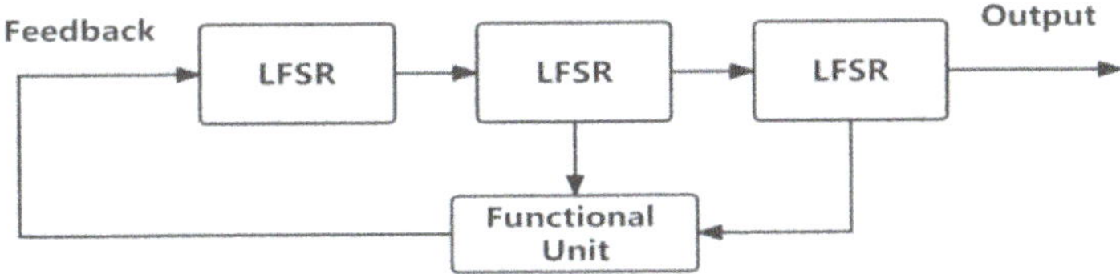

Fig. 8. Block diagram of cascaded designed key-stream generator.

A simulated result for cascaded step generator is shown in Fig. 9.

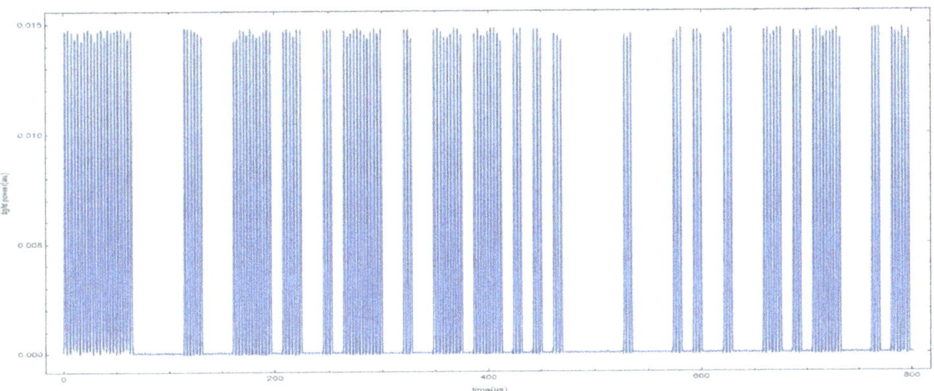

Fig. 9. Simulated result for cascaded step generator at 320 Gb/s using the 15th and 21th unit.

In Fig. 10, the schematic of an alternating step generator is shown.

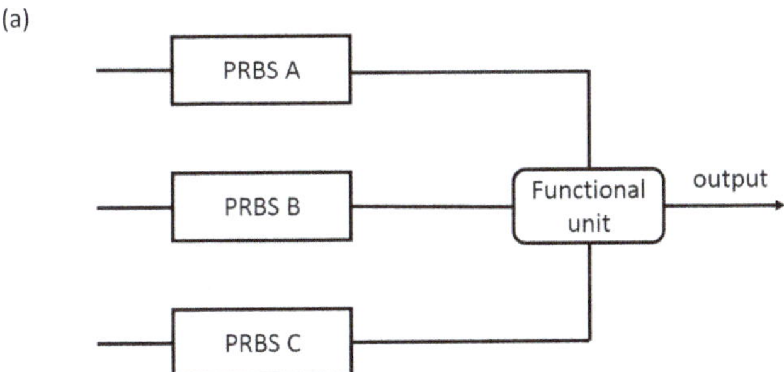

Fig. 10. Design of the alternating step generator (ASG). (a) Schematic diagram of ASG; (b) schematic of the functional unit in (a).

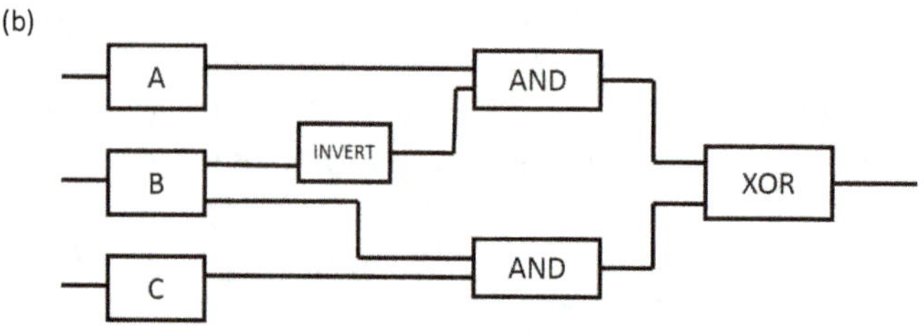

Fig. 10. (*Continued*).

A simulated result for alternating step generator is shown in Fig. 11. We provide three different initial inputs and using 4th and 7th unit for the three PRBS in this design.

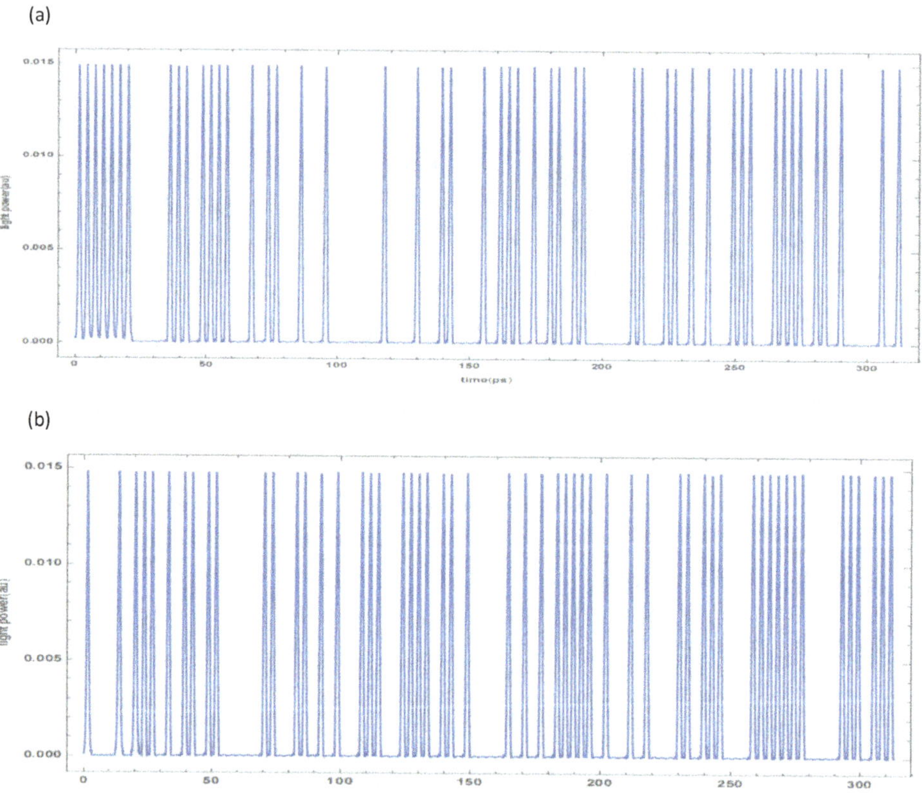

Fig. 11. Simulation of the alternating step generator at 320 Gb/s. (a) Simulation results of PRBS A sequence with input "1111111"; (b) Simulation result of PRBS B with input "10100001"; (c) Simulation result of PRBS C with input "0010110"; (d) Simulation results of the output sequence of ASG.

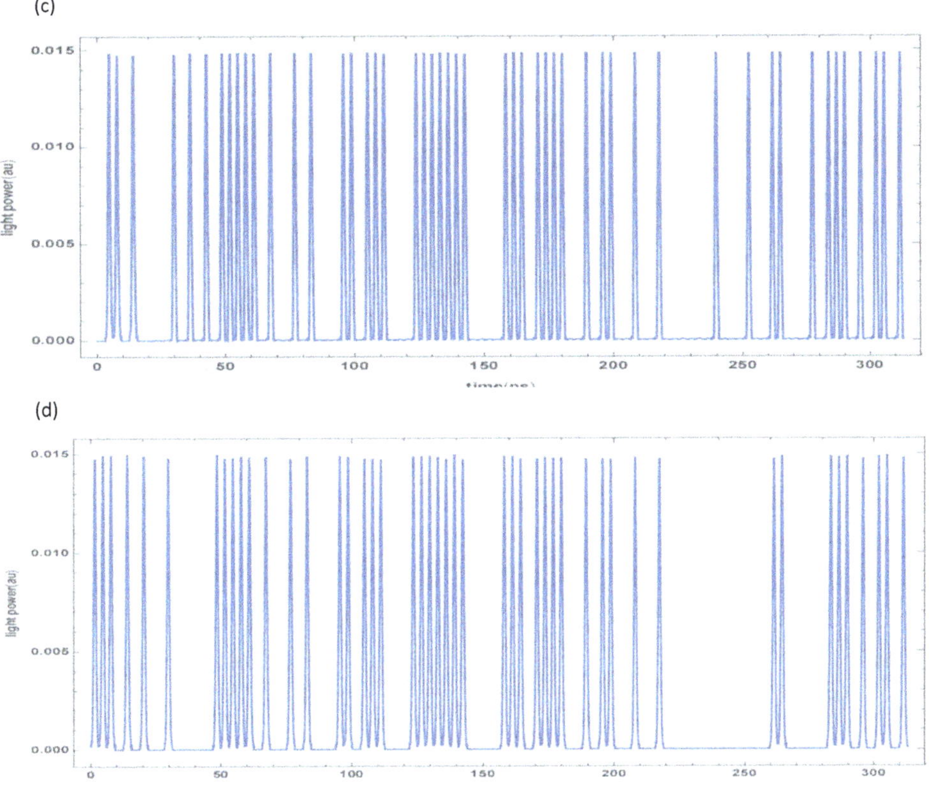

Fig. 11. (*Continued*).

In Fig. 12, a parallel designed key-stream generator is shown and Fig. 13 shows the results.

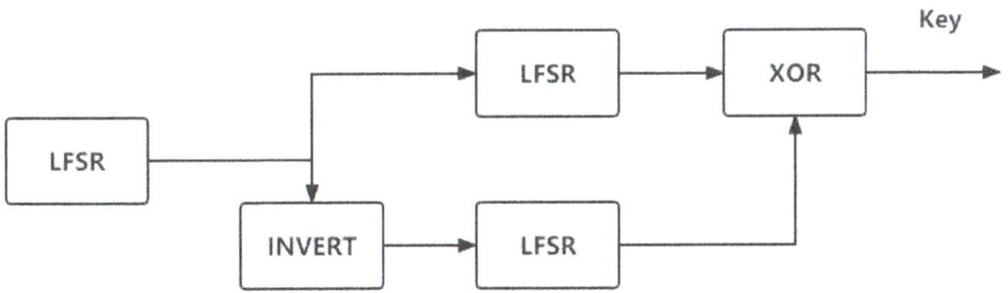

Fig. 12. Block diagram of parallel designed key-stream generator with its functional unit.

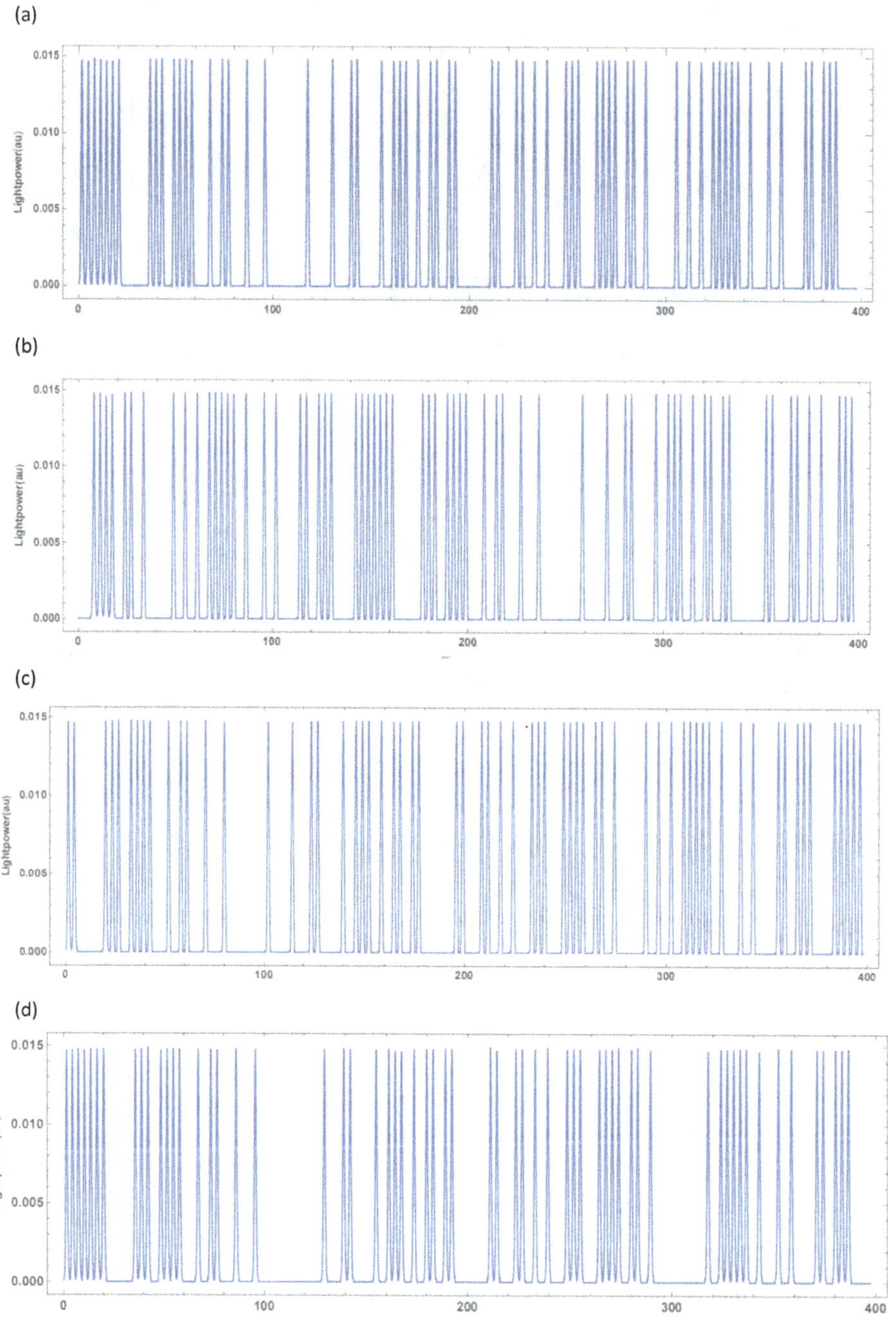

Fig. 13. Simulation of parallel designed key-stream generator at 320 Gb/s. (a) Simulation result of first LFSR, initial input is "1111111". (b) Simulation result of second LFSR in the design, input is "15th–21st" of the first LFSR. (c) Simulation result of third LFSR in the design, input is inverse of "15th–21st" of the first LFSR. (d) Simulation result of Output (XOR between second and third LFSR.).

6. Conclusion

The optical logic XOR, AND gates based on quantum dot-semiconductor optical amplifier Mach–Zehnder interferometer (QD–SOA–MZI) can operate at 320 Gb/s and can be used in the design of encryption circuits. The designs we propose can function as stable and complex key-stream generators at a high data rate.

References

1. K. E. Zoiros, T. Houbavlis and M. Kalyvas, "Ultra-high speed all-optical shift registers and their applications in OTDM networks," Invited paper, Optical and Quantum Electronics 36(11), 1005–1053 (2004).
2. G. P. Agrawal, Fiber-Optic Communication Systems, John Wiley, New York (2002).
3. N. K. Dutta, "Semiconductor Optical Amplifiers," Chapter 9, Second Edition, World Scientific (2012).
4. T. Akiyama, M. Sugawara and Y. Arakawa, "Quantum-dot semiconductor optical amplifiers," Proceedings of the IEEE 95, 1757–1766 (2007).
5. A. A. Krylov, S. G. Sazonkin, V. A. Lazarev, D. A. Dvoretskiy, S. O. Leonov, A. B. Pnev, V. E. Karasik, V. V. Grebenyukov, A. S. Pozharov and E. D. Obraztsova, "Ultra-short pulse generation in the hybridly mode-locked erbium-doped all-fiber ring laser with a distributed polarizer," Laser Physics Letters 12, 065001 (2015).
6. H. Ju, A. Uskov, R. Nötzel, Z. Li, J. M. Vázquez, D. Lenstra, G. Khoe and H. Dorren, "Effects of two-photon absorption on carrier dynamics in quantum-dot optical amplifiers," Applied Physics B 82, 615–620 (2006).
7. T. W. Berg, S. Bischoff, I. Magnusdottir and J. Mork, "Ultrafast gain recovery and modulation limitations in self-assembled quantum-dot devices," IEEE Photonics Technology Letters 13, 541–543 (2001).
8. X. Zhang and N. K. Dutta, "Effects of two-photon absorption on all optical logic operation based on quantum-dot semiconductor optical amplifiers," Journal of Modern Optics 65, 166–173 (2018).
9. P. Borri, W. Langbein, J. M. Hvam, F. Heinrichsdorff, M. H. Mao and D. Bimberg, "Spectral Hole-Burning and Carrier-Heating Dynamics in Quantum-Dot Amplifiers: Comparison with Bulk Amplifiers," Physica Status Solidi (B) 224, 419–423 (2001).
10. T. Akiyama, H. Kuwatsuka, T. Simoyama, Y. Nakata, K. Mukai, M. Sugawara, O. Wada and H. Ishikawa, "Application of spectral-hole burning in the inhomogeneously broadened gain of self-assembled quantum dots to a multiwavelength-channel nonlinear optical device," IEEE Photonics Technology Letters 12, 1301–1303 (2000).
11. W. Li, H. Hu, X. Zhang and N. K. Dutta, "High speed all optical logic gates using binary phase shift keyed signal based On QD-SOA," International Journal of High Speed Electronics and Systems 24, 1550005 (2015).
12. J. Vazquez, H. Nilsson, J.-Z. Zhang and I. Galbraith, "Linewidth enhancement factor of quantum-dot optical amplifiers," IEEE Journal of Quantum Electronics 42, 986–993 (2006).
13. K. E. Zoiros, T. Houbavlis and M. Kalyvas, "Ultra-high speed all-optical shift registers and their application in OTDM networks," Optical and Quantum Electronics 36, 1005–1053 (2004).

14. A. Kotb and K. E. Zoiros, "Performance of all-optical XOR gate based on two-photon absorption in semiconductor optical amplifier assisted Mach-Zehnder interferometer with effect of amplified spontaneous emission," Optical and Quantum Electronics 46, 935–944 (2014).
15. S. W. Golomb, Shift Register Sequence, Holden-Day, San Francisco (1967).
16. M. Sugawara, H. Ebe, N. Hatori, M. Ishida, Y. Arakawa, T. Akiyama, K. Otsubo and Y. Nakata, "Theory of optical signal amplification and processing by quantum-dot semiconductor optical amplifiers," Physical Review B 69, 235332 (2004).

Behavioral Modeling of the Pinched Hysteresis Loop of a Pt/TiO$_2$/Pt Memristor

Aalvee Asad Kausani[*] and Mehdi Anwar[†]

*Department of Electrical and Computer Engineering, University of Connecticut,
Storrs, CT 06269, USA*
**aalvee_asad.kausani@uconn.edu*
†a.anwar@uconn.edu

The fourth fundamental circuit element, the memristor, has become a promising candidate to substantially improve the energy and area efficiencies of circuits as traditional complementary metal-oxide-semiconductor (CMOS) technology is approaching its physical limit. However, a mathematical representation of the experimentally obtained current-voltage characteristic of the memristor is necessary to develop and test memristor-based circuitry in electrical design simulators. Here we have developed a behavioral model for the I-V trace of a Pt/TiO$_2$/Pt memristor that can relate the fitting equations with the physical processes associated with the device in response to applied electrical excitation. Multiple conduction mechanisms are involved in memristor that depend upon its latest state. Therefore, the I-V has distinct segments that altogether form a hysteresis loop pinched at the center. In accordance with the predominant conduction mechanisms at each segment, our model defines the form of the equations. The behavioral model can adequately represent the experimental I-V retrieved from existing work.

Keywords: Memristor; pinched hysteresis; behavioral modeling.

1. Introduction

The fourth fundamental circuit element, the memristor, was theoretically introduced in 1971 by Leon L. Chua to define the relationship between two of the four fundamental circuit variables – the charge q and flux-linkage φ [1]. Chua demonstrated active circuit realization of the memristor. However, the physical realization of a two terminal memristor without internal power supply could not be realized at that time. It was foreseen by Chua that the simple charge-flux relationship of the theoretical memristor could yield a hysteric current-voltage curve. Followed by the conceptual understanding of memristor properties, the theoretical possibilities of its novel applications were discussed. In 2008, Strukov *et al.* from the HP labs [2] experimentally demonstrated the memristive property of a Pt/TiO$_2$/Pt assembly that showed hysteric I-V for the application of a sinusoidal voltage. The I-V graph is pinched at the center confirming no flow of current in the absence of external bias. Strukov *et al.* also presented an intuitive model for the device to capture the experimental

---

[†]Corresponding author.

I-V behavior in an analytical expression. The analysis is based on the proposition that hysteresis needs an atomic rearrangement that can modify the electronic current. The mathematical expression models this behavior with the help of a state variable which is physically restricted by boundary function in compliance to the device geometry. The model developed by Strukov *et al.* could successfully feature characteristics found in other material systems such as organic films, chalcogenides, metal oxides, perovskites [3–6] etc. otherwise categorized as bipolar switching. After the physical realization of memristor and its mathematical modeling, research on memristive structures and their novel applications have been greatly expedited. The two distinct resistive regions in the hysteric I-V characteristic of memristors and their retention in absence of power make them suitable as non-volatile memory devices for digital applications. The revolutionary concept of organizing the two-terminal memristor devices at the cross points of horizontal (word line) and vertical (bit line) wires has resulted in highly dense memory storage. In combination with memristors' non-volatility property, the crossbar structure has significantly facilitated in-memory processing of data especially required by artificial neuromorphic architectures for energy-efficient and fast computation [7, 8].

However, the understanding of the underlying physical mechanism of a memristor and developing a circuit model based on the device physics are equally significant for the exact representation of the devices in circuit simulators. In general, experimentally obtained I-V data along with the conduction mechanism are available although the analytical expressions are yet to be reported. Some PSpice models of memristor have been developed following the formulation by HP lab, the difference being the design of the window function to restrict the state variable in some physical limit [9–11]. These models need to solve the time-dependent coupled equations of charged dopant and electrons at the run time and therefore are complicated approaches. Moreover, the HP lab developed the model based on the assumption of the drifting charged dopants in response to applied electrical excitation which makes the device homogenously conduct current at the device's cross-sections. Later it was discovered that conduction in metal/oxide/metal-thin film memristors is dictated by native conductive channels [12]. The detailed physics of $Pt/TiO_2/Pt$ memristor was reported in [13] along with the analytical formulation of the growth of conductive paths known as high conductive filaments. The discovery of filamentary conduction requires a new model for the I-V trace of the device. Mazady *et al.* [14] developed a procedure to model any memristor device in circuit simulators such as PSpice. Instead of tracing the current for a complete cycle of applied voltage by solving coupled equations at each point, this method separately models each significant current segment and defines them as voltage dependent current sources. The switching between the currents is controlled by a series of SPST switches. The pinched hysteric I-V characteristic of a 50nm long $Pt/TiO_2/Pt$ memristor having a 50nm × 50nm cross-sectional area has been reproduced from the work of Mazady *et al.* and is illustrated in Fig. 1. Mazady's method is an ingenious idea to model any memristor since it is able to separately capture each current region eliminating the need of window functions which are more mathematical than physical. However, Mazady *et al.* adapted polynomial fitting to the experimental device data to develop the voltage dependent current equations. The polynomial-based best-fit equations have the constant coefficients that physically resemble offset current without any voltage dependency violating the pinched hysteric behavior. Therefore, while it is a comprehensive algorithm for emulating the I-V trace of a memristor, it is also imperative to incorporate the physical behavior in determining the expressions of current to accurately emulate the processes observed during conduction.

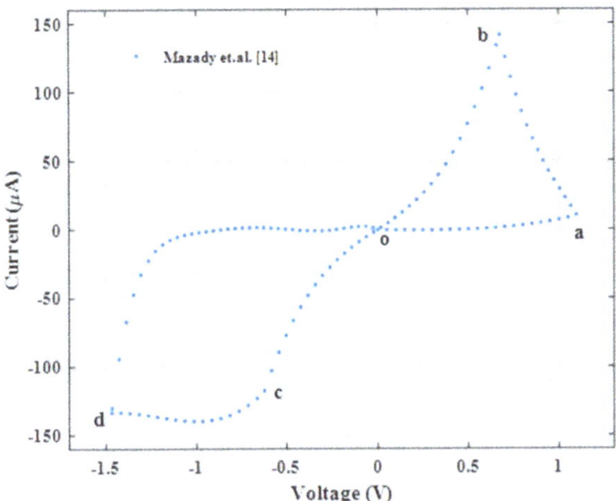

Fig. 1. The Current vs Voltage (I-V) trace obtained from the polynomial fitting equations reported in the DC circuit model of memristor [14].

Here we present a set of voltage dependent expressions of memristor currents inspired by the conduction mechanisms of the device. The conduction is dictated by gradually growing or dissolving filaments depending on the polarity of the applied electric field [13]. The partially developed filaments initiate the tunneling mechanism while the complete filamentary paths establish Schottky contact at the interfaces with the electrodes. A careful design of the parameters is important to prioritize a conduction mechanism over the others depending on the magnitude of applied voltage.

2. Behavioral Model

2.1. *Formulation of Current*

The hysteric I-V trace in Fig. 1 comprises of five distinct sections denoted as *oa*, *ab*, *boc*, *cd*, and *do*. Each segment is dictated by the physical mechanism of conduction described in detail by Mazady *et al.* [13]. The fundamental mechanism of this metal/oxide/metal structure involves the formation and dissolution of conductive filaments in the high resistance oxide layer in response to the applied electric field. The development and rupture of conductive channels are dependent upon the polarity of applied voltage. Besides the filamentary conduction, the resistive bulk oxide and low-conductive filaments formed during the electroformation process modulate the conduction. The behavioral model approximates the expressions of the current based on the dominant conduction methods.

The development of filaments for the application of a voltage to the memristor is depicted in Fig. 2(a). The applied electric field causes the oxidation of TiO_2 to $Ti_4O_5^{+2}$ near the anode, makes it drift to the cathode and reduces it to Ti_2O_3 particles at the cathode. The filament tip gradually progresses from the cathode to the anode as the Ti_2O_3 aggregates. The incrementing filaments initiate a tunneling current the value of which is gradually increasing as the gap between the electrodes is diminishing.

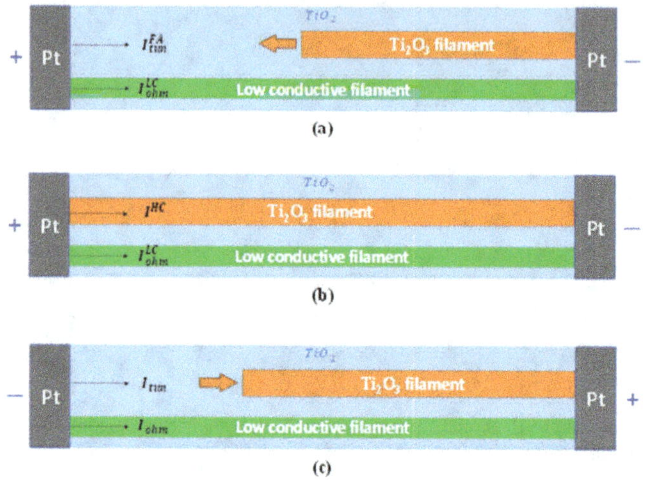

Fig. 2. The formation and rupture of high conductive filaments in $Pt/TiO_2/Pt$ memristor in relation to the applied voltage. (a) Growth of filament, (b) Bridging the gap between the electrodes by complete filament, and (c) Dissolution of filament for the reversal of voltage polarity.

The tunneling current has exponential dependence on the applied voltage [15] and does not appear to be significant for smaller voltages since the tunneling distance is large. Therefore, the ohmic contribution of the low conductive filaments is dominant over the tunneling current for smaller applied voltage. The bulk TiO_2 in between the electrodes can contribute to the conduction with a bulk tunneling current. However, the tunneling distance for the TiO_2 remains unchanged for the application of electrical excitation and is large enough to safely discard it from the behavioral modeling. The *oa* region in the Fig. 1 represents the combination of tunneling and ohmic currents during the applied voltage from 0 volt to 1.2 volts. The mathematical behavior in this region is as follows.

$$I_{oa} = I^{FA}_{oa,tun} + I^{LC}_{oa,ohm} + I^{bulk}_{oa,tun}$$

$$I_{oa}(\mu A) \approx I^{FA}_{oa,tun} + I^{LC}_{oa,ohm} = A_1(e^{B_1 v} - 1) + (v/R_{LC}) \times 10^6 \qquad (1)$$

$I^{FA}_{oa,tun}$ and $I^{LC}_{oa,ohm}$ are the filament-assisted tunneling current and the ohmic current, respectively. The resistance of the low conductive filaments has been reported as $637K\Omega$

[13]. The parameter values suitable for the device in interest are found to be $A_1 = 1, B_1 = 2.4537$.

The *ab* region represents positive applied voltage with decreasing magnitude yielding an increasing current. This negative differential resistance behavior is attributed to the gradually growing filament length that decreases the tunneling distance. Therefore, the filament-assisted tunneling current will dominate the overall conduction and is expressed as follows:

$$I_{ab}(\mu A) = A_2(1/v)e^{1-B_2 v}, \quad A_2 = 200.52 \text{ and } B_2 = 2.6074 \tag{2}$$

The filaments are completely developed by the time voltage reaches around V_b and bridge the gap between the electrodes as depicted in Fig. 2(b). The difference between the work functions of Pt and Ti_2O_3 makes their junctions Schottky contacts. Therefore, the $Pt/Ti_2O_3/Pt$ structure resembles two Schottky diodes connected back-to-back. Hence, the current along *boc* is dictated by an ohmic current through the high conductive filament modulated by the Schottky contacts and the ohmic current through the low conductive filaments. The conductivity of the low conductive filaments is smaller by a magnitude than that of the high conductive filaments. Therefore, while the high conductive filaments are bridging the gap, the contribution of the low conductive filaments has been ignored to keep the model simple. Combining these ideas, the *boc* region has been modelled as follows:

$$I_{bo} = I_{bo}^{HC} + I_{bo,ohm}^{LC} = \left(I_{bo,sch}^{HC} + I_{bo,ohm}^{HC}\right) + I_{bo,ohm}^{LC}$$

$$I_{bo}(\mu A) \approx I_{bo,sch}^{HC} + I_{bo,ohm}^{HC} = A_3(e^{B_3 v} - 1) + (v/R_{HC}) \times 10^6 \tag{3}$$

$$I_{oc}(\mu A) \approx I_{oc,sch}^{HC} + I_{oc,ohm}^{HC} = -A_3\left(e^{B_3 |v|} - 1\right) - (|v|/R_{HC}) \times 10^6 \tag{4}$$

The expressions for *ob* and *oc* conductions are similar except the sign adjustments along *oc* to replicate the negative current for negative voltage region in the graph. Here, $A_3 = 11.33, B_3 = 3.78$ and R_{HC} is the resistance of the high conductive filaments reported as $63.7K\Omega$ [13].

With the reversed polarity and increasing magnitude of applied voltage, V_c is the threshold voltage when filaments start to rupture. Disconnected filaments initiate tunneling current which would decrease with increasing magnitude of voltage in relation to the increasing tunneling distance. However, inspecting the behavior of the I-V along the *cd* segment, we assume that neither all of the filaments of a device with multiple conductive paths rupture at V_c nor do any of them abruptly disconnect from an electrode. Therefore, the conduction along *cd* is a combination of tunneling current through the ruptured filament and ohmic current through the intact filaments.

$$I_{cd} = I_{cd,tun}^{FA} + I_{cd,ohm}^{HC}$$

$$I_{cd}(\mu A) = A_4(e^{B_4 v} - 1) + (v/R_{HC}) \times 10^6 \tag{5}$$

Here, $A_4 = 110.4849, B_4 = 5.4575$. A dominant tunneling current begins to flow past the location d, which is the maximum value of applied voltage with reversed polarity. The decreasing exponential behavior along do can be attributed to the complete separation of the filaments from the electrode at voltage V_d as illustrated in Fig. 2(c) and their gradual dissolution due to negative applied voltage that increases the tunneling distance. However, the length of the filaments is sufficiently small at d' and the conduction is only dictated by the low conductive filaments along $d'o$. The conductions along dd' and $d'o$ are modelled as follows:

$$I_{do} = I^{FA}_{do,tun} + I^{LC}_{do,ohm}$$

$$I_{dd'} \approx I^{FA}_{dd',tun} \text{ and } I_{d'o} \approx I^{LC}_{d'o,ohm}$$

$$I_{dd'}(\mu A) = -A_5\left(e^{B_5|v|} - 1\right), A_5 = 8.3296 \times 10^{-5}, B_5 = 9.5273 \tag{6}$$

$$I_{d'o}(\mu A) = A_6(v/R_{LC}) \times 10^6, A_6 = 0.730002 \tag{7}$$

The I-V expressions in this model are developed observing the physical behavior of the device and based on the expected mathematical interpretation of the behavior. The parameter values have been estimated by solving consecutive equations of currents at their intersections to avoid any abrupt change in the I-V. The corresponding quantities of current and voltage at the intersections are extracted from the polynomial fitting-based model reported in [14]. Figure 3 illustrates the behavioral model along with the baseline model [14] to demonstrate the closeness of our model to the existing work.

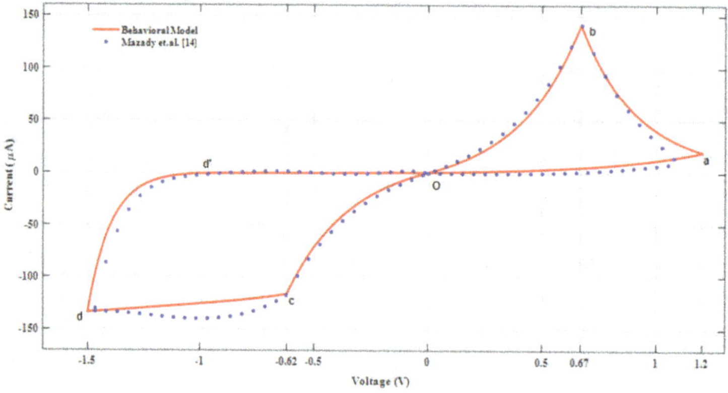

Fig. 3. Illustration of the behavioral model and the DC circuit model [14].

2.2. *Formulation of Voltage*

The I-V modeling is associated with the generation of a pseudo-DC voltage that could traverse a complete cycle comprising of positive and negative regions to adequately represent the gradual changes in voltage along with the reversal of polarity. The voltage waveform is expected to have a positive and negative maximum magnitude that coincide with the voltage values at the locations a and d of Fig. 3, respectively. The b, o and c

points stay on the decreasing direction of the voltage while the d' point resides on the increasing direction after the negative maximum. Combining the conditions, the voltage waveform is designed similar to the 50 Hz triangular waveform in [14] and is illustrated in Fig. 4(a). The corresponding equations of the time dependent voltage waveform are as follows:

$$v(t) = \begin{cases} \dfrac{1.2}{t_1} t & t_s \leq t \leq t_1 \\[2ex] \dfrac{2.7t - (1.5t_1 + 1.2t_2)}{t_1 - t_2} & t_1 < t \leq t_2 \\[2ex] \dfrac{1.5(t - t_f)}{t_f - t_2} & t_2 < t \leq t_f \end{cases} \tag{8}$$

$t_s = 0$ ms, $t_f = 20$ ms, $t_1 = 5$ ms and $t_2 = 15$ ms

$v_{max} = v(t_1) = 1.2$ volt, $v_{min} = v(t_2) = -1.5$ volt

Intermediate voltage points t_b, t_o, t_c and $t_{d'}$ partition the voltage waveform according to the distinct current regions. The corresponding values are: $t_b = 6.96$ ms, $t_o = 9.44$ ms, $t_c = 11.74$ ms and $t_d = 16.67$ ms. The behavioral modeling of the current in relation to the voltage waveshape is depicted in Fig. 4(b).

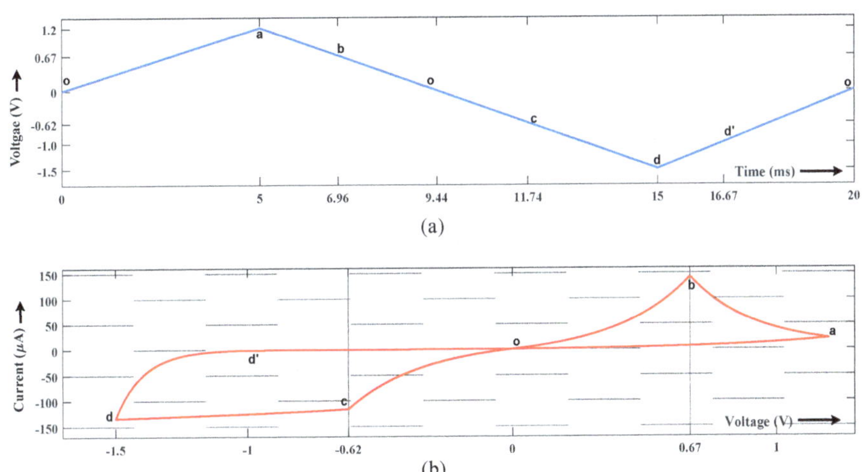

Fig. 4. The complete illustration of the behavioral model presented in this work. (a) The voltage waveform, and (b) The I-V characteristics.

3. Conclusion

A set of behavioral expressions for the I-V trace of a Pt/TiO₂/Pt memristor had been developed in relation to the conduction mechanisms of the device in response to electrical excitation. The model is in close proximity with the existing DC circuit model [14] derived from the experimental data. By exclusively calibrating the model parameters, the generalized expressions of currents are expected to comply with any memristor of metal/oxide/metal configuration that results in filamentary conduction. However, the contributions of the insignificant conduction mechanisms are eventually ignored in this model from the exact general expressions of different current segments. If a memristor structure demonstrates significant contribution from such mechanisms, this model might not be able to accurately capture the experimental I-V trace. On that note, an important extension to this work would be to retain all conduction mechanisms irrespective of their contributions and calibrate the related parameters accordingly.

Acknowledgments

One of the authors (AAK) was supported by a Department of Education GAANN funding.

References

1. L. Chua, *IEEE Transactions on Circuit Theory* **18**, 507–519 (1971).
2. D. B. Strukov, G. S. Snider, D. R. Stewart and R. S. Williams, *Nature* **453**, 80–83 (2008).
3. J. C. Scott and L. D. Bozano, *Advanced Materials* **19**, 1452–1463 (2007).
4. M. N. Kozicki, Mira Park and M. Mitkova, *IEEE Transactions on Nanotechnology* **4**, 331–338 (2005).
5. R. Waser and M. Aono, *Nature Mater* **6**, 833–840 (2007).
6. R. Oligschlaeger, R. Waser, R. Meyer, S. Karthäuser and R. Dittmann, *Applied Physics Letters* **88**, 042901 (2006).
7. T. Van Nguyen, J. An and K.-S. Min, Comparative study on quantization-aware training of memristor crossbars for reducing inference power of neural networks at the edge, in *2021 International Joint Conference on Neural Networks (IJCNN)*, Shenzhen, China, 18–22 July 2021.
8. J. Chen, Y. Wu, Y. Yang, S. Wen, K. Shi, A. Bermak and T. Huang, *IEEE Transactions on Neural Networks and Learning Systems* **33**, 1779–1790 (2022).
9. J. Zha, H. Huang and Y. Liu, *IEEE Transactions on Circuits and Systems II: Express Briefs* **63**, 423–427 (2016).
10. T. Prodromakis, B. P. Peh, C. Papavassiliou and C. Toumazou, *IEEE Transactions on Electron Devices* **58**, 3099–3105 (2011).
11. Z. Biolek, D. Biolek and V. Biolková, *Radioengineering* **18**, 210–214 (2009).
12. J. J. Yang, M. D. Pickett, X. Li, D. A. A. Ohlberg, D. R. Stewart and R. Stanley Williams, *Nature Nanotechnology* **3**, 429–433 (2008).
13. A. Mazady and M. Anwar, *IEEE Transactions on Electron Devices* **61**, 1054–1061 (2014).
14. A. Mazady and M. Anwar, DC circuit model of a memristor, in *2011 International Semiconductor Device Research Symposium (ISDRS)*, College Park, MD, USA, 07–09 December 2011.
15. S. Sze and K. K. Ng, *Physics of Semiconductor Devices* (Hoboken, NJ, USA: Wiley, 2007).

Design and Simulation of Multi-State D-Latch Circuit Using QDC-SWS FETs

A. Almalki[*,†], B. Saman[*,†], R. H. Gudlavalleti[*], J. Chandy[*], E. Heller[‡] and F. C. Jain[*,§]

[*]*Department of Electrical and Computer Engineering, University of Connecticut, CT, USA*
[†]*Department of Electrical Engineering, College of Engineering, Taif University, Saudi Arabia*
[‡]*Synopsys Inc., Ossining, NY 10562, USA*
[§]*faquir.jain@uconn.edu*

This paper presents a novel D-latch circuit using multi-state quantum dot channel (QDC) spatial wavefunction-switched (SWS) field-effect transistors (FET). The SWS-FET has two or more vertically stacked quantum-well or quantum dot (QD) layers where the magnitude of the gate voltage determines the location of carriers in each channel. Spatial location is used to encode multiple logic states along with the carrier transport in mini-energy bands formed in GeO_x-Ge/ SiO_x-Si quantum dot superlattice (QDSL), and to obtain 8-states operation. The design is based on the 8-state inverter using QDC SWS-FETs in CMOS-X configuration. This could be a new paradigm for designing flip-flops and registering more complex sequential circuits. The proposed design leads to reduced propagation delay and a smaller Si footprint.

Keywords: Multistate D-Latch; quantum dot channel (QDC) FET; SWS-FET; multi-state/multi-bit logic.

1. Introduction

Over the past few decades, integrated circuits (IC) have been able to achieve better speed, higher density, and lower power consumption due to the scaling down of MOSFETs. The requirement for greater information density and faster processing speed created motivation towards multi-bit logic architectures as device scaling reached its limits. Unlike conventional FETs, spatial wavefunction switched (SWS)-FETs are composed of two or more vertically stacked coupled quantum well or quantum dot channels. Features of multi-state QDC-FETs using QDSL as a transport channel (7 nm × 7 nm) for more than 8 dots per channel were reported by Jain *et al.* [1].

Figure 1 shows QDC-QDG FET structure and experimental I-V characteristics for Si quantum dot channel (QDC) FET [2] with quantum-dot gate. The steps are due to mini-energy band transitions. Figure 2 shows the schematic of (SWS)-QDC FET device and experimental ID-VG characteristics at various drain voltage VD. The drain current ID is due to conduction in lower quantum-dot channel (QD2) as the gate is increased, the drain current decreases in QD2 and increases in QD1. This results from the spatial wavefunction

[§]Corresponding author.

switching (SWS) between the lower channel QD2 and upper channel QD1 [3]. The drain current is accessed through drain D1 for upper channel QD1 and deep drain D2 for lower channel QD2.

The 3D schematic in Fig. 3 illustrates a Ge-O_X cladded Ge quantum dot field-effect transistor (QDC FET) consisting of two Ge QD channels, namely QD2 (the lower channel) and QD1 (the upper channel), with each channel composed of two layers of self-assembled Ge quantum dots. The subsequent section details the QDC-SWS FET 8-states operation.

2. 8-State QDC SWS-FET and Multi-Bit Logic

Jain *et al.* reported [1] an 8-state QDC-SWS FET using two Ge quantum dot channels (QD2 and QD1) shown in Fig. 3. The 8-state operation, listed in Table 1 is based on the drain current, which depends on the number of mini-energy bands in transport quantum channel (QC) and transfer via tunneling between lower (QD2) and upper (QD1) channels. The 8-state operation is listed in Table 1.

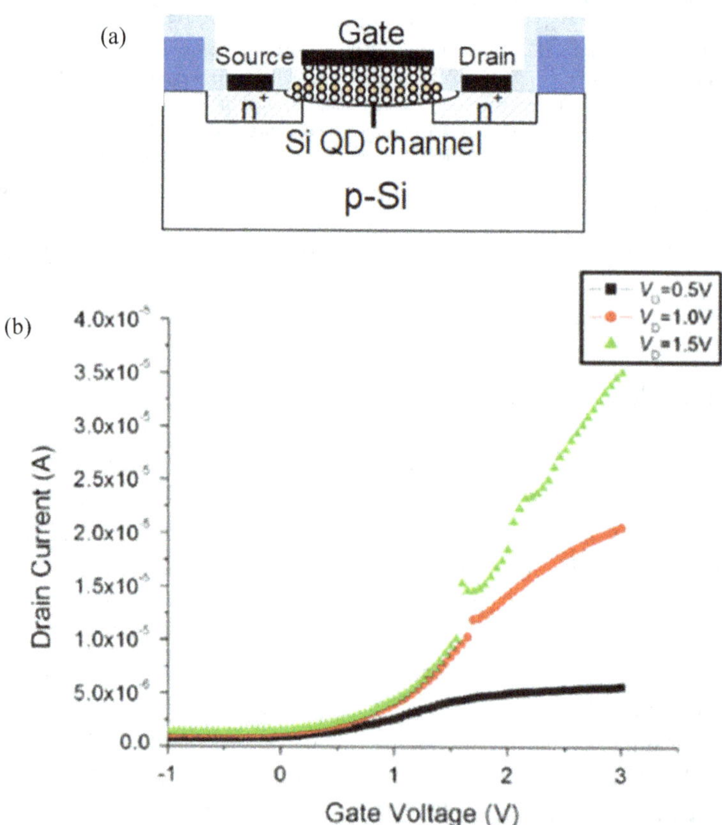

Fig. 1. QDC-QDG FET (a) Schematic (Top) (b) ID-VG at VD (black line at 0.5V, red line at 1.0V and green line at 1.5V). (Bottom).

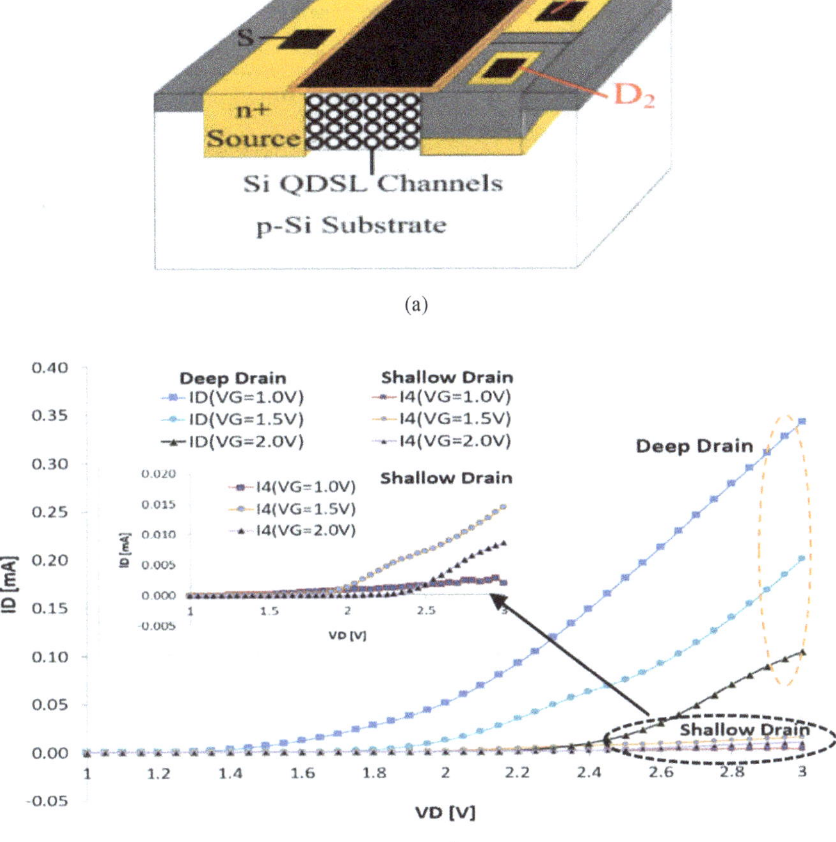

Fig. 2. SWS-QDC-FET (a) schematic (b) ID-VG characteristics for deep and shallow (shown in inset) drain voltage at different gate voltages, VG.

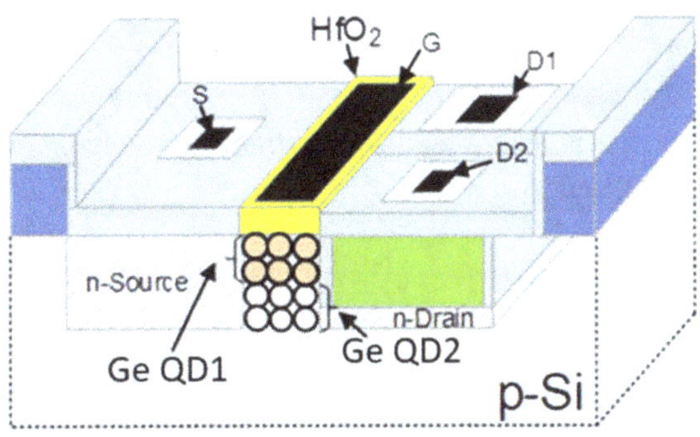

Fig. 3. 3D schematic view of Ge QDC-FET.

Table 1. QDC SWS-FET 8-state operation.

State	Electrons position
000	No electrons in any channel
001	Electron in first mini-energy band of QD2 (0_B1)
010	Electrons in first and second mini-energy bands of QD2 (0_B2B1)
011	Electrons in three mini-bands (0_B3B2B1)
100	Electrons in first mini-energy band of QD1 and two mini-energy bands of QD2 (B1_B2B1)
101	Electrons in first and second mini-energy bands of QD1 and first mini-energy band of QD2 (B2B1_B1)
110	Electrons in three mini-energy bands of QD1 and first mini-energy band of QD2 (B3B2B1_B1),
111	Electrons in three mini-energy bands of QD1 and none in QD2 (B3B2B1-0)

3. QDC-SWS-FET Analog Behavioral Model (ABM)

In the ABM model, the SWS-FET drain currents for the wells behave very similarly to that of a conventional FET, with Well 1's drain current matching that of its conventional equivalent. However, when charges are transferred from Well 2 to Well 1 after a certain threshold, the equation changes for Well 2, and the drain current is adjusted accordingly [1, 6–8]:

$$I_D = \begin{cases} 0 & V_{GS} > V'_{TH} \\ \mu_n C_{ox} \left(\frac{W}{L}\right) \left[\left(V_{GS} - V'_{TH}\right) V_{DS} - \frac{V_{DS}^2}{2} \right] & V_{GS} \leq V'_{TH} \quad \& \, V_{DS} > V_{GS} - V'_{TH} \\ \mu_n C_{ox} \left(\frac{W}{L}\right) \frac{\left(V_{GS} - V'_{TH}\right)}{2} & V_{GS} \leq V'_{TH} \quad \& \, V_{DS} \leq V_{GS} - V'_{TH} \end{cases} \tag{1}$$

where the adjusted V'_{TH} is defined as

$$V'_{TH} = V_{TH2} + \alpha \tag{2}$$

The matching parameter α is determined by the transition voltage (V_{UL}), the threshold voltage in well 1 (VTH1), and V_{GS}:

$$\alpha \equiv matching\ parameter = \begin{cases} \dfrac{(V_{GS} - V_{UL})^2}{(V_{TH1} - V_{UL})} & V_{GS} > V_{UL} \\ 0 & elsewhere \end{cases} \tag{3}$$

In the QDC-SWS FET model, four additional ABM blocks are used to realize drain currents' variations in each channel which depends on the number of mini-energy bands in transport quantum channel (QC). Figure 4 shows simulation of I_{DS}-V_{GS} characteristics for n- and p-channel QDC-SWS FETs.

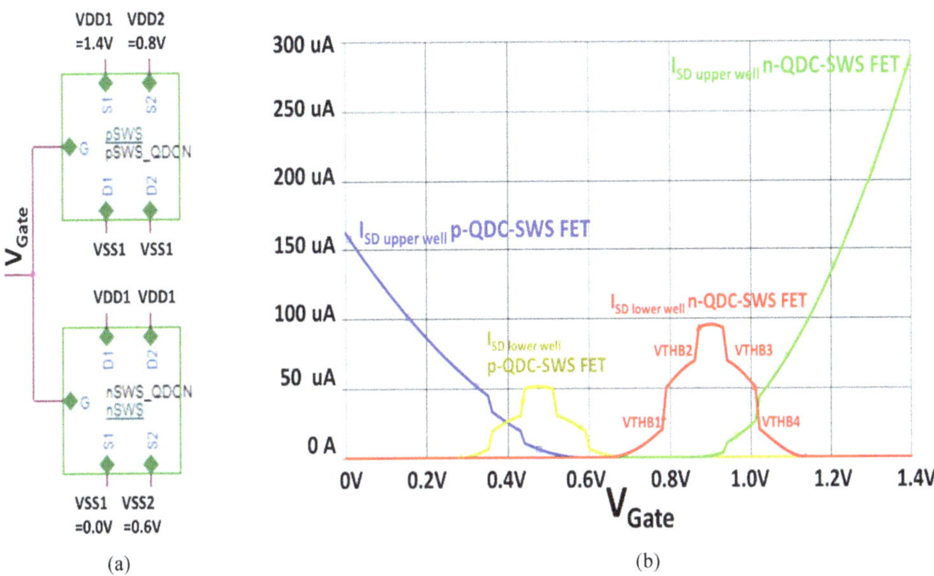

Fig. 4. (a) Schematic (left) (b) IDS-VGS characteristics (right) for n- and p-channel QDC-SWS FETs.

4. 8-State QDC-SWS FET-Based Inverter Model

The model is based on an integration of the Berkeley Short-channel IGFET Model (BSIM4.6) and the Analog Behavioral Model (ABM) [8]. Each part of the Complementary QDC-SWS FET-based inverter is modeled by two transistors using 180 nm technology [7]. The transistors represent upper wells and lower wells of the device with different threshold voltages. Additional ABM blocks are also used to produce drain current step-like characteristics in each channel caused by the mini-band transitions in the QDC [1]. To achieve the inverter logic, drains D_2 and D_1 are connected as the output and both gates are connected as the input. In addition, the source of the two QC1 and QC2 in the n-channel FET is connected to VSS_1 = Ground and VSS_2 = 0.6 V and the source of QC1 and QC2 in the p-channel FET is connected to VDD_1 = 1.4 V and VDD_2 = 0.8 V supplies, as shown in Fig. 5.

Simulation of the inverter using 8-state QDC-SWS FETs in CMOS-X configuration is presented in Fig. 6. It also shows the device schematic in Cadence Simulator and the parameters used for the inverter input (top panel in green line) and output (dashed red line) as a function of time.

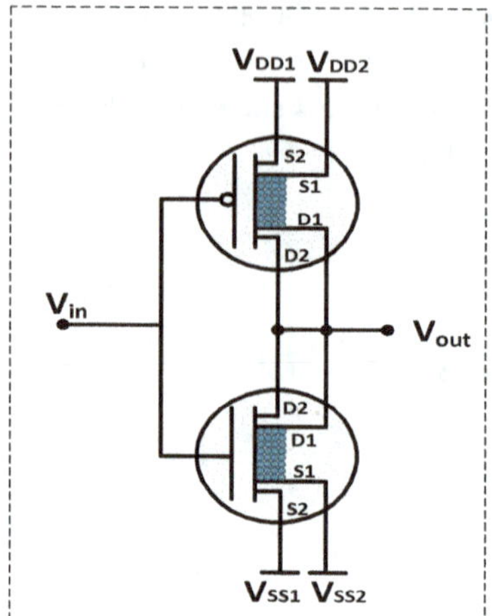

Fig. 5. 8-state QDC-SWS FET-based inverter schematic.

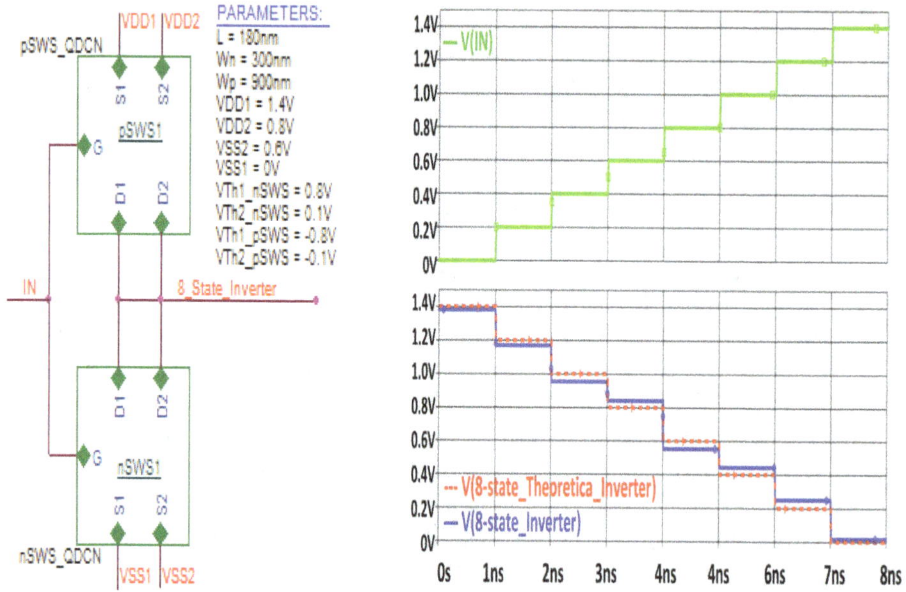

Fig. 6. Inverter schematic, parameters and simulation results showing the 8 states using n- and p-QDC-SWS FETs.

5. D Latch Circuit Based on the 8-State Inverter Using QDC-SWS FETs

The proposed D latch circuit shown in Fig. 7 is made up of two 8-state QDC-SWS FET-based inverters and two transmission gates. Table 2 shows the transmission gate truth table. Figure 8 shows the operation of the D-latch circuit, when Clk is high, transmission gate 1

(TG1) is ON and transmission gate 2 (TG2) is OFF, so output (Q) directly follows the input (D). When Clk is low (0) TG1 is OFF and TG2 is ON, now new data entering the latch are stopped and we get only previously stored data at the output.

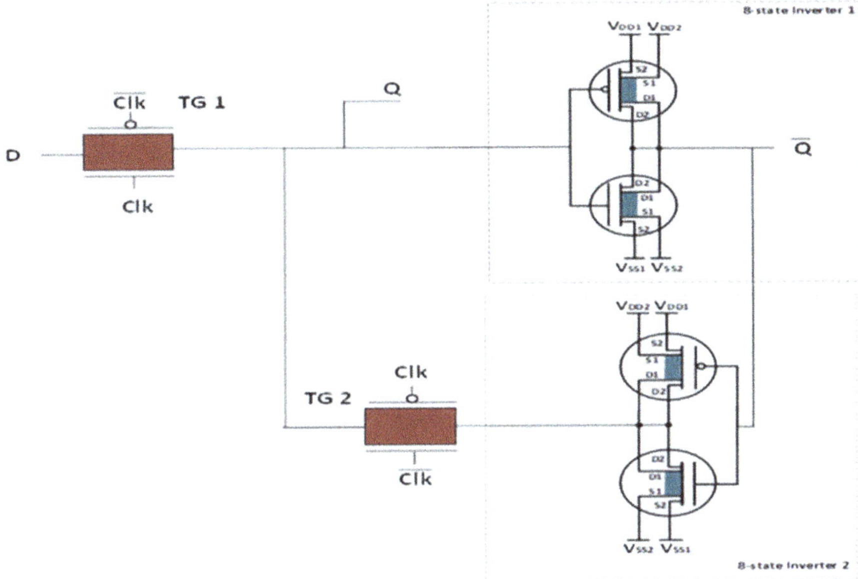

Fig. 7. Proposed D latch circuit schematic.

Table 2. Transmission gate truth table.

Control (Clk)	TG-PMOS	TG-NMOS	IN	OUT
1	0	1	0	0
1	0	1	1	1
1	0	1	2	2
1	0	1	3	3
1	0	1	4	4
1	0	1	5	5
1	0	1	6	6
1	0	1	7	7
0	1	0	0	Z
0	1	0	1	Z
0	1	0	2	Z
0	1	0	3	Z
0	1	0	4	Z
0	1	0	5	Z

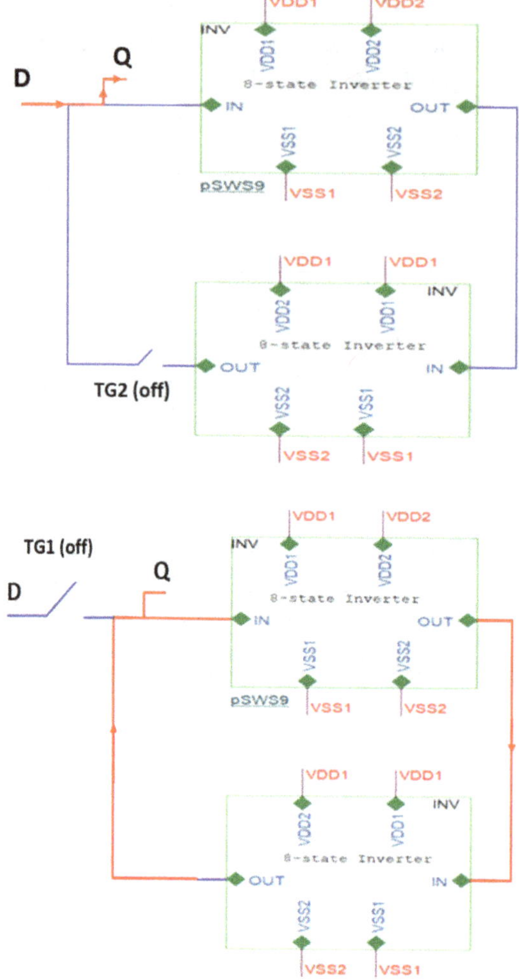

Fig. 8. Output (Q) when Clk is high (Top) and output (Q) when Clk is low (Bottom).

6. D-Latch Circuit Simulation Results

Figure 9 shows the simulated circuit schematic of the D-Latch using two QDC-SWS FET-based inverters and two transmission gates TG1 and TG2. The 8-state D-latch transient analysis waveform has been depicted in Fig. 10. The top panel (green line) shows five cycles of 8-state data (state 0–7). If the enable signal (Clk) given in the middle panel (blue line) is high, the output follows the input, and if the enable signal is low, the latch retains one of eight states previously stored inside the latch. The red line in the bottom panel shows the latch's output (Q) where each logic state of the latch is tested during the five cycles of DATA depending on the enable signal (Clk).

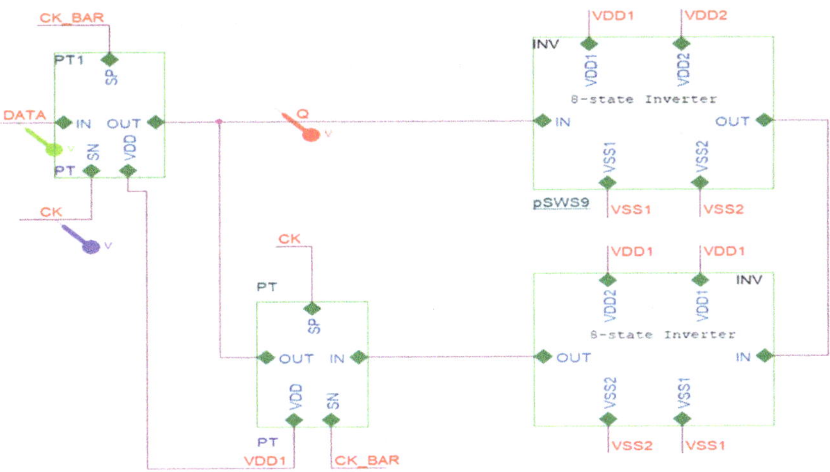

Fig. 9. Schematic of a D-Latch circuit based on 3-bit/8-state inverters.

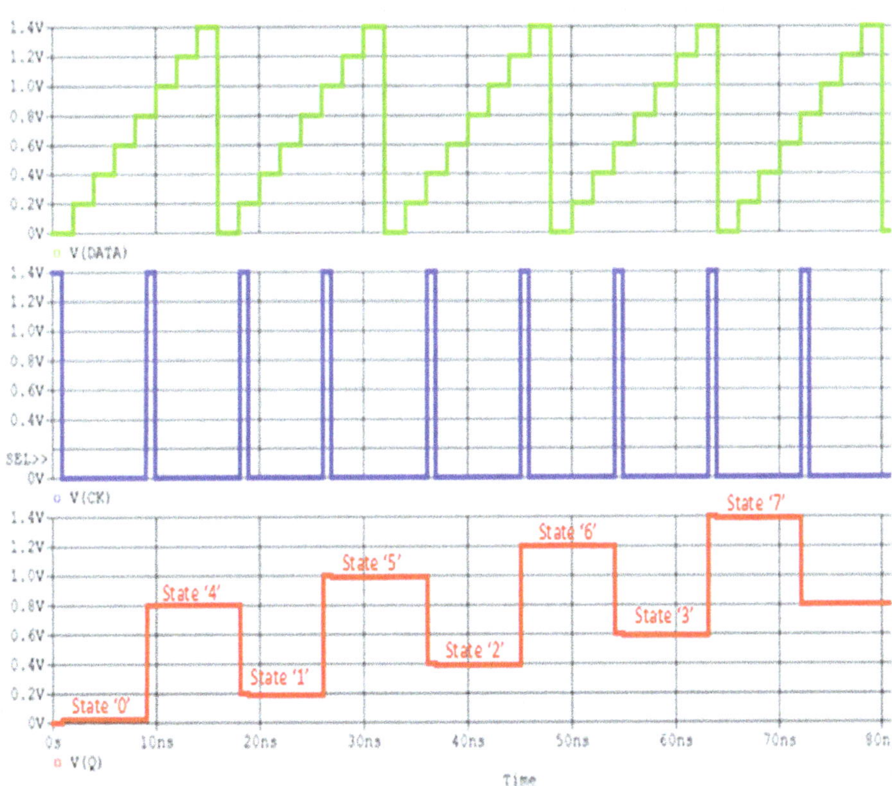

Fig. 10. Simulation results for proposed SWS-CMOS D-Latch based on the 8-state/3-bit SWS-CMOS inverter.

7. Conclusion

This paper presented a novel D-latch circuit using 8-state QDC SWS-FET inverters. The simulations were done in Cadence using ABM and BSIM4.6 models. The encoding of multiple logic states is achieved using spatial location of wavefunctions and carrier transport in different mini-energy bands formed in GeOx-Ge QDSL. The use of 8-state inverters in the implementation of D-latch allows advanced multibit sequential circuits. Additionally, the latch provides faster switching between states, resulting in reduced power consumption and delay. Overall, this work provides a new pathway towards compact multistate microprocessors.

References

1. F. Jain, R. Gudlavalleti, R. Mays, B. Saman, P-Y. Chan, J. Chandy, M. Lingalugari and E. Heller, "Quantum Dot Channel FETs Harnessing Mini-Energy Band Transitions in GeO$_x$-Ge and Si QDSL for Multi-Bit Computing," *International Journal of High Speed Electronics and Systems*, 31(01n04), 2240012 (2022).
2. F. Jain, S. Karmakar, P.-Y. Chan, E. Suarez, M. Gogna, J. Chandy and E. Heller, "Quantum Dot Channel (QDC) Field-Effect Transistors (FETs) using II-VI Barrier Layers," *Journal of Electronic Materials*, 41, 2775 (2012).
3. F. Jain, M. Lingalugari, B. Saman, P.-Y. Chan, P. Gogna, E.-S. Hasaneen, J. Chandy and E. Heller, "Multi-State Sub-9 nm QDC-SWS FETs for Compact Memory Circuits," *46th IEEE SISC*, Dec 2–5 (2015).
4. F. Jain, B. Saman, R. Gudlavalleti, R. Mays, J. Chandy and E. Heller, "Low-Threshold II–VI Lattice-Matched SWS-FETs for Multivalued Low-Power Logic," *Journal of Electronic Materials*, 50, 2618 (2021).
5. P.-Y. Chan, M. Lingalugari, E. Heller and F. Jain, "An Investigation on Quantum Dot Super-lattice (QDSL) Diode," *International Journal of High Speed Electronics and Systems*, 23(01n02), 1420004 (2014).
6. F. Jain B. Saman, R. H. Gudlavalleti, R. Mays, J. Chandy and E. Heller, "Multi-Bit SRAMs, Registers, and Logic Using Quantum Well Channel SWS-FETs for Low-Power," *High-Speed Computing International Journal of High Speed Electronics and Systems*, 28, 03n04 (2019).
7. B. Saman, P. Mirdha, M. Lingalugari, P. Gogna, F. C. Jain, E.-S. Hasaneen and E. Heller, "Logic Gates Design and Simulation Using Spatial Wave-function Switched (SWS) FETs," *International Journal of High Speed Electronics and Systems*, 24, 03n04 (2015).
8. F. Jain, B. Saman, R. H. Gudlavalleti, J. Chandy and E. Heller, *International Journal of High Speed Electronics and Systems*, 27(3 & 4), 1840020 (2018).
9. S. K. Saw, M. Maiti, P. Meher and S. K. Chakraborty, "Design and Implementation of TG based D Flip Flop for Clock and Data Recovery Application," *International Conference on Recent Trends in Engineering, Science & Technology - (ICRTEST 2016)*, Hyderabad, 2016, pp. 1–3, doi:10.1049/cp.2016.1491.

https://doi.org/10.1142/9789811283765_0007

Hybrid Mode-Locked Fiber Ring Laser Using Graphene Saturable Absorbers to Generate 20 and 50-GHz Pulse Trains

A. Rahman[*], S. Fan and N. K. Dutta

Department of Physics, University of Connecticut, Storrs, Connecticut 06269, USA
ashiq.rahman@uconn.edu

Optical pulses at high repetition rates are generated using rational harmonic mode locking and saturable absorber made of graphene nanoparticles in a fiber laser. The pulse generation from the fiber laser is modeled by solving the Generalized Nonlinear Schrodinger Equation. The computation involved varying the various saturable absorption parameters, such as linear and nonlinear absorption coefficients. Experimentally stable pulse trains at 20 GHz and 50 GHz are generated with a pulse width of ~ 2.7 ps. This result agrees with the simulation.

Keywords: Fiber optics; fiber lasers; mode locking; graphene saturable absorbers; RHML.

1. Introduction

The advancement of high-speed transmission fibers has greatly benefited global internet connectivity and long-distance fiber-optic communications. Recently, a new design of fiber optic technology has achieved data rates as high as 178-Tbps over a distance of 40-kilometers, which is a significant achievement [1]. In order to achieve optical transmission of data at exceptionally high rates with minimal errors, it is necessary for the pulses to be extremely narrow and for the pulse trains to have very high frequencies [2–4]. Over the years, various techniques have been shown to produce narrow width pulses with high frequency. One such method, known as hybrid mode locking, which combines active and passive mode locking, was detailed in [5]. Many fiber laser designs have incorporated graphene as saturable absorbers, as demonstrated in various studies [6–14]. A Thulium-(Tm) doped mode-locked fiber laser was created using layers of graphene/PMMA (poly methyl methacrylate) saturable absorber, which was synthesized through chemical vapor deposition (CVD) [15]. Sheng *et al.* also exhibited a fiber laser that was mode-locked and had adjustable power output [16]. In addition to CVD, trituration is a feasible technique for obtaining graphene or charcoal that can be utilized for saturable absorption aims, as demonstrated in a study by Lin *et al.* [17]. Here, trituration was used on several carbon-based materials to generate pulses for Erbium-Doped Fiber lasers.

---

[*]Corresponding author.

This paper describes how Rational Harmonic Mode Locking (RHML) and carbon-based saturable absorbers were jointly integrated into a fiber ring laser system, as depicted in Fig. 1. The incorporation of graphene nanoparticle layer into the laser cavity resulted in a stable output of 20-GHz pulse trains, with the pulse width reduced from 5.3-ps to 2.8-ps. When the cavity frequency was adjusted, a 50-GHz pulse train with a pulse width of 2.7-ps was produced. The researchers also used numerical simulations to investigate the effects of altering saturable absorption parameters, such as saturation intensity and the saturable and non-saturable absorption components of the graphene layers. The simulation results for minimum pulse-widths were consistent with the experimental results obtained under the same parameters.

2. Experimental Setup

The technique of imprinting-exfoliation-wiping process is utilized to incorporate graphene saturable absorbers within the fiber ring laser cavity [18]. To start the process, the pencil nibs are finely ground, and the resulting powder is applied to one end face of a regular single mode fiber (SMF) cable using a brush. A patch-cord connector is utilized to sandwich graphene layers between the two fiber ends, which are then separated to imprint graphene on both faces. To achieve a more uniform distribution of graphene and decrease cluster size, the process is repeated multiple times, reducing the absorption loss from self-aggregation of graphene nanoparticles [19]. Figure 2 illustrates the steps of this process.

Figure 1 demonstrates the integration of the graphene nanoparticles sandwiched in the fiber laser circuit. The 980-nm diode laser powers the fiber ring laser setup, and the pump laser is connected to the circuit through a 980/1550-nm wavelength-division multiplexer (WDM). Gain is achieved in the ring laser by utilizing a homemade 23-m long Erbium doped fiber amplifier (EDFA). The output coupler (OC) allows only 10% of the light in the ring to be output and 90% to be fed back into the system. An optical isolator is also used in the circuit to prevent feedback and ensure unidirectional flow. Passive mode-

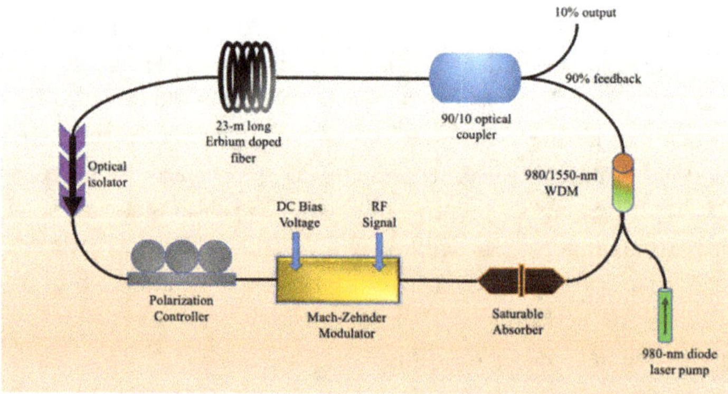

Fig. 1. Er-doped fiber ring laser setup, incorporating graphene saturable absorber in the cavity, pumped by a 980 nm diode laser.

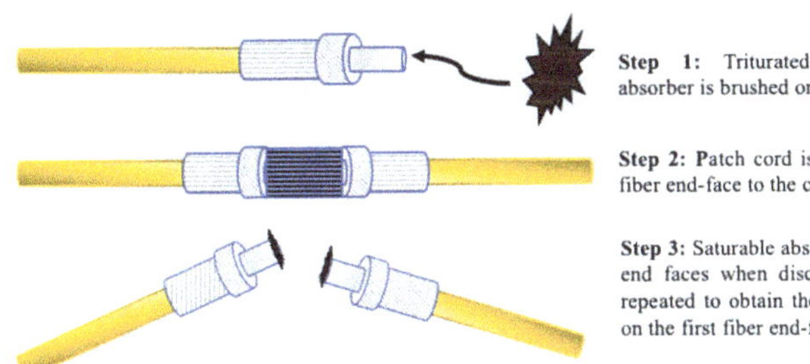

Fig. 2. Schematic of the imprinting-exfoliation-wiping process step by step.

locking is achieved using the patch-cord sandwiched graphene nanoparticle, while active mode-locking is accomplished through the use of a Mach-Zehnder Modulator (MZM). The MZM, a $LiNbO_3$ modulator, is driven by a radio frequency of around 10-GHz, provided by a synthesizer. The MZM is polarization-sensitive, requiring a polarization controller (PC) to be connected before the input end of the modulator. Standard 1550-nm single mode fiber (SMF) wires are used to connect all equipment mentioned.

3. Active and Passive Mode-Locking

3.1. *Active mode-locking*

The MZM is employed for active mode-locking, driven by a synthesizer's RF frequency, and carefully adjusted by the polarization controllers. The conditions necessary to obtain this phenomenon involve the modulation frequency of the laser cavity, which when it satisfies the condition $f_m = (n + 1/p)f_c$, the laser pulses that are output from the ring laser cavity is p times the modulation frequency f_m. Here, n and p are integers and f_c is the cavity mode separation ($f_c = c/n_i L$, where c is the speed of light in vacuum, n_i is the average index of the fiber and L is the length of fiber loop). In the experiment, $n_i = 1.5, c = 2.997 \times 10^8$ m/s, and $L = 50$ m, yielding $f_c = 4.29$ MHz. Therefore, for $f_m \sim 10$ GHz, $n = 2329$. By detuning f_c to f_c/p from nf_c, the laser travels around the ring cavity p times more before the pulses stabilize and the output coupler can emit them, resulting in the output pulses having a frequency that is a factor p higher than f_m. This process is called Rational Harmonic Mode-locking (RHML), which is the active mode-locking scheme in the experiment. The output of the 20-GHz pulse train is obtained when $p = 2$. Similarly, to get the 50-GHz pulse train, $f_c = 2.80$ MHz and $n = 3571$, with the value of $p = 5$ for this experiment.

3.2. *Passive mode-locking*

The researchers achieved saturable absorption in the laser cavity by incorporating graphene nanoparticles, which were reduced in size from 500-nm to 300-nm through an imprinting-exfoliation-wiping process. Graphene, which is the main component of graphite, exhibits optical absorption properties governed by the Pauli blocking effect [20]. When laser light passes through graphene layers, electronic transitions occur, but the absorption of incoming photons ceases when the final states are filled. This saturation effect makes graphene-based materials effective as saturable absorbers. The absorption of saturable absorbers is described using the following equation: [21].

$$\alpha = \alpha_{lin} + \alpha_{non}\left(1 + \frac{I}{I_{sat}}\right)^{-1}.$$

(1)

The given equation shows that the absorption has two components: a linear component that does not saturate (α_{lin}), and a nonlinear component that saturates (α_{non}). The saturation of the nonlinear term is determined by both the intensity of the incoming laser (I) and the saturation intensity of the material (I_{sat}).

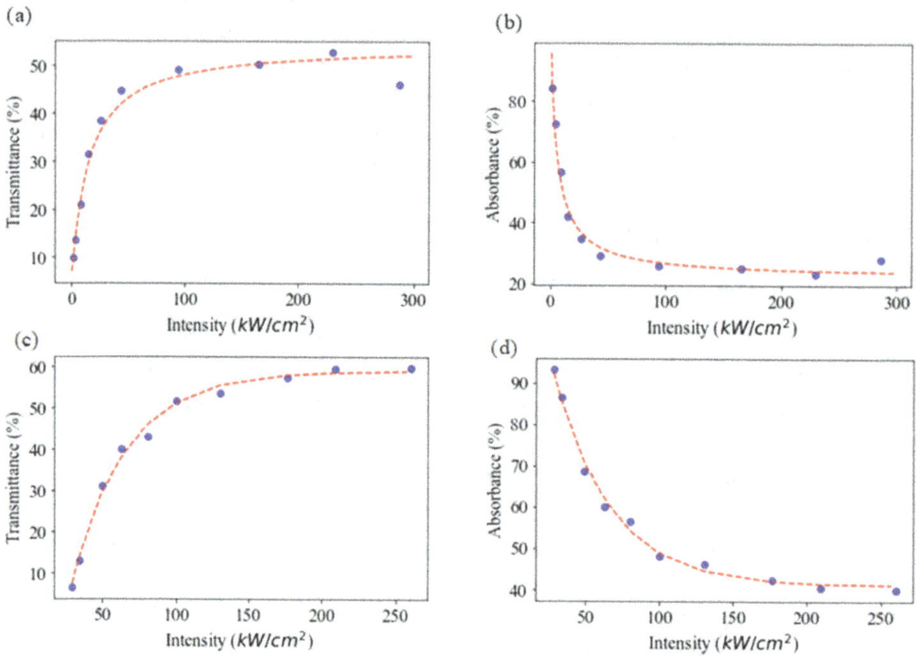

Fig. 3. (a) Transmittance of graphene saturable absorbers for the 20-GHz experiment, at a saturation intensity of ~100 kW/cm² (b) Corresponding modulation depth of ~50% (c) Transmittance of graphene for 50-GHz experiment, saturation intensity ~150 kW/cm² (d) modulation depth ~60%.

To prepare for the fiber laser experiment, the team first needed to determine the modulation depth of the saturable absorber, which measures the difference in absorption at high and low applied intensities for a particular wavelength. A high modulation depth is necessary for achieving strong pulse shaping, narrower pulse-widths and reliable self-starting. To determine this quantity, the team passed the mode-locked laser signal through a graphene sample, followed by a variable optical attenuator (VOA). The team found that the modulation depth of the graphene nanoparticles used in this experiment was approximately 50%, as shown in Fig. 3. Raising the input intensity from 20-kW/cm² to 250-

kW/cm^2, the saturable absorption was found to decrease, and the saturable transmission increased from 0 to 0.58. Curve fitting analysis indicated that the linear absorption (α_{lin}) was 0.6, the nonlinear absorption (α_{non}) was 0.5, and the saturation intensity (I_{sat}) was approximately 100-kW/cm^2. After determining these values, the saturable absorber was used in the main fiber ring laser circuit.

4. Numerical Simulation

4.1. *Laser pulse propagation mathematical model*

The electric field in an optical pulse can be written as

$$E(x,y,z,t) = \frac{1}{2}\left[A(z,t)\psi(x,y)e^{i(\omega_0 t - \beta_0 z)} + cc \right],$$

where $A(z,t)$ represents the slowly varying complex envelope of the pulse, $\psi(x,y)$ represents the transverse electric field distribution of the mode, ω_0 represents center frequency and β_0 represents the center propagation constant at ω_0. In the presence of optical loss or gain, second-order dispersion and third-order nonlinearity, (by using Maxwell's electromagnetic wave equations) the complex envelope $A(z,t)$ can be shown to satisfy the following equation [22]:

$$\frac{\partial A(z,t)}{\partial z} + \frac{\alpha}{2}A(z,\tau) + \sum_{k\geq 2}\frac{i^{k-1}}{k!}\beta_k\frac{\partial^k A(z,\tau)}{\partial \tau^k}$$

$$= \frac{g}{2}A(z,\tau) + \frac{g}{2\Omega_g^2}\frac{\partial^2 A(z,\tau)}{\partial \tau^2} + i\gamma|A(z,\tau)|^2 A(z,\tau). \tag{2}$$

Expanding and keeping up to the $k = 3$ term results in Eq. (3) which is used to model the pulse propagation in this paper. The loss term (α) is neglected here.

$$\frac{\partial A(z,t)}{\partial z} + \frac{i}{2}\beta_2\frac{\partial^2 A(z,\tau)}{\partial \tau^2} - \frac{1}{6}\beta_3\frac{\partial^3 A(z,\tau)}{\partial \tau^3}$$

$$= \frac{g}{2}A(z,\tau) + \frac{g}{2\Omega_g^2}\frac{\partial^2 A(z,\tau)}{\partial \tau^2} + i\gamma|A(z,\tau)|^2 A(z,\tau). \tag{3}$$

When the Split Step Fourier Method (SSFM) is applied to solve this equation, the resulting pulse propagation can be described by several parameters, including the slowly varying pulse envelope's amplitude $A(z,\tau)$ in relation to both the propagation distance z and time-delay parameter τ, as well as the propagation constants β_k and the nonlinear parameter γ. The nonlinear parameter is proportional to the nonlinear susceptibility χ^3. This nonlinear susceptibility arises from dependence of polarization on cube of the electric field. Equation (4) provides the relationship between the gain g of the Erbium doped fiber and the gain bandwidth Ω_g, energy of the pulse envelope E, and the small signal gain g_0. As the energy of the pulse envelope E approaches E_{sat}, the gain becomes saturated.

$$g = g_0 \left(1 + \frac{E}{E_{sat}}\right)^{-1}. \tag{4}$$

Simulating pulse propagation in the single mode fiber (SMF) is a simple process since in this case, the gain g is set to zero. To simulate RHML, the transmission properties of the Mach–Zehnder Modulator are modeled. The output amplitude of the MZM is determined mathematically by [23].

$$A_{\text{out}}(z, \tau) = A_{\text{in}}(z, \tau) \cdot \cos\left(\frac{\phi(t)}{2}\right) \text{ with } \phi(t) = \frac{\pi V(t)}{V_\pi}. \tag{5}$$

The modulator imparts a sinusoidal shape onto the incoming laser. A phase shift of π is attained when the voltage between the two arms of the modulator equals V_π. The modulator is activated by the applied voltage $V(t)$, which is determined by the bias voltage V_b and the amplitude of the sinusoidal RF signal $V_m \sin(\omega_m t)$, i.e., $V(t) = V_b + V_m \sin(\omega_m t)$.

Equation (6) gives the output amplitude of the pulse envelope after it goes through the graphene saturable absorber, by subtracting Eq. (1) from 1 to find the percent transmitted.

$$A(z, \tau)_{\text{out}} = A_{\text{in}}(z, \tau) \cdot \left(1 - \alpha_{lin} - \frac{\alpha_{non}}{1 + I/I_{sat}}\right). \tag{6}$$

The output coupler is simulated by Eq. (7), which is the amplitude of the 10% stable output pulses.

$$A(z, \tau)_{out} = A_{in}(z, \tau) \cdot R, \tag{7}$$

where R is obtained by comparing Eqs. (6) and (7). As mentioned earlier, the laser pulse travels through different optical components within the laser cavity. To simulate this, the MATLAB program uses different parameters that represent these components. Table 1 provides a summary of the values used for all the various parameters needed to solve the generalized nonlinear Schrödinger equation (GNLSE – Eq. (3)).

4.2. *Variation of pulse-width with absorption parameters*

In this section, the focus is on studying the impact of varying the saturable absorption, non-saturable absorption and saturation intensity on pulse-width using the model described in the previous section. The goal was to determine the optimal parameter values that would result in strong pulse shaping and the narrowest possible pulse-width. Figure 4 illustrates the effect of changing the intensity at which absorption gets saturated on pulse-width. The simulation was performed for two values of non-saturable absorption. Both cases show that pulse-width can be halved in a range of $I_{sat} \sim 30$–40 kW/cm^2. Figures 5 and 6 demonstrate how pulse-width changes with linear and nonlinear absorption components of the saturable absorber, respectively. Both simulations were carried out for constant I_{sat} values of 40 kW/cm^2 and 100 kW/cm^2. Without using a saturable absorber, the pulse-width of a rationally harmonic mode-locked pulse was found to be 5.3-ps. However, after incorporating the graphene nanoparticle saturable absorber, the pulse-width was compressed by up to approximately 50%. Figure 7 shows the impact of using graphene saturable absorbers on the roundtrip time required to stabilize the pulses. Besides a lower FWHM, the output requires less roundtrips to stabilize due to stronger pulse shaping.

Table 1. Modulator parameters used for generating the two hybrid mode-locked pulse trains.

Component	Parameter	Value
Erbium	β_2	-0.13×10^{-3} ps^{2}m^{-1}
Doped	β_3	0.135×10^{-3} ps^{3}m^{-1}
Fiber	γ	3.69 W^{-1}km^{-1}
	g_0	10 dBm^{-1}
Single	β_2	-22.1×10^{-3} ps^{2}m^{-1}
Mode	β_3	0.171×10^{-3} ps^{3}m^{-1}
Fiber	γ	1.20 W^{-1}km^{-1}
	g_0	0 dBm^{-1}
Graphene	α_{lin}	0.4
Saturable	α_{non}	0.4
Absorber	I_{sat}	30 kWcm^{-2}
Coupler	R	90 %
Modulator	V_π	6 V
	V_m	6 V
	V_b	0 V
	f_m	~ 10 GHz
	f_c	4.29 MHz (20 GHz)
	f_c	2.80 MHz (50 GHz)
	n	2329 (20 GHz)
	n	3571 (50 GHz)
	p	2 (20 GHz)
	p	5 (50 GHz)

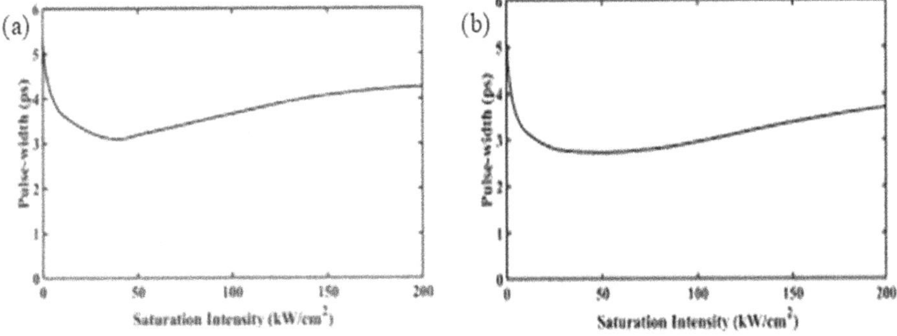

Fig. 4. Keeping the saturable absorption $\alpha_{non} = 0.5$ for both cases, the pulse-width of the mode-locked fiber laser is graphed against the saturation intensity, with the calculation done for two different non-saturable absorption: (a) $\alpha_{lin} = 0.2$ and (b) $\alpha_{lin} = 0.4$.

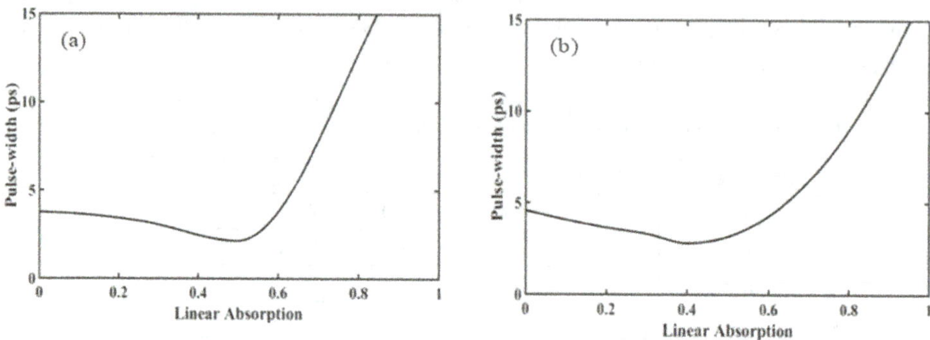

Fig. 5. The pulse-width of the mode-locked fiber laser is graphed against the linear (non-saturable) absorption α_{lin} for graphene-SA while keeping the saturable portion of the absorption constant. (a) For a saturation intensity of 40-kW/cm^2, and (b) for a saturation intensity of 100-kW/cm^2.

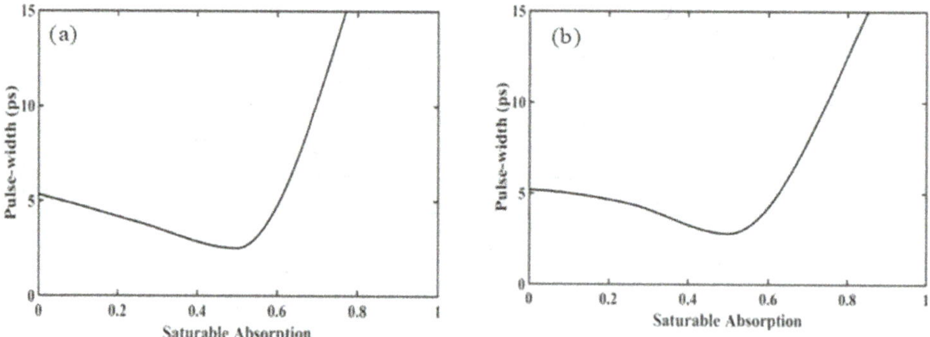

Fig. 6. The pulse-width of the mode-locked fiber laser is graphed against the nonlinear (saturable) absorption α_{non} while keeping the non-saturable portion of the absorption constant. (a) For saturation intensity of 40 kW/cm^2 (b) For saturation intensity of 100 kW/cm^2.

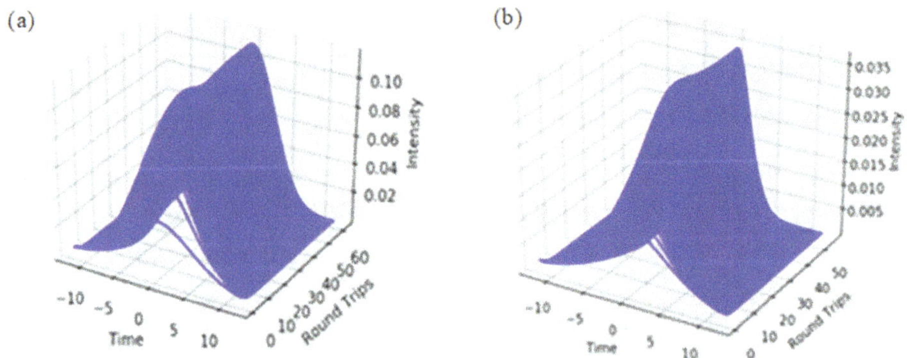

Fig. 7. Intensity vs. time vs. number of roundtrips required to stabilize the pulses simulated plotted for (a) without the presence of a saturable absorber and (b) with a saturable absorber included. Saturable absorber yields a reduction of output pulse-width, and a decrease in the number of roundtrips required to stabilize the pulses.

5. Experiment Results of Hybrid Mode-Locking

In this section, the results of using graphene nanoparticles as saturable absorbers in the optical cavity are presented. The output of the optical spectrum analyzer (OSA) is displayed in Fig. 8. Incorporating the saturable absorber removes noise and results in a smoother output pulse. Furthermore, the 3-dB bandwidth is increased, expanding to 1.4 nm from 0.8-nm for the 20-GHz pulse train and to 0.68-nm from 0.36-nm for the 50-GHz pulse train, as shown in Fig. 9. The autocorrelator trace of the output pulses is shown in Fig.10, indicating that the FWHM pulse width is shortened from 5.3-ps to 2.8-ps and 2.7-ps, respectively, for the two output pulse trains, after introducing the graphene saturable absorber. There is a good agreement between the simulation and experiment outcomes.

6. Conclusion

In conclusion, this research paper demonstrates a technique to generate high frequency pulses using rational harmonic mode locking and a saturable absorber made of graphene nanoparticles. Two separate experiments resulted in pulse trains with frequencies of 20-GHz and 50-GHz, with respective pulse widths of 2.8 ps and 2.7 ps. The study also simulated the effects of different absorbance properties, such as the linear and nonlinear absorbance components and the saturation intensity, on the pulse-width. The short pulses at high repetition rates are important for high data rate optical network studies.

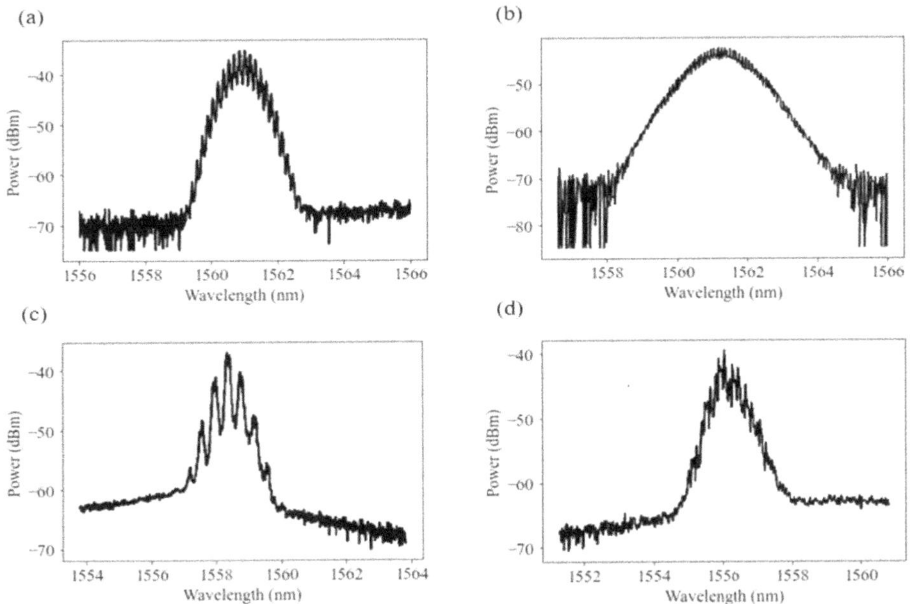

Fig. 8. OSA spectra for 20-GHz pulse train: (a) without graphene, bandwidth is 0.80-nm, (b) with graphene, 1.38-nm, and for 50-GHz pulse train: (c) without graphene, bandwidth is 0.68-nm and (d) with graphene, 0.36-nm.

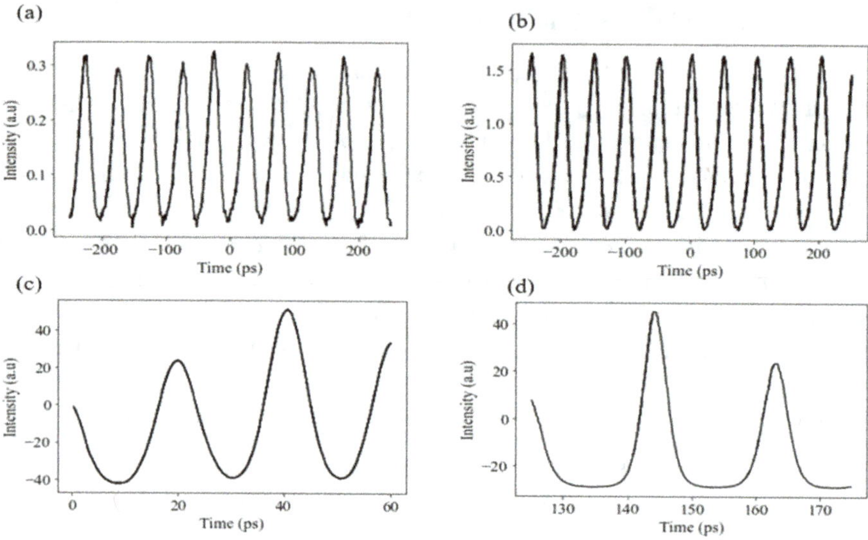

Fig. 9. Oscillator traces for 20-GHz pulse train (a) without graphene and (b) with graphene. Autocorrelator traces of 50-GHz pulse train (c) without graphene and (d) with graphene.

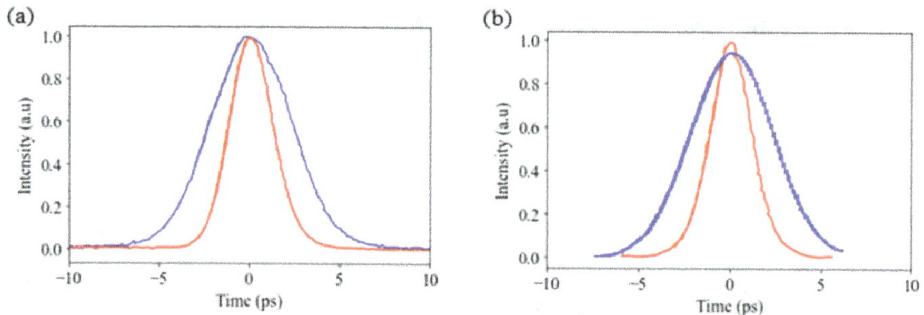

Fig. 10. FWHM pulse-widths of (a) 20-GHz and (b) 50-GHz pulse trains, decreased from 5.3-ps (RHML) to 2.8-ps and 2.7-ps, respectively (using hybrid mode-locking).

Disclosures

The authors declare no conflicts of interest.

References

1. L. Galdino *et al.*, "Optical fibre capacity optimisation via continuous bandwidth amplification and geometric shaping," IEEE Photon. Technol. Lett. **32**, 1021–1024 (2020), doi:10.1109/LPT.2020.3007591.
2. Z. Ahmed, and N. Onodera, "High repetition rate optical pulse generation by frequency multiplication in actively mode locked fiber ring laser," Electron. Lett. **32**, 455–457 (1996).
3. C. Wu, and N.K. Dutta, "High-repetition-rate optical pulse generation using a rational harmonic mode-locked fiber laser," IEEE J. Quantum Electron. **36**, 145–150 (2000).

4. A.O. Wiberg, C.S. Bres, B.P. Kuo, J.X. Zhao, N. Alic and S. Radic, "Pedestal-free pulse source for high data rate optical time-division multiplexing based on fiber-optical parametric process," IEEE J. Quantum Electron. **45**, 1325–1330 (2009).

5. W. Li, "Different methods to achieve hybrid mode locking," Cogent Phys. **6**, 1707624 (2019), doi:10.1080/23311940.2019.1707624.

6. Q. Bao, H. Zhang, Y. Wang, Z. Ni, Y. Yan, Z.X. Shen, K.P. Loh and D.Y. Tang, "Atomic-layer graphene as saturable absorber for ultrafast pulsed lasers," Adv. Funct. Mater. **19**, 3077–3083 (2009).

7. Q. Bao, H. Hang, Z. Ni, Y. Wang, L. Polavarapu, Z. Shen, Q. Xu, D. Tang and K.P. Loh, "Monolayer graphene as a saturable absorber in a mode-locked laser," Nano Res. **4**, 297–307 (2011).

8. G. Sobon, J. Sotor and K.M. Abramski, "All-polarization maintain femtosecond Er-doped fiber laser mode-locked by graphene saturable absorber," Laser Phys. Lett. **9**, 581–586 (2012).

9. W. Li, H. Hu, X. Zhang, S. Zhao, K. Fu and Niloy K. Dutta, "High-speed ultrashort pulse fiber ring laser using charcoal nanoparticles," Appl. Opt. **55**, 2149–2154 (2016).

10. J.D. Zapata, D. Steinberg, L.A.M. Saito, R.E.P. de Oliveira, A.M. Cardenas and E.A. Thoroh de Souza, "Efficient graphene saturable absorbers on D-shaped optical fiber for ultrashort pulse generation," Sci. Rep. **6**, 20644 (2016).

11. Y.W. Song, S.Y. Jang, W.S. Han and M.K. Bae, "Graphene mode-lockers for fiber lasers functioned with evanescent field interaction," Appl. Phys. Lett. **96**, 051122-1–051122-3 (2010).

12. S. Thapa, A. Rahman and N.K. Dutta, "Mode-locked fiber ring laser using graphene nano-particles as saturable absorbers," Int. J. High-Speed Electron. Syst. **31** (01n04), 2240002 (2022), https://doi.org/10.1142/s012915642240002x

13. Y.H. Lin, C.Y. Yang, J.H. Liou, C.P. Yu and G.R. Lin, "Using graphene nano-particle embedded in photonic crystal fiber for evanescent wave mode locking of fiber laser," Opt. Exp. **21**, 16763 (2013).

14. H. Hu, X. Zhang, W. Li and N.K. Dutta, "Hybrid mode-locked fiber ring laser using graphene and charcoal nanoparticle as saturable absorbers," Proc. SPIE **9836**, 983630 (2016).

15. G. Sobon, J. Sotor, I. Pasternak, A. Krajewska, W. Strupinski and K. Abramski, "Thulium-doped all-fiber laser mode-locked by CVD-graphene/PMMA saturable absorber," Opt. Exp. **21**, 12797–12802 (2013), doi:10.1364/OE.21.012797.

16. Q. Sheng, M. Feng, W. Xin, H. Guo, T. Han, Y.-G. Li, Y.-G. Liu, F. Gao, F. Song, Z.-B. Liu and J. Tian, "Tunable graphene saturable absorber with cross absorption modulation for mode-locking in fiber laser," Appl. Phys. Lett. **105**, 041901 (2014), doi:10.1063/1.4891645.

17. Y.-H. Lin, C.-Y. Yang, S.-F. Lin and G.-R. Lin, "Triturating versatile carbon materials as saturable absorptive nano powders for ultrafast pulsating of erbium-doped fiber lasers," Opt. Mater. Exp. **5**, 236 (2015), doi:10.1364/OME.5.000236.

18. Y.H. Lin and G.-R. Lin, "Kelly sideband variation and self-four-wave-mixing in femtosecond fiber soliton laser mode-locked by multiple exfoliated graphite nano-particles," Laser Phys. Lett. **10**, 045109 (2013).

19. Y.H. Lin, Y.C. Chi and G.-R. Lin, "Nanoscale charcoal powder induced saturable absorption and mode-locking of a low-gain erbium-doped fiber-ring laser," Laser Phys. Lett. **10**, 055105 (2013).

20. Z.Q. Li, C.J. Lu, Z.P. Xia, Y. Zhou and Z. Luo, "X-ray diffraction patterns of graphite and turbostratic carbon," Carbon **45**, 1686–1695 (2007).

21. U. Keller, K.J. Weingarten, F.X. Kartner, D. Kopf, B. Braun, I.D. Jung, R. Fluck, C. Honninger, N. Matuschek and J. Aus der Au, "Semiconductor saturable absorber mirrors (SESAM's) for femtosecond to nanosecond pulse generation in solid-state lasers," IEEE J. Sel. Top. Quantum Electron. **2**, 435–453 (1996).
22. G.P. Agrawal, Nonlinear Fiber Optics, 4th edn. (Elsevier, 2007); N. K. Dutta and X. Zhang "Optoelectronic Devices" (World Scientific, 2018), Appendix 3.
23. H.M. Chen, "A study of high repetition rate pulse generation and all-optical add/drop multiplexing," Ph.D dissertation (University of Connecticut, 2002), Chap. 4, p. 68.

Low Noise Gain and Index Tailored External Cavity Laser Operating at 1310 nm for Performance Enhancements of IMDD Photonic Links

R. Dougenik[*], R. Lacomb and F. Jain

ECE, University of Connecticut, Storrs, CT 06269-4157, USA
[*]robert.dougenik@uconn.edu

A novel gain and index tailored (GIT) external cavity laser (ECL) is presented. The single frequency laser demonstrates high power operation in excess of 250 mW and shot noise limited relative intensity noise (RIN) at 100 mW. This laser source is an ideal source for intensity-modulated direct detection (IMDD) photonic links. Link budget analysis is completed based on measured RIN and optical power. The analysis demonstrates that there are significant radio frequency performance advantages to operating at higher optical power while maintaining low RIN.

Keywords: Gain and index tailored (GIT); external cavity laser (ECL); microwave photonic link; intensity-modulated direct detection (IMDD); photonic links; shot noise limited relative intensity noise (RIN).

1. Background

Microwave photonic systems involve the conversion and transmission of radio frequency (RF)/microwave and millimeter wave information optical signals characterized by amplitude and phase. A typical intensity-modulated direct detection (IMDD) photonic link is illustrated in Fig. 1.

The link includes a low noise single mode laser, a balanced Mach Zehnder modulator (MZM) and a high speed photo-detector. The advantages of photonic links over conventional electronic transmission architectures (coaxial cable or waveguides) include reduced size and weight (1.7 kg/km vs 567 kg/km), reduced loss (0.5 dB/km vs 360 dB/km @ 2 Ghz) [1], insusceptibility to electromagnetic interference and large instantaneous bandwidths. Short distance links are required for wide band signal processing and manipulation of RF signals for radar and electronic warfare applications. The applicability of RF-over-Fiber links is directly related to system-level performance quantified by the sensitivity and dynamic range. Both of these system-level performance metrics are greatly dependent upon the performance of the laser source.

---

[*]Corresponding author.

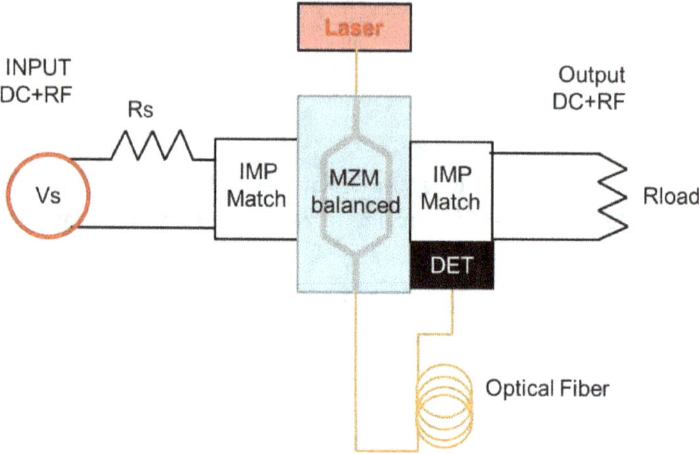

Fig. 1. Schematic of intensity-modulated direction detection (IMDD) photonic link.

The IMDD photonic link, as illustrated in Fig. 1, utilizes a balanced Mach Zehnder modulator (MZM) to convert RF inputs to an optical signal provided by the laser source. The linearity of the response is directly related to the bias control (Vdc) of the MZM, which maintains the device in quadrature and the amplitude of the RF signal (Vrf) [2, 3]. The IMDD acts as an RF power amplifier, when Vrf<<Vdc linearity is maintained and the small signal gain is given by

$$G_{rf} = \left(\frac{\Re \alpha \pi P_{Laser}}{V_{\pi}} \right)^2 Z_{in} Z_{out} \, , \tag{1}$$

where R is the responsivity of the photodetector, α is the combined link and MZM loss, V_{π} is the dc bias control voltage for quadrature, Z_{in} is the input link impedance, Z_{out} is the output link impedance and P_{Laser} is the optical power of the laser source. Maximizing the link gain requires optimization of the laser power, minimization of V_{π} and insertion loss and optimization of photodetector responsibility and power handling.

In addition to link gain, photonic links are quantified by noise via the RF noise figure and linearity-based metrics expressed by the compression dynamic range (CDR) and spurious-free dynamic range (SFDR). Three noise sources dominating the photonic link performance are the thermal noise N_{th}, detector shot noise N_{sh} and laser noise N_{Laser}. Noise levels are often expressed in terms of relative intensity noise (RIN), relative to the dc electrical power at the link output (i.e. the photodiode terminals) $Power_{dc\text{-}out} = I_{dc}^2 Z_{out}$. The dc photocurrent is directly related to the laser power $I_{dc} = 0.5 R \alpha P_{Laser}$. The RIN of the laser is expressed as follows:

$$RIN_{Laser} = \frac{N_{Laser}}{I_{dc}^2 Z_{out}} = \frac{4 N_{Laser}}{\Re^2 \alpha^2 P_{Laser}^2} \, . \tag{2}$$

The sources of laser noise are spontaneous emission, relaxation oscillations (interchange between electron/photon dynamics), mode suppression ratio, modal stability, thermal instability, pump instability and gain non-uniformities. Likewise, the output-referenced relative intensity noise of the photonic link is given by

$$\mathrm{RIN}_{Link} = \frac{N_{total}}{I_{dc}^2 Z_{out}} = \frac{N_{Laser} + N_{th} + N_{sh}}{I_{dc}^2 Z_{out}}. \tag{3}$$

The RIN of the link is thus ultimately limited by the laser source. When the laser performance is sufficient such that RIN_{Link} is shot noise limited by the noise figure (NF) is given in Eq. (8). The compression dynamic range (CDR) and the spurious-free dynamic range (SFDR) are both related to the total noise N_{total} and therefore are dependent upon the link Noise and ultimately the laser noise N_{Laser}. The 1 dB CDR is defined as the range of input powers over which the output powers are out of the noise and compressed less than 1 dB. The CDR is given by [2]

$$CDR_{linear-1HzBW} = \frac{10^{.1} P_{rf-out}^{1dB}}{N_{total}} --> CDR_{\log}(dB \cdot Hz) = P_{rf-out}^{1dB} + 1 - N_{total}(dBm / Hz), \tag{4}$$

and the SFDR is given by

$$SFDR_{linear-1Hz-BW} = \frac{OIPN^{(n-1)/n}}{N_{total}} --> SFDR_{\log}(dB \cdot Hz^{(n-1)/n}) = \frac{n-1}{n} OIPn^{(n-1)/n} - N_{total}(dBm / Hz). \tag{5}$$

All of these performance metrics can be improved by optimization of the laser source, which is accomplished by increasing the laser power P_{Laser} while maintaining shot noise limited performance provided the MZM and the photodetector can handle the increased power. The goal of this project is to optimize the laser source.

2. Identification and Significance of the Innovation

High performance laser sources for RF analog photonic applications require high power single frequency operation. The RIN of these laser sources is ultimately limited by cavity stability, Q of the cavity and output power. Continuous wave operation of a laser produces random fluctuations in carrier and photon distributions. Random carrier and photon recombination and generation events produce instantaneous time variations in the carrier and photon densities. This leads to instantaneous variations in the output power establishing a noise floor and a minimum linewidth. The coupling between the carrier density and photon density establishes a resonant oscillation peak associated with intensity noise, the RIN peak. The characteristics of the RIN spectrum are shown in Fig. 2 [3], the RIN spectrum is seen to be a function of laser output power. The RIN peak at resonance is demonstrated to decrease 9 dB with every doubling of output power. Beyond the resonant peak, the RIN noise levels out to the quantum limit. The quantum limit drops 3 dB for every doubling of the laser power, so even at powers at which the RIN peak has diminished;

further increases in power continue to reduce the laser noise. The dependency of the RIN on current density also necessitates the utilization of low noise injection sources. For shot noise limited current sources, the RIN noise floor reduces to a constant ($2e/I_{Laser}$) at a given intensity level, where e is the electron charge.

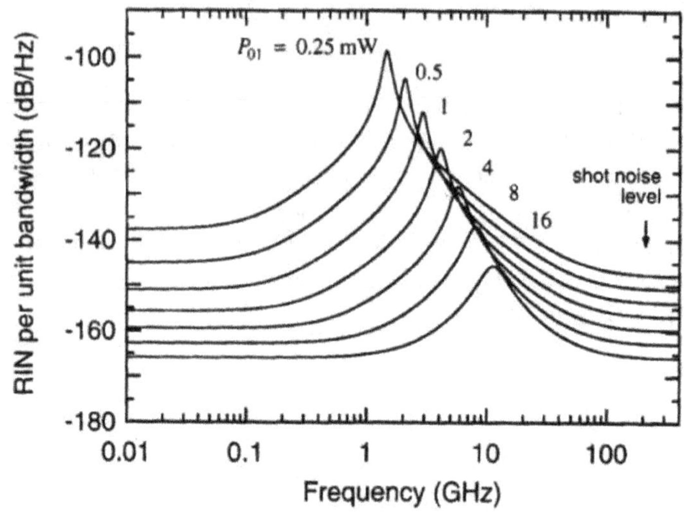

Fig. 2. Calculated RIN for different output power levels for a semiconductor in-plane laser source [3].

Conventional laser architectures (fiber lasers and solid state lasers) employed for microwave photonics utilize solid state gain media which exhibit superior noise performance compared to conventional semiconductor laser architectures. Solid state gain media inherently express reduced gain-refractive index dependencies but suffer from poor thermal quality and inefficient optical pumping leading to low-power conversion efficiencies. Solid state lasers are larger, typically more expensive and less robust than their semiconductor laser counterparts.

We have recently developed a semiconductor external cavity laser (ECL) exhibiting both high power and low RIN. The laser architecture utilizes a custom Gain and Index Tailored (GIT) gain chip and a quantum dot active layer epitaxial design. The chip is designed to support wide spatial single-mode operation; when coupled to an ECL optical circuit single frequency operation is observed. Noise factors arise from photonic and electronic-induced effects inside the semiconductor media due to gain/refractive index coupling and temperature variations. The quantum dot active layer provides improved quantum efficiency, increased thermal stability and lower thresholds over conventional quantum well structures. ECLs act to reduce phase and RIN noise by decreasing the percentage of cavity round-trip time that a photon spends inside the semiconductor media (effectively reducing the strength of this coupling) and by elongating the overall cavity length (known to be inversely proportional to the linewidth).

3. Gain & Index Tailored (GIT) Ridge Waveguide Laser

The novel Gain and Index Tailored semiconductor gain chip design is based upon a novel GIT-Ridge Waveguide (RWG) laser diode structure shown in Fig. 3.

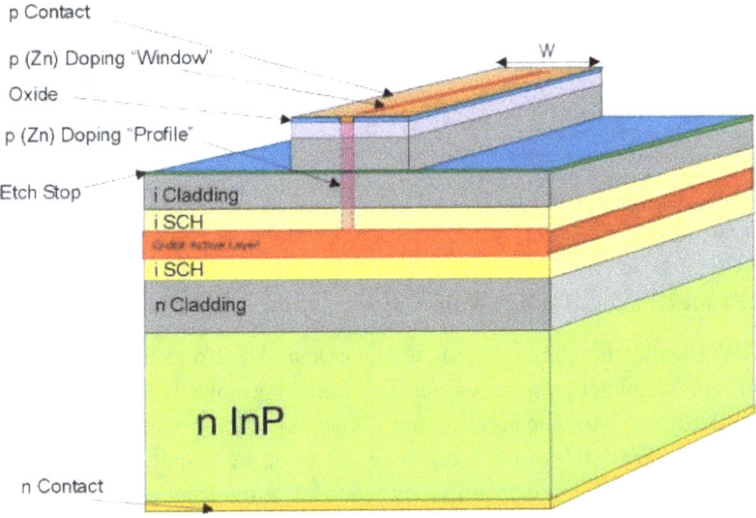

Fig. 3. Schematic gain and Index RWG semiconductor laser with quantum dot epitaxial structure.

The key design aspects of the GIT-RWG laser diode feature a quantum dot epitaxial structure and a multi-mode ridge waveguide with selective p-type doping channels to tailor the injected carrier distribution and thus the optical gain profile in the active layer to favor single mode operation. By selective p-type doping, the injected carrier profile is designed to overlap with the central portion of the fundamental mode where the intensity is greatest, thus favoring stimulated recombination (stimulated emission probability is directly related to the photon intensity) while simultaneously reducing the effects of spatial hole burning (where the gain regions overlapping with high modal intensity are depressed) thus improving noise performance. Single spatial mode widths 3–4 times wider than conventional single mode structures are realizable using this approach. Semiconductor peak power is ultimately limited by catastrophic optical damage (COD) at the facet (defined by a critical power density at which the facet melts). Wide mode laser operation enables higher output powers before the onset of COD and improved thermal performance. High-power operation is required for low RIN performance.

In conventional structures, the entire ridge is contacted and current spreading extends injected carriers to regions with low-modal intensity leading to higher probability of spontaneous photon generation along the outside edges of the ridge during lasing. Figure 4 illustrates a comparison of the modal-gain overlap for the GIT structure compared to a conventional RWG structure.

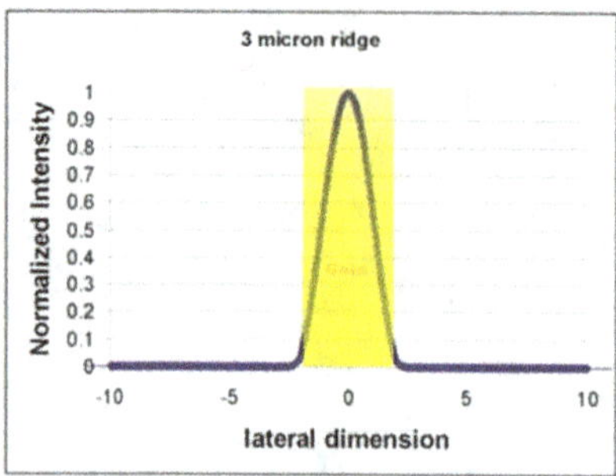

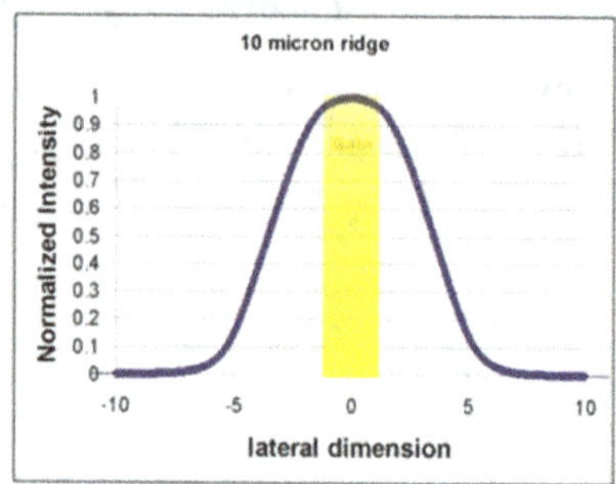

Fig. 4. Gain (yellow) modal intensity (blue) overlap for (a) conventional RWG laser structure "left" and (b) for GIT lase structure "right".

4. Gain & Index Tailored (GIT) Gain Chip

The GIT-RWG semiconductor laser provides an essential platform for the development of a low-noise semiconductor gain chip supporting both high-power operation and low noise essential to microwave photonic performance. The main difference between the diode laser device and a gain chip resides in the reflective nature of the output coupler "front facet." For a diode laser, the chip requires sufficient feedback at the front facet to support lasing. A gain chip is formed by the reduction of the front facet reflectivity (with application of an anti-reflection (AR) coating in conjunction with a curved waveguide) to reduce the optical feedback such that the laser threshold is not achieved. In this configuration, the semiconductor chip acts as a photonic amplifier, thus providing gain. A schematic of a curved waveguide GIT gain chip is given in Fig. 5. As illustrated in the schematic, the p-type dopant is offset from the center of the ridge toward the outside edge of the ridge structure. This design ensures that the carrier injection is centered at the peak of the mode which is pushed to the outside edge for a curved waveguide. The semiconductor epitaxial layers employ band gap engineering to confine carrier recombination (photon generation) to an active layer while providing a perpendicular waveguide for modal confinement.

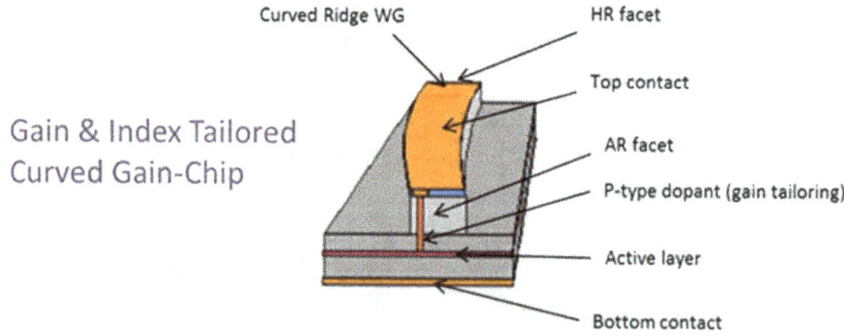

Fig. 5. Semiconductor curved ridge Gain & Index Tailored Gain Chip, HR (high reflection coating) and AR (antireflection coating).

Quantum confinement is employed by the active layer to increase the density of states for higher dynamic gain and improved linewidth.

5. Semiconductor Gain Chip External Cavity Laser

An external cavity laser consists of a gain chip housed in an optical resonator. The optical resonator is formed between the HR coating of the semiconductor gain chip and a frequency selective external mirror (typically a Fiber Bragg grating or Volumetric Bragg grating) acting as the output coupler, the AR coating and curved facet of the gain chip is design to diminish any reflections between the HR and output coupler semi-reflective mirror. In Fig. 6, a schematic of a semiconductor ECL package is shown. The package features a semiconductor gain chip coupled to a lenticulated polarization maintain (PM) single-mode fiber housing a Bragg grating. The high-reflectivity (HR) back mirror of the gain chip and the Bragg grating within the optical fiber form the laser cavity.

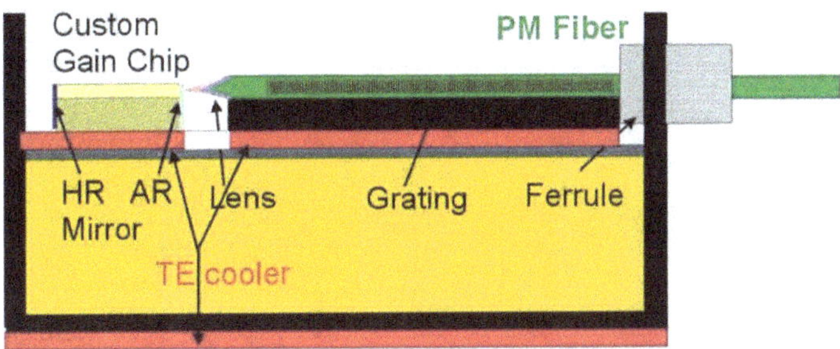

Fig. 6. External cavity laser architectures fiber Bragg grating configuration.

6. GIT-Gain Chip External Cavity Laser Prototype Results

A prototype GIT-ECL featuring a FBG, 14-pin butterfly package, PM fiber pig-tail was developed. The GIT gain chip used in this prototype featured a 5 μm wide ridge (with refractive index tailored to provide a lateral mode width > 5 μm $1/e^2$ width) with a 3 mm cavity length. Performance data are provided in Fig. 7 (power vs current vs voltage, i.e. light-current-voltage (LIV)) and demonstrates the 1310 nm operation as well as power vs current profile. Figure 8 shows the turn-on time of the laser demonstrating very quick stabilization.

7. Relative Intensity Noise Measurement

The relative intensity noise of the GIT-ECL laser was measured at NRL. The measurement involves analyzing the spectrum produced at the photodetector with an electric spectrum analyzer. The setup must be calibrated for loss as a function of frequency. Measured results for the GIT-ECL laser are given in Fig. 9. This illustrates that the RIN data for the GIT-ECL taken at 100 mW are shot noise limited. When the laser is shot noise limited, the NF is inversely related to the laser power (i.e. drops by 3 dB for every doubling of laser power). Since the RIN data shown in Fig. 9 demonstrates shot noise limited RIN @ 100 mW of laser power, the RIN is expected to decrease as the laser power is increased provided it remains shot noise limited. The projected RIN is given in Fig. 9 for the GIT-ECL operating at max power. Actual data were not taken at this level for a low noise current source and a higher power detector was not available. Low noise current sources are required for RIN measurement for current source noise can add to the laser RIN.

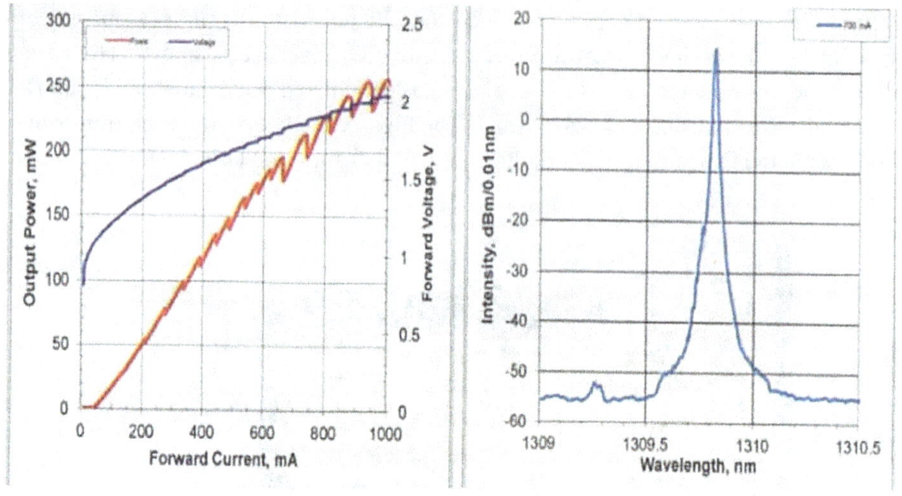

Fig. 7. GIT-ECL proto-type results, Left (Power (cw) vs Current & Voltage vs Current. Right) wavelength spectrum.

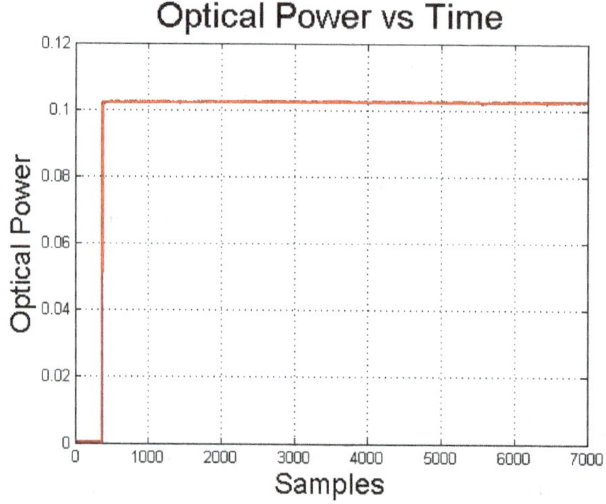

Fig. 8. Output power vs time for GIT-ECL prototype (1 sample = 0.2 s).

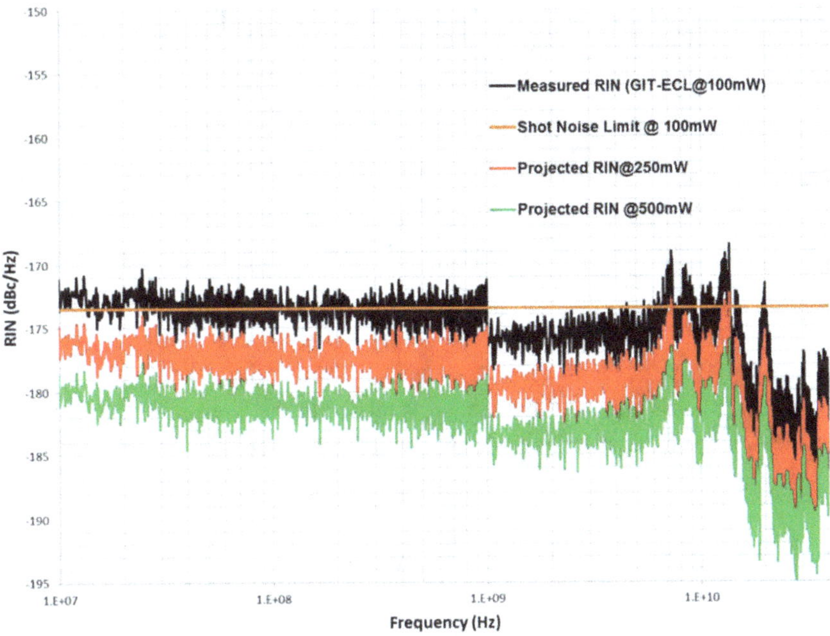

Fig. 9. Relative Intensity Noise Black measured for the GIT-ECL @ 1310 nm @ 100 mW (green). (Red) RIN projection for operation @ 250 mW, (green) RIN projected @ 400 mW.

8. IMDD Link Analysis–Projection

8.1. *Link Gain*

Link projections for an intensity-modulated direct detection photonic link are made as a function of laser power. A link performance analysis is made for the GIT-ECL assuming a commercially available modulator slope efficiency sm and photodetector responsivity R in conjunction with measured laser performance data. Utilizing these terms, the link gain is given by [5, 6]

$$g_{link} = \frac{Pload}{P_{source}} = s_m^2 \Re^2,$$

$$G_{Link} = 10\log(g_{link}). \tag{6}$$

The link gain as a function of optical power is provided in Fig. 10.

8.2. *Noise Figure*

A standard quantitative metric describing the effect of noise on a circuit is the noise factor F, the noise figure is 10Log(F). Considering the IMDD link as a 2-port network, the noise factor is the ratio of the output noise power per unit bandwidth to the input noise generated by the input source @ room temp [6]. The noise figure is given by

$$NF = 10\log\frac{SIG_{in} / Noise_{in}}{SIG_{out} Noise_{out}}. \tag{7}$$

For high input power, the IMDD link can be estimated to be shot noise limited, for the laser RIN has diminished below the shot noise of the detector (see Fig. 9). For a shot noise limited IMDD link, the noise figure can be estimated by [6]

$$NF = 10Log\left(2 + \frac{2qI_D R_{load}}{s_m^2 \Re kT} \right). \tag{8}$$

Looking at Eq. (8), it is apparent that the NF is related to the laser power and wavelength through the slope efficiency of the MZM. Projections of the NF for the GIT-ECL are provided in Fig. 10.

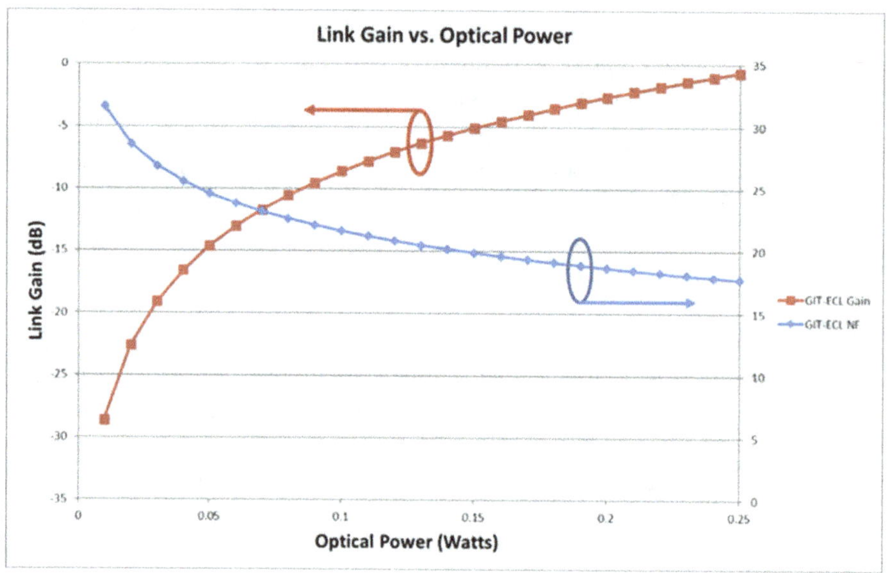

Fig. 10. Link Gain and Noise Figure Projections for an ideal Shot Noise limited IMDD photonic link as a function of laser power for GIT-ECL prototype @ 1310 nm.

9. Conclusion

A novel gain and index tailored (GIT) structure was presented incorporating quantum dot epitaxy for low-noise continuous wave operation 1310 nm. The gain and index tailored structure optimizes stimulated emission by confining the optical gain to regions of high lateral modal intensity. The quantum dot epitaxy harnesses the one-dimensional density of states limiting the linewidth, increasing the characteristic temperature to and reducing the threshold compared to quantum-well structures. The combination of these two factors works together to reduce noise sources that contribute to relative intensity noise in semiconductor lasers. The 1310 nm operating wavelength is also advantageous over 1550 nm since the modulator efficiency is inversely proportional to the square of the laser wavelength ($1/\lambda 2$) [6]. This relationship can be understood by considering the mode size as a function of wavelength. Considering a Mach-Zehnder modulator, the separation in modulation electrodes is related to the optical mode size, smaller modes allow for narrower electrode spacing and therefore increased electric field for a given voltage. Given that the radio frequency performance metrics of intensity-modulated direct detection links are related to laser power, the relative intensity noise projected performance improvements were presented based on prototype laser results and commercially available components with the exception of photodetector power handling.

References

1. A. Agarwal, T. Banwell, T. K. Woodward, "Optically Filtered Microwave Photonic Links for RF Signal Processing Applications," J. Lightwave Technol. 29, 16, 2394–2401 (2011).
2. V. Urick, P. Devgan, J. McKinney, J. Dexter, "Laser Noise and Its Impact on the Performance
3. of Intensity-Modulation with Direct-Detection Analog Photonic Links," NRL/MR/5652--07-9065, 2007.
4. L. A. Coldren, S. Corzine, M. Mashanovitch, Diode Lasers and Photonic Integrated Circuits, Second Edition, 1995.
5. K. Kikuchi, T. Okoshi, "Estimation of linewidth enhancement factor of AlGaAs laser by correlation measurement between FM and AM noises," IEEE J. Quantum Electron. 21, 6, 669–673 (1985).
6. V. Urick Jr, J. McKinney, K. Williams, "Fundamental of Microwave Photonics," ISBN978-1-118-29320-1, 2015.
7. C. Cox, "Analog Optical Links," Cambridge University Press, ISBN 0521621631.
8. R. Neuhaus, Diode Laser Locking and Linewidth Narrowing Appl-1012, TOPTICA Photonics AG.
9. Application Note: Introduction to Laser Frequency Stabilization, https://www.newport.com/n/introduction-to-laser-frequency-stabilization.
10. R. W. P. Drever, J. L. Hall, F. V. Kowalski, J. Hough, G. M. Ford, A. J. Munley, H. Ward, "Laser Phase and Frequency Stabilization Using an Optical Resonator," Appl. Phys. B 31, 97–105 (1983).
11. Estimating Laser Diode Lifetimes and Activation Energy, https://assets.newport.com/webDocumentsN/images/AN33_Laser_Diode_Activation_IX.pdf.

Compute-in-Memory SRAM Cell Using Multistate Spatial Wavefunction Switched (SWS)-Quantum Dot Channel (QDC) FET

Raja Hari Gudlavalleti[*], Evan Heller[†], John Chandy[*] and Faquir Jain[*,‡]

[*]*University of Connecticut, CT, USA*
[†]*Synopysis Inc., Ossining, NY, USA*
[‡]*faquir.jain@uconn.edu*

This paper presents multistate spatial wavefunction switched (SWS)-quantum dot channel (QDC) field-effect transistor (FET) static random access memory (SRAM)-based Compute-in-Memory (CIM) cell. The SWS-QDC FETs have two or more vertically stacked coupled quantum dot channels, and the spatial location of carriers within these channels is governed by the applied gate voltage. The location of the carriers can be utilized to encode multiple logic levels within a single device. The utilization of SWS-QDC FETs in CIM cell increases the data storage and energy-efficient computation in the memory. CIM reduces the data access time and improves performance for energy-efficient artificial intelligence (AI) edge devices.

Keywords: Compute-in-memory (CIM); multistate SRAM; SWS-FET; quantum-dot channel (QDC); artificial intelligence (AI).

1. Introduction

Compute-in-memory (CIM) technology is one of the emerging solutions towards efficient data transfer between logic processor and memory by computing within the memory macro [1]. Static random access memory (SRAM)-based CIM has attracted the attention of researchers for its access speed, write energy and compatibility with CMOS technologies. Architectures and techniques for SRAM-based CIM were reported [2, 3] using conventional MOSFETs in CMOS technology. The efficiency of the in-memory computation can be increased by handling multiple bits in the same clock cycle [4]. Earlier, in-memory computing using various NVRAMs has also been reported [5, 9]. Quantum dot (QD) NVRAMs have been suggested for multi-bit in-memory computing [6, 8]. In this work, we present spatial wavefuncion switched (SWS)-quantum dot channel (QDC) field effect transistor (FET) based memory structure for CIM applications. Jain *et al.* proposed SWS-QDC-FET that can handle multibit for computation in logic circuits and multibit storage in volatile/non-volatile memory cells [5–8]. Our approach in this paper is to increase memory density and computation ability in SRAM-based CIM cell using SWS-QDC-FETs.

[‡]Corresponding author.

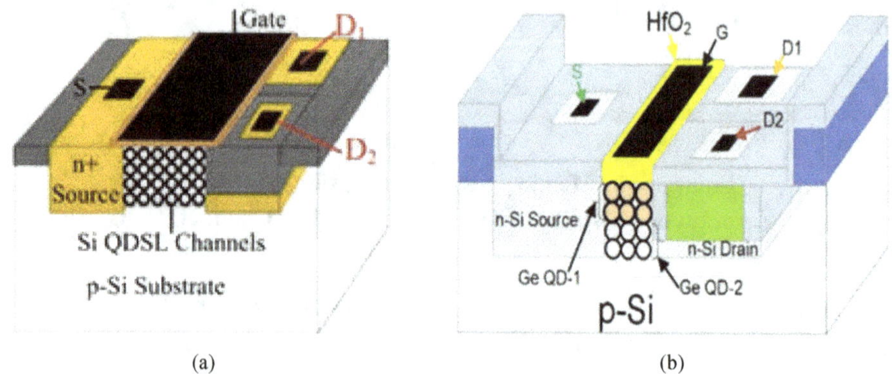

Fig. 1. Four state 2-channel spatial wavefunction switched (SWS) QDC FET with (a) SiOx-Cladded Si QDs (b) in the channel region.

SWS-QDC-FET has four layers of SiOx-cladded Si QDs and GeOx-cladded Ge quantum dots, as shown in Figs. 1(a) and 1(b), respectively, self-assembled in the recessed channel region. The upper two QD layers act as the transport channel from S and shallow drain D_1 and the lower two QD layers act as the transport channel from source, S and deep drain, D_2. The gate voltage, V_G, determines the spatial location of the charges in the upper or lower quantum-wells/-dots in the channel region. This property of the device allows encoding logic levels at each threshold level of the device for the multistate logic operation. The location of electrons in lower, upper, both and neither in the channels encodes the logic states <01>, <10>, <11> and <00>, respectively, which provides a 4-state/2-bit FET operation in logic circuits. In this paper, a conventional 8T CMOS SRAM cell [3] is modified to utilize the read-port as the dot-product multiplication cell. However, this implementation is done using SWS-QDC-FET which performs multibit storage and dot-product computation.

2. SRAM-Based CIM Circuit

The block diagram representation of SRAM-based CIM unit cell is shown in Fig. 2, which consists of a 4-state/2-bit SRAM cell and analog dot-product computation unit. The circuit is implemented using 4-state/2-bit SWS-QDC-FET/SWSFET. The circuit model of SWSFET, implementation of 2-bit SRAM and dot-product computation unit cell are presented in the following sections.

2.1. *SWSFET-based complementary 2-bit/4-state SRAM*

2.1.1. *2-bit/4-state SWSFET-based inverter*

The SWS-CMOS-based inverter, as shown in Fig. 3(a), exhibits four logic states '0', '1', '2' and '3' compared to a conventional CMOS-based inverter which exhibits two logic states '0' and '1'. Figures 3(b) and 3(c) show the DC transfer and transient characteristics

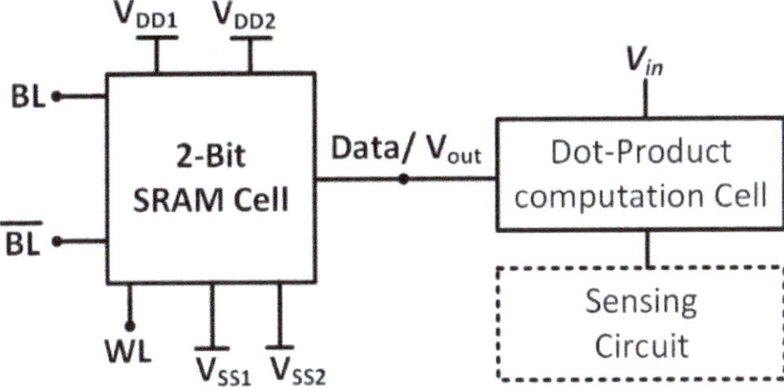

Fig. 2. Dot-product multiplier cell using four-state SRAM and two SWS-QDC transistors.

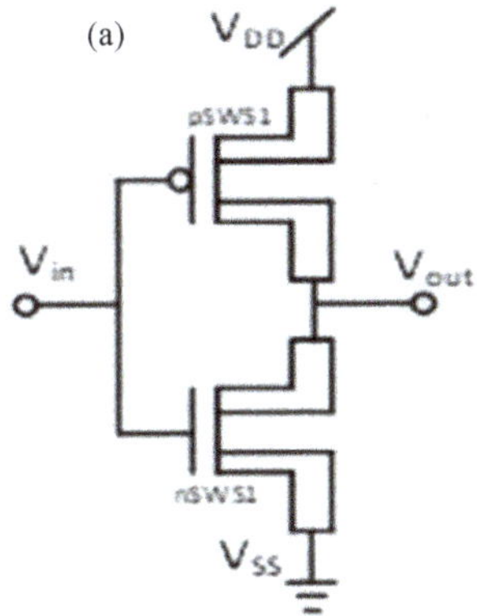

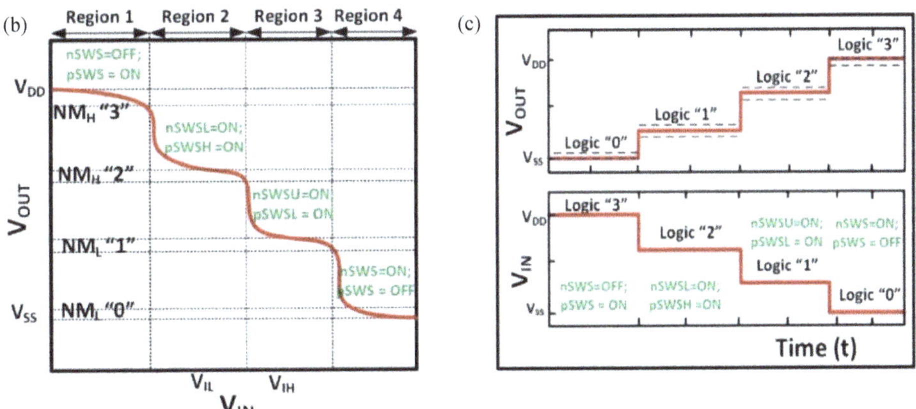

Fig. 3. SWS-CMOS inverter (a) Schematic (b) DC transfer characteristics (c) Input-output waveforms for pulse inputs.

of SWS-CMOS inverter, respectively. The characteristics indicate that an SWS-CMOS inverter can process multiple logic levels with the same number of devices, as compared to conventional CMOS-based inverter. Therefore, SWS-CMOS-based logic circuit design can perform multistate logic computation, and multistate storage in SRAM and DRAM memories. This would increase computation capability in processors and storage density in memories.

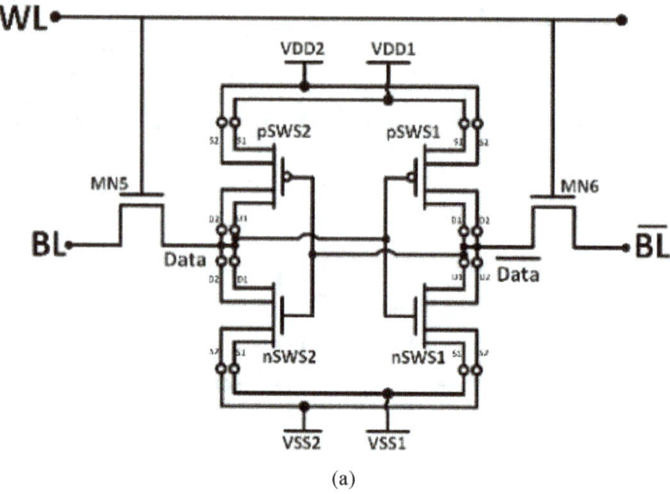

(a)

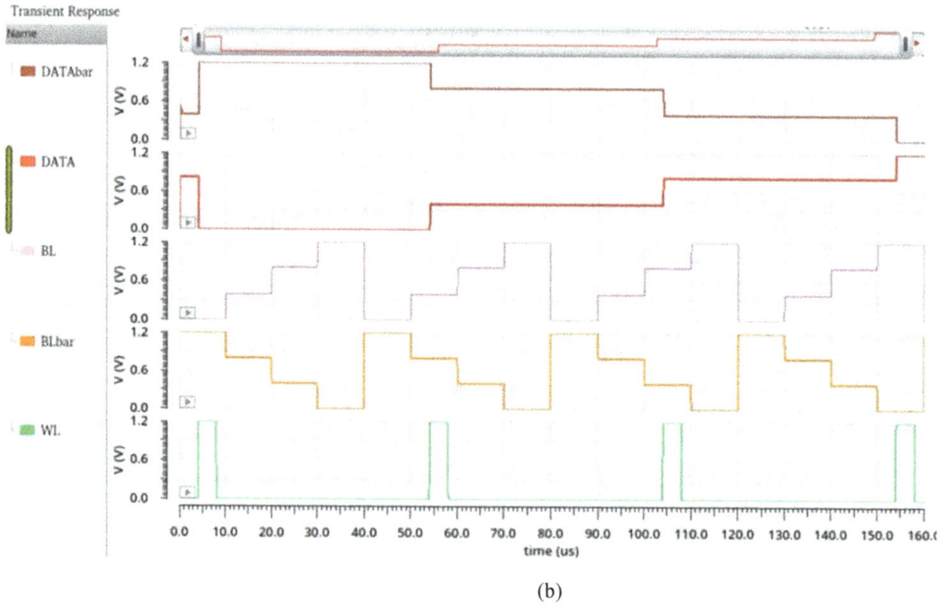

(b)

Fig. 4. (a) Circuit schematic of CMOS spatial wavefunction switching (SWS)-Static Random-Access Memory (SRAM) unit cell. (b) DATA (top panel), write pulse (middle panel) and stored DATA bar (bottom panel).

2.1.2. *2-bit/4-state SWSFET-based SRAM unit cell*

The conventional six-transistor (6T) SRAM cell consists of two cross-coupled inverters and two word-line-enabled pass transistors to read/write data from memory cell. As shown in Fig. 4(a) [7], the SRAM cell consists of SWS-CMOS inverters as the cross-coupled pair that can store multiple supply levels encoded as logic bits. Here, quaternary logic states '0' - '3' are stored in the memory cell corresponding to 0 V, 0.4 V, 0.8 V and 1.2 V voltage signals, respectively. The simulation results are shown in Fig. 4(b). When the wordline (WL) signal (green solid line) goes high, the corresponding bitline (BL) (pink solid line) and (bitline) (BL) (orange solid line) signals are transferred to the SRAM DATA (solid light red color) and (DATA) (solid dark red color), respectively.

2.2. **SWSFET-based SRAM compute-in-memory cell**

A circuit simulation model for n-SWSFET (p-SWSFET) has been implemented using two conventional n-MOS (p-MOS) transistors and a Verilog-A/analog behavior model (ABM) compact model. Each transistor represents the transport of charges from source (S2) to Deep drain (D2), represented as nmos1 and source (S1) to shallow drain (D1), represented as nmos2 governed by the gate voltage. The Verilog-A compact model accounts for the spatial location of the electron (hole) wavefunction which depends on the input gate voltage. The n-SWSFET symbol and circuit simulation model are shown in Figs. 5(a) and 5(b) respectively. Figure 5(c) shows the Cadence I_D-V_G characteristics for shallow drain D1 (green solid line) and deep drain D2 (dotted pink line). Multiple thresholds associated within the single SWSFET can be utilized to encode each state at different thresholds as indicated with state '0' to state '3'.

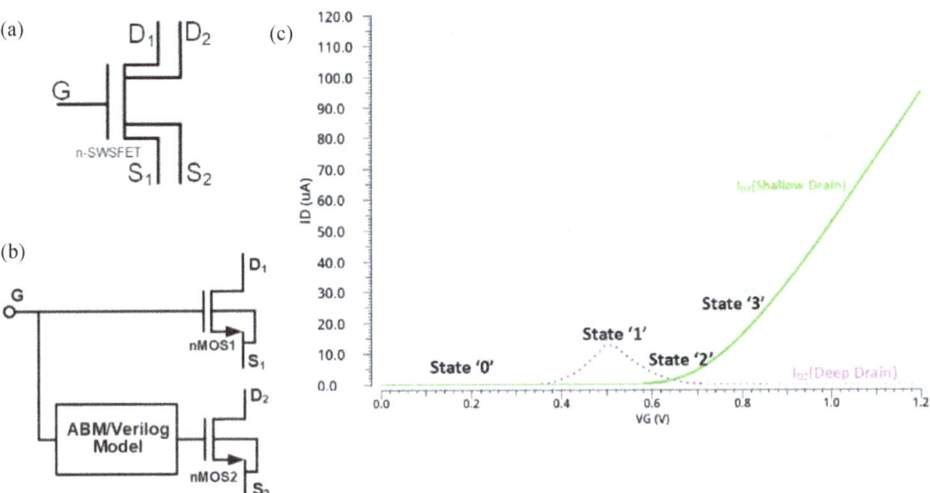

Fig. 5. An n-SWSFET (a) symbol (b) two-transistor with ABM/Verilog model (c) SWSFET IDS-VGS characteristics.

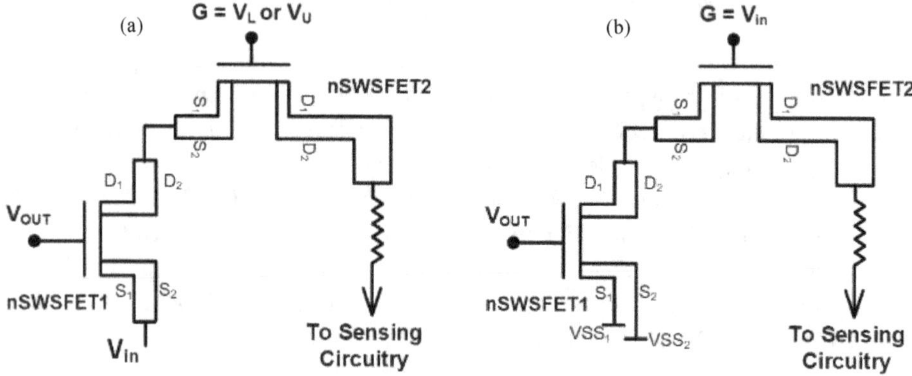

Fig. 6. SWS-QDC FET-based analog dot-product computation using two different configurations (a) and (b).

SWS-QDC FET-based analog dot-product computation using two different configurations is shown in Figs. 4(a) and 4(b). The output of the SRAM cell (V_{OUT}) has four states and the carrier transport in nSWSFET2 is based on the applied gate voltage (G). Due to multiple thresholds associated with nSWSFET, the output current to the sensing circuitry can be weighted based on the dot-product multiplication of applied input voltage and conductance associated with each input voltage to nSWSFET1 and nSWSFET2 transistors. In Fig. 6(a), the output current to the sensing is a dot product based on the applied input voltage on the source of nSWSFET1 and equivalent conductance nSWSFET1 and nSWSFET2. In Fig. 6(b), dot-product output current is the multiplication of the applied input on the gate of nSWSFET2 and equivalent conductance nSWSFET1 and nSWSFET2.

Figure 7 shows the current flow in each state for 4-state outputs (V_{out}) of CMOS SWS-QDC SRAM and input voltage (G2) applied to the dot-product cell when the dot-product multiplication cell is configured, as shown in Fig. 4(a). Similarly, Fig. 4(b) can be dealt.

2.3. *Simulation results*

The DC simulation for the I_D-V_{in} characteristics when $V_G = V_L$ and $V_G = V_U$ are shown in Figs. 8(a) and 8(b), respectively. The dotted region of the curves shows the linear output current variation with input voltage. The linearity can be improved by adding additional output analog circuitry which will be dealt with in future contributions.

3. Conclusion

This work shows compute-in-memory analog dot-product multiplication using 4-state/ 2-bit SRAM cell and two SWSFET devices. In comparison to conventional MOS-based design, this approach increases the storage density of SRAM and multibit compute-in-memory which increases the overall efficiency of the system. Other QDC-FET and QDC-QDG-FET and QD-NVRAMs may be used for CIM and In-Memory computing [8].

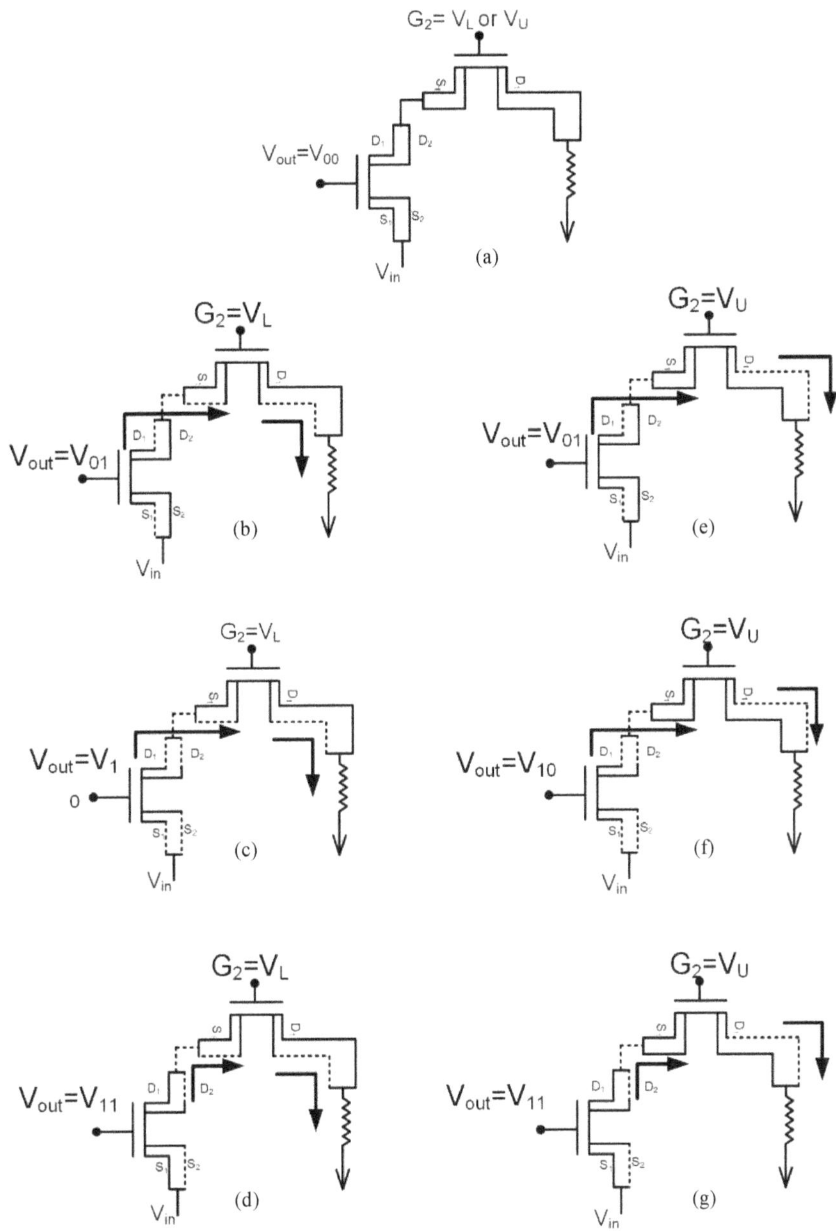

Fig. 7. The current flow in each state for 4-state outputs (V$_{out}$) of CMOS SWS-QDC SRAM and input voltage (G2) applied to the dot-product cell. [a] When SRAM output is at '00' state, the resistor output is 0V. [b-d] The schematic shows the current for SRAM output states '01', '10' and '11' when input voltage G$_2$ = V$_L$, where the charge transport happens between S$_2$ and D$_2$. [e-f] The schematic shows the current for SRAM output states '01', '10' and '11' when input voltage G$_2$ = V$_U$, where the charge transport happens between S$_1$ and D$_1$.

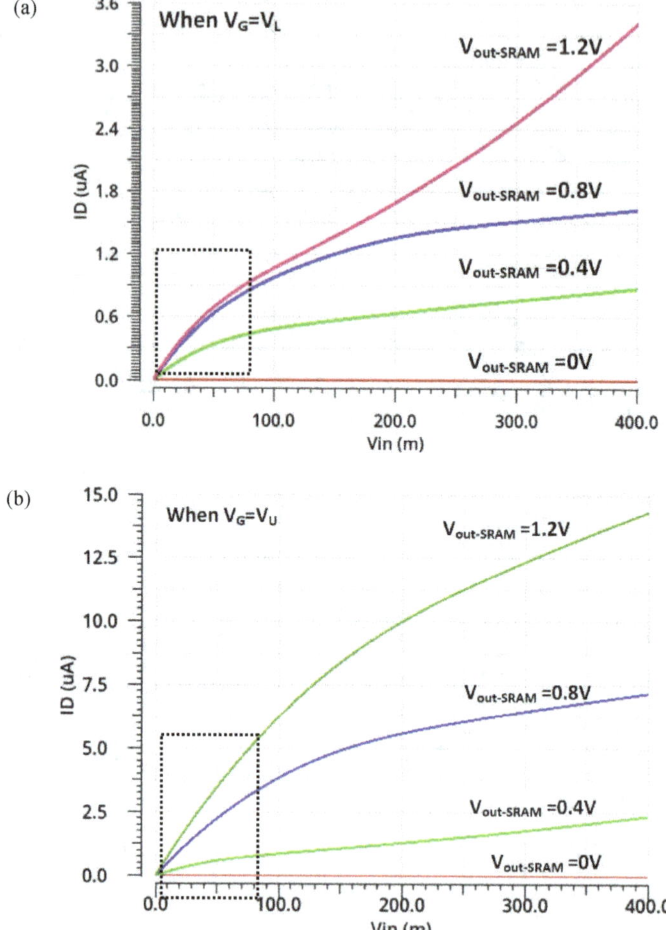

Fig. 8. The output current I_D flowing through the resistor (a) when $V_G = V_L$ (b) when $V_G = V_u$. The dotted region shows the linear region of the I_D-V_{in} curve.

References

1. H. S. Stone, A Logic-in-Memory Computer, *IEEE Trans. Comput.*, **C-19**, 73-78 (1970).
2. C.-J. Jhang, C.-X. Xue, J.-M. Hung, F.-C. Chang and M.-F. Chang, Challenges and Trends of SRAM-Based Computing-In-Memory for AI Edge Devices, *IEEE Trans. Circuits Syst. I, Reg. Papers*, **68**, 1773-1786 (2021).
3. A. Jaiswal, I. Chakraborty, A. Agrawal and K. Roy, 8T SRAM Cell as a Multibit Dot-Product Engine for Beyond von Neumann Computing, *IEEE Trans. Very Large Scale Integr. (VLSI) Syst.* **27**, 2556-2567 (2019).
4. R. Khaddam-Aljameh, P.-A. Francese, L. Benini and E. Eleftheriou, An SRAM-Based Multibit In-Memory Matrix-Vector Multiplier with a Precision That Scales Linearly in Area, Time, and Power, *IEEE Trans. on Very Large Scale Integr. (VLSI) Syst.* **29**, 372-385 (2021).

5. F. Jain, R. Gudlavalleti, R. Mays, B. Saman, P.-Y. Chan, J. Chandy, M. Lingalugari and E. Heller, Quantum Dot Channel FETs Harnessing Mini-Energy Band Transitions in GeOx-Ge and Si QDSL for Multi-Bit Computing, *Int. J. of High Speed Electron. Syst.* **31**, 2240012 (2022).

6. F Jain, R Gudlavalleti, R Mays, B Saman, J Chandy and E Heller, Integrating QD-NVRAMs and QDC-SWS FET-Based Logic for Multi-Bit Computing *Int. J. of High Speed Electron. Syst.* **31**, 2240020 (2022).

7. R. H. Gudlavalleti, B. Saman, R. Mays, Evan Heller, J. Chandy and F. Jain, A Novel Peripheral Circuit for SWSFET Based Multivalued Static Random-Access Memory, *Int. J. of High Speed Electron. Syst.* **29**, 2040010 (2020).

8. F. Jain, B. Saman, R. Gudlavalleti, R. Mays, J. Chandy and E. Heller, Low-Threshold II–VI Lattice-Matched SWS-FETs for Multivalued Low-Power Logic. *J. Electron. Mater.* **50**, 2618-2629 (2021).

9. D. Lelmini and H.-S. P. Wong, In-memory computing with resistive switching devices, *Nat. Electron.* **1**, 333-337 (2018).

High Speed 1550 nm Indium Gallium Arsenide-Indium Phosphide Photodetector

Erik Perez, Ronald LaComb[*] and Faquir Jain[†]

*Department of Electrical and Computer Engineering,
University of Connecticut, Storrs, CT, USA*
[*]ronald.lacomb@uconn.edu
[†]faquir.jain@uconn.edu

This paper presents preliminary results of a high speed 1550 nm indium gallium arsenide (InGaAs)-based mesa-type modified uni-traveling carrier photodiode (M-UTC-PD) structure. Conventional UTC-PD refers to P-I-N type photodiodes which selectively use electrons as active carriers. Photons absorbed in the relatively thin P-type absorber create minority carriers which are field accelerated toward a depleted collector thereby establishing high velocity ballistic transport, making these structures applicable for high speed applications. The M-UTC-PD structure presented uses spatially tailored P-type absorber regions to limit minority carrier generation both in the lateral and axial dimensions. Utilizing an otherwise conventional UTC-PD epitaxial structure where the top P-type layers are undoped, the spatially tailored P-type regions are defined by closed ampoule Zinc diffusion techniques. The M-UTC-PD structure presented utilizes a series of nested p-doped rings within a mesa structure to limit dark current and reduce overall capacitance to improve high speed operation. Two photodiode structures will be investigated for this research project, a conventional UTC-PD structure and a modified structure, utilizing similar device designs, epitaxial designs and fabrication processes. The conventional structure will be utilized for fabrication process development, verification of epi quality and development of rapid prototyping approach toward chip-based testing and subsequent high speed RF testing procedures. Conventional UTC-PD device results will be used as a comparison to quantify the performance of the M-UTC-PD structure utilizing Zn-doped defined p-type absorber regions. Results are given for chip tests of UTC-PD chips verifying epitaxial quality and fabrication process, subsequent testing of packaged devices and RF analysis remains. Process development of the Zn-doped devices is underway, once completed, these devices will be compared to the base design to quantify performance enhancement associated with the modified design.

Keywords: Uni-traveling photodiode; modified uni-traveling photodiode; rapid prototyping.

1. Introduction

A photodiode's main purpose is to accurately measure the intensity of incident light upon it [1, 3] converting optical signals to electrical signals enabling measuring of encoded information. Photodiodes are used in various applications including sensing, optical communications, fiber optic systems like radio frequency over fiber and high speed architectures and photonic wave generators, photomixers and receivers [2–4, 6, 7]. In

---

[*]Corresponding author.

general, bandwidth and efficiency dictate the function capabilities of a high speed photo-detector due to the tradeoff between the two characteristics [4, 5].

The efficiency and responsiveness of a diode are dependent on the characteristics it demonstrates based on the current and capacitance of the device [3]. The total current of a photodiode is dependent on the current generated by incident light on the device creating electron–hole pairs and the current that is generated in the absence of incident light, also known as dark current.

2. Conventional Photodiode Structure and Operation

The standard photodiode follows a P-I-N structure where there is a p-type material, an intrinsic material, and an n-type material [5]. Here, the p-type layer and the n-type layer are separated by the intrinsic layer where photogenerated carriers are generated, these carriers are immediately swept out of the layer by the depletion electric field, whereby the electrons are drawn to the cathode and holes move towards the anode creating a photocurrent [5]. High speed operation of conventional P-I-N photodiode structures is inherently limited by the hole mobility limiting the velocity upon which the carriers are swept out of the depleted region.

3. Uni-traveling Carrier Photodiode Structure and Operation

Uni-traveling photodiodes (UTC-PD) are P-I-N type photodiodes which selectively use electrons as active carriers [5]. This is due to the holes being the majority carriers in the absorption region that have a relatively small hole energy relaxation time [3]. This means the photoresponse is mainly dictated by the minority carriers, electrons. In the device fabrication during the epitaxial growth there is a physical separation of the absorber region and the collector region [3–5]. By separating these two regions the overall time it takes for the electron to travel is lessened because the electric field generated by absorption is very small so the electron current is mainly generated by diffusion. Once light is incident on the n-type material, photons are absorbed through the relatively thin P-type absorber to create minority carriers which are field accelerated toward a depleted collector establishing high velocity ballistic transport. This occurs due to the reverse biased voltage that is applied so it sweeps only the electrons out of the absorber region and into the collector region. Ballistic transport is needed for superior photoresponse performance due to having electrons travel at high speeds in any given medium without scattering or colliding with other electrons.

3.1. *Research Approach*

Two photodiode structures will be fabricated, a conventional UTC-PD featuring a p-mesa type device structure defining the absorber area and a modified structure utilizing tailored p-type absorbing regions within mesa-defined structures. Both device designs will utilize very similar epitaxial designs and device geometries to enable performance comparison.

The conventional structure will be used to develop the fabrication process and establish a baseline performance metric upon which the modified structure will be compared.

The conventional UTC-PD device structure is illustrated in the bottom left of Fig. 1. The device features a mesa structure to be formed extending into the epitaxial layers to expose the buried N-type layers. Several different designs were made varying the p-mesa area for comparison. The conventional UTC-photodiode epitaxial structure consists of 14 layers as shown in Fig. 1 top left. The epitaxial structure is tailored for $1.55\mu m$ operation by selecting appropriate mole fractions for In (x) and As (y) while maintaining the appropriate lattice constants for an InP substrate, as illustrated in Fig. 1 [8]. Selective dopants of the layers, p-type with Be and n-type with Si establish the PIN photodiode device structure. The device structure utilizes a mesa etch to isolate the current injection into the absorbing layers and to allow contacts to be made to the buried n-type layers, a schematic is also shown in Fig. 1.

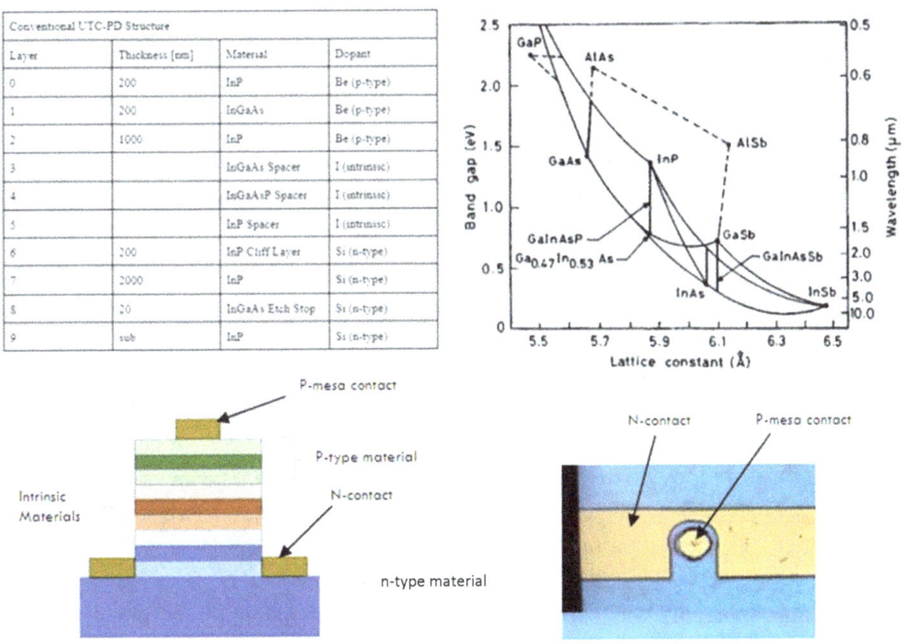

Layer	Thickness [nm]	Material	Dopant
0	200	InP	Be (p-type)
1	200	InGaAs	Be (p-type)
2	1000	InP	Be (p-type)
3		InGaAs Spacer	I (intrinsic)
4		InGaAsP Spacer	I (intrinsic)
5		InP Spacer	I (intrinsic)
6	200	InP Cliff Layer	Si (n-type)
7	2000	InP	Si (n-type)
8	20	InGaAs Etch Stop	Si (n-type)
9	sub	InP	Si (n-type)

Fig. 1. Top Left: Epitaxial Layer Design for UTC-PD. Top Right: Band Gap as a function of lattice constants for III-V Compounds [8]. Bottom Left: Schematic of the device structure. Bottom Right: Picture of a photodetector prototype.

Fabrication of the device began with definition of the anode −ve p-type contacts by a liftoff process of sputtered metals consisting of sputtered Ti, Pt and Au. Then an overlapping mesa structure was established using reactive ion etching (RIE) as a dry etch down to the n-type with 6-7 layers in the epitaxial growth shown in Fig. 1 followed by a wet etch finish, exposing the buried etch stop layer number 8 used for contacting the buried

N-type layers. Cathode contacts were formed by patterning 300 nm of PECVD silicon oxide utilizing exposing contact windows by etching with a wet etch process (Buffered Oxide Etch (BOE)) as the etchant. A lift-off process was utilized to form the n-type contacts which consisted of a multi-layer stack of: Ge, Au, Ni and Au. An example of a prototype device is shown in the bottom right of Fig. 1. The device samples were then annealed for contact alloy formation at 420 degrees for 30 seconds utilizing a Rapid Thermal Anneal process, this was followed by substrate lapping and polishing to 200μm thickness to enable backside illumination of the devices.

4. Experimental Testing and Setup

For rapid prototyping, a chip-level test set was devised to verify chip functionality and select chips for subsequent packaging and testing. To accomplish chip-level testing, a 3D printed holder was designed to hold the sample on a glass slide as shown in Fig. 2. Below the glass slide was a mirror mounted to a 45° platform where a 1064 nm laser was used in lieu of a 1550 nm laser for chip testing for lack of availability of a collimated 1550 nm laser. 0 nm laser was aimed and reflected onto the backside of the diode in order to achieve a back-illuminated device. The two probes were placed on the *p*-type and *n*-type gold contacts, respectively, as shown in Fig. 2. Before illumination testing began, an I-V Characteristic was generated utilizing a LabView virtual interface. Here, a reverse voltage bias was also obtained by switching the leads on the probes touching the contacts. Once the IV characteristics verified diode operation, known good chips were rested for photo-conduction. This was accomplished by connecting the device to a benchtop photodiode amplifier (Thorlabs PDA 200C) utilizing two probe connectors while illuminating the backside of the sample. Backside illumination was accomplished by directing a collimated 1064 nm laser to reflect off a mirror positioned beneath the transparent device mount, photocurrent was measured as a function of laser current. The 1064 nm laser that was applied to the device was current-controlled and varied from 0 to 750 mA and the response current was measured to two decimal places. The data points were taken with brief breaks in between in order to ensure that the photodiode was not saturating from constant exposure to the laser. The chip test measurement was accomplished at zero bias to avoid damaging the chips.

The testing apparatus was used to solely determine functionality of the devices for future packaging and RF testing. Figure 3 shows the I-V characteristics of a typical photodiode under zero illumination depicting the forward and reverse bias regions along with the breakdown region. The device shows the breakdown voltage occurring to the left of −2.5 V, the linear portion between −2.5 V and 0 V is the reverse biased and any voltages above 0 show the forward biased region. Once the device was back-illuminated, the response current could be measured based on the different intensities created by the varied current source for the laser with a zero biased voltage. In Fig. 3, as the light intensity increases so does the current response, meaning the device was detecting a change in light and responded appropriately as more power was applied. The device tested between 0 mA and 550 mA showed that no saturation had occurred as evidenced by no plateau in Fig. 3.

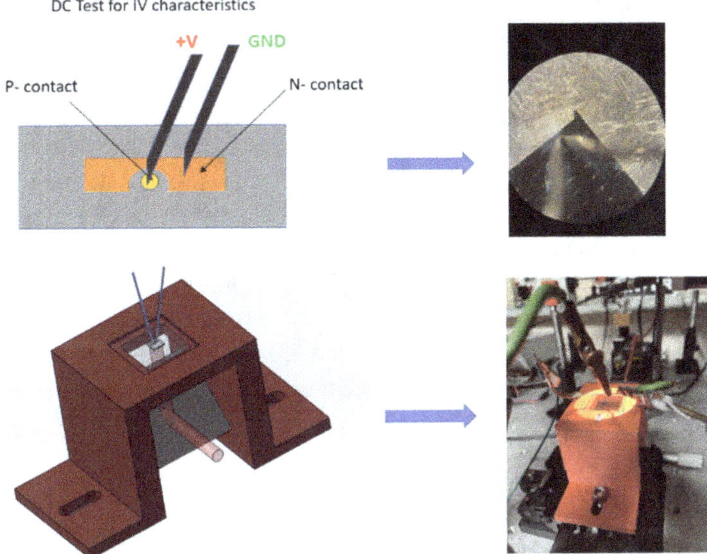

Fig. 2. Top Left: Schematic of photodetector chip with p mesa and n contact with probes. Top Right: Picture of probes shown touching the p mesa and n contact. Bottom Left: Schematic model of back-illumination test for rapid prototyping. Bottom Right: Picture of 3D Printed back-illuminated model with mirror and probes.

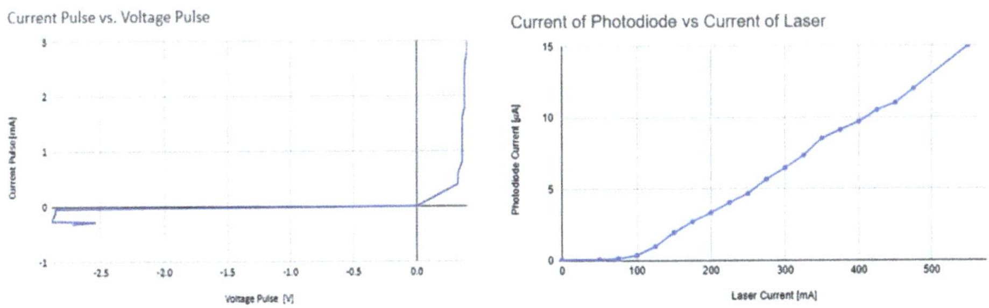

Fig. 3. Left: I-V Characteristics of a photodiode under no illumination depicting the forward and reverse bias along with the breakdown region. Right: Current of photodiode vs. the current of the laser.

5. Future Work

A change from the conventional UTC-PD would be the rapid prototyping of a modified UTC-PD that has an inclusion of zinc guard bands that reduces capacitance. By being able to lower the capacitance, the tailored p-type regions are going to be doped into the intrinsic layer creating sectioned-off portions of p-type material that will be separated by intrinsic materials. [Refer Fig. 4 for epitaxial layers]. By looking at the equation for capacitance,

$$C = A * d , \tag{1}$$

where C is capacitance, ε is permittivity of dielectric, A is area and d is the distance between parallel plates, we can determine that as the area decreases so will the overall capacitance. The spatially tailored p-type materials that were diffused into an intrinsic material reduce the area of the p-type material since it only exists in smaller regions as seen in Fig. 4. The smaller areas of Zn will appear as a series of concentric rings ranging in size so that different variations could be observed simultaneously. This design will still allow for the same number of photons to create electron–hole pairs in the depletion region as in a conventional UTC-PD that are then swept to the collector region. One way to measure the bandwidth and speed of the photodetector is to perform RF testing and use the cut-off frequency. Cut-off frequency is the measure of the photodiode response to incident light and is the frequency at which the output current decreases by 3 dB. The planned RF testing will test the conventional UTC-PDs that were already fabricated and compare them to Zinc guard band UTC-PDs.

Zinc Diffused UTC-PD Structure			
Layer	Thickness [nm]	Material	Dopant
0	200	InP	I (intrinsic)
1	200	InGaAs	I (intrinsic)
2	1000	InP	I (intrinsic)
3		InGaAs Spacer	I (intrinsic)
4		InGaAsP Spacer	I (intrinsic)
5		InP Spacer	I (intrinsic)
6	200	InP Cliff Layer	Si (n-type)
7	2000	InP	Si (n-type)
8	20	InGaAs Etch Stop	Si (n-type)
9	sub	InP	Si (n-type)

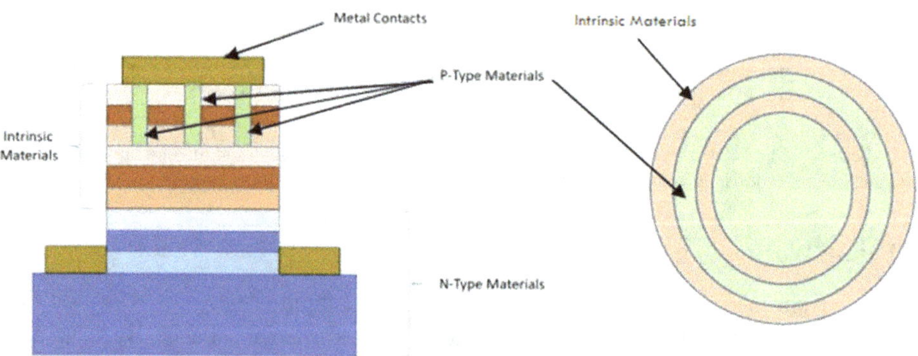

Fig. 4. Top: Epitaxial layer design for Zinc Diffused UTC-PD Structure. Bottom: Schematic of device structure (left: cross section and right: top view).

6. Conclusion

A UTC-PD epitaxial design was grown to fabricate prototype devices. A fabrication process was developed verifying the applicability of the contact lithography mask set and all subsequent process steps required to make prototype devices. A novel chip-testbed was devised to verify the functionality of devices on a pre-packaged format enabling cherry-picking of good devices for subsequent packaging and RF testing. Overall, the fabrication and testing of the UTC-PD using the process described, yielded samples housing an array of functioning devices exhibiting I-V diode characteristics and photoconductivity thereby verifying the quality of the epitaxial wafer and the fabrication process used to make the prototype devices. Future work includes RF testing of the UTC-PD devices to establish a performance baseline, fabrication of the M-UTC-PD devices, and completing a device performance analysis comparing the performance of two structures.

References

1. A. Beling, "Periodic Travelling wave photodetectors with serial and parallel optical feed based on InP", (2006).
2. J.-W. Shi, C. B. Huang and C. L. Pan "Millimeter-wave photonic wireless links for very high data rate communication", NPG Asia Mater., Vol. 3, no. 4, 41–48 (2011).
3. L. Guo, Y. Huang, X. Duan, X. Ren, Q. Wang and X. Zhang, "High-speed modified uni-traveling-carrier photodiode with a new absorber design", Chin. Opt. Lett. Vol. 10, no. S1, S12301–S12304 (2012).
4. R. Zhang, B. Hraimel, X. Li, P. Zhang and X. Zhang, "Design of broadband and high-output power uni-traveling-carrier photodiodes", Opt. Exp. Vol. 21, 6943–6954 (2013).
5. S. Srivastava and K. P. Roenker, "Numerical modeling study of the InP/InGaAs uni-traveling carrier photodiode", Solid-State Electron. Vol. 48, no. 3, 461–470 (2004).
6. T. Ishibashi, N. Shimizu, S. Kodama, H. Ito, T. Nagatsuma and T. Furuta, UTC PD in Tech.Dig.Ultrafast electron, Optoelectron, Vol. 3 (1997).
7. T. Yoshimatsu, Y. Muramoto, S. Kodama, T. Furuta, N. Shigekawa, H. Yokoyama and T. Ishibashi, Electron. Lett. Vol. 46 (2010).
8. P. D. Dutta, H. L. Bhat and V. Kumar, The physics and technology of gallium antimonide: An emerging optoelectronic material. Available from: https://www.researchgate.net/figure/Band-gap-as-a-function-of-lattice-constant-for-III-V-compounds-and-their-ternary-and_fig3_224488301.

1D and 2D Chaotic Time Series Prediction Using Hierarchical Reservoir Computing System

Md. Razuan Hossain, Anurag Dhungel, Maisha Sadia, Partha Sarathi Paul and Md. Sakib Hasan*

*Electrical and Computer Engineering Department,
University of Mississippi, Oxford, Mississippi, USA*
**mhasan5@olemiss.edu*

Reservoir Computing (RC) is a type of machine learning inspired by neural processes, which excels at handling complex and time-dependent data while maintaining low training costs. RC systems generate diverse reservoir states by extracting features from raw input and projecting them into a high-dimensional space. One key advantage of RC networks is that only the readout layer needs training, reducing overall training expenses. Memristors have gained popularity due to their similarities to biological synapses and compatibility with hardware implementation using various devices and systems. Chaotic events, which are highly sensitive to initial conditions, undergo drastic changes with minor adjustments. Cascade chaotic maps, in particular, possess greater chaotic properties, making them difficult to predict with memoryless devices. This study aims to predict 1D and 2D cascade chaotic time series using a memristor-based hierarchical RC system.

Keywords: Reservoir computing; chaotic time series; cascade chaotic maps; biomolecular memristor; logistic map; Hénon map.

1. Introduction

Neuromorphic Computing, a novel architecture that diverges from the traditional Von Neumann system, has recently gained traction in numerous applications. First proposed by Dr. Carver Mead in 1980, Neuromorphic Computing emulates biological processes through highly parallel computing architectures [1]. An Artificial Neural Network (ANN) is a computational system designed to imitate the function of biological neurons. Like the human nervous system, ANNs connect artificial neurons to one another via synapses [2]. The most prevalent types of ANNs are Feedforward Neural Networks (FNN) and Recurrent Neural Networks (RNN). FNNs have no internal loops, with connections only directed from input to output layers, capturing

---

*Corresponding author.

data at a specific moment [3]. Conversely, RNNs possess internal loops and recurrent connections, enabling data to remain in the network for a certain duration [4], allowing them to process spatial and temporal data. Although RNN connections can hold past and present data, the training algorithm's complexity increases with minor changes in recurrent connections. As a result, training such networks entails intricate processes and is considerably slow [5].

Reservoir Computing, a burgeoning neural-inspired computing approach, employs RNNs as high-dimensional reservoirs for processing intricate training data [6]. This concept is rooted in the information processing principles of the human brain, which are highly dimensional and complex [7]. A standard reservoir features sparse, extensive, and either undirected or directed recurrent nodes. The current states of the reservoir's node, which serve as the output, are determined by the input and previous states. The output node is trained using straightforward linear regression [8].

Recent studies have demonstrated that memristor-based RC systems can effectively address complex temporal problems with outstanding performance [8]. Generally, memristors are two-terminal resistive devices with memory effects. The device's current state relies on one or more internal state variables and can be altered based on the history of its previous state [9–12]. Memristors also have widespread applications in Neuromorphic computing [13–15]. Neuromorphic computing systems, which consist of artificial neurons and synapses, provide a more efficient platform for implementing neural network algorithms. Neuromemristive systems, a subset of neuromorphic computing systems, employ memristors to emulate the synaptic plasticity found in biological neurons, where the memristor conductance or current functions as synaptic weights [16, 17].

Various types of memristors are currently being incorporated into physical RC systems [18–21]. For instance, Huaqiang Wu et al. developed a solid-state memristor using a stacked cross-point structure of $Ti/TiO_x/TaO_y/Pt$ [22]. Najem et al. recently reported a biomolecular memristor with composition, structure, switching mechanism, and ionic transport similar to bio-synapses [18]. This device exhibits generic memristive properties and can perform complex computations while consuming significantly less power [8]. In this study, we utilize the **solid state memristor** model from Ref. 22 and the **biomolecular memristor** model from Ref. 18 to develop a physical RC system within a simulation platform.

Predicting chaotic nonlinear dynamics has garnered significant interest across various research domains. Chaos refers to a phenomenon in which the temporal development of a deterministic nonlinear dynamic system becomes aperiodic and highly sensitive to its initial condition [23]. Chaotic maps exhibit ergodicity and unpredictability, allowing them to generate diverse chaotic sequences with different initial conditions. Cascade Chaotic Maps (CCM), which combine two chaotic maps (seed maps), display enhanced chaotic performance and more complex chaotic properties than traditional seed maps [24, 25].

In this study, our objective is to predict cascade chaotic time series sequences using various memristor-based RC systems. We have selected the cascade 1D logistic map and the cascade 2D Hénon map for our analysis. Additionally, we present two different hierarchical RC systems as enhancements to the system. The paper is structured as follows: Section 2 explores the two distinct dimensions of Cascade chaotic time series maps; Sec. 3 outlines conventional RC architectures; Sec. 4 delves into the Hierarchical RC system; Sec. 5 presents the simulation results and an analysis of the output work; and finally, Sec. 6 concludes the paper with a summary.

2. Cascade Chaotic Maps

Cascade chaotic maps are formed by connecting two seed maps in a series, resulting in more chaotic properties than individual seed maps. The chaotic performance of CCM is assessed using bifurcation plots and the Lyapunov exponent (LE) [26]. A bifurcation diagram [27] generates and plots several steady-state sequences relative to a controlled parameter, essentially depicting the chaotic region of the corresponding map [28]. In a chaotic sequence, even slight changes in the initial condition cause two neighboring trajectories to diverge exponentially fast. LE, represented by λ, is a commonly employed metric to measure the sensitivity of a chaotic circuit to its initial condition. Negative LE values signify a periodic region [29], while positive values indicate a chaotic region [30].

Figure 1 illustrates the seed and cascaded maps. For 1D and 2D seed maps, each seed map is connected to another to construct the cascaded map.

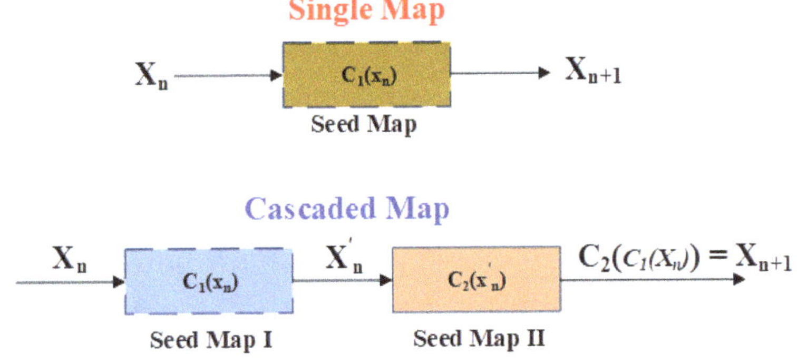

Fig. 1. Schematic of cascading scheme

2.1. *Cascade 1D Logistic Map*

The logistic map is a well-known 1D chaotic map with strong performance [31]. For input values within the range [0,1], this map generates an output sequence within the same bounds. The seed logistic map is mathematically defined as:

$$C_1(x_n) = 4ax_n(1 - x_n). \tag{1}$$

After series with another seed map, the cascaded equation becomes as follows:

$$C_2(C_1(x_n)) = x_{n+1}. \tag{2}$$

Here, a represents the bifurcation parameter between the range of 0 and 1. Figure 2 displays the bifurcation, and LE diagram [32] for both the seed and cascaded logistic maps. The LE value in the cascaded map is nearly double, indicating that it is more chaotic than the seed map. In this study, we have chosen the bifurcation parameter, $a = 1$.

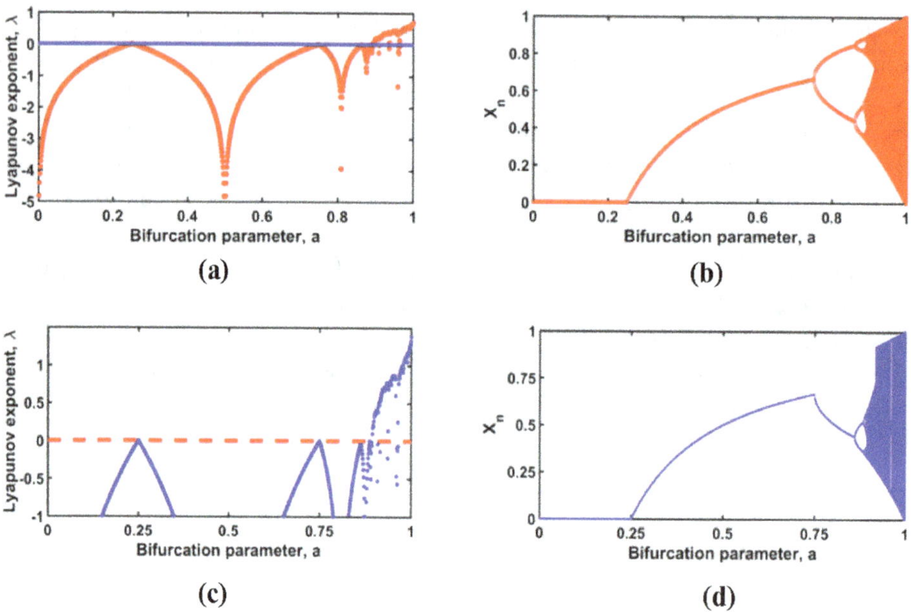

Fig. 2. Seed and Cascaded Logistic map characteristics analysis. (a) LE plot of Logistic map (seed map). (b) Bifurcation plot of Logistic map (seed map). (c) LE plot of cascaded Logistic map. (d) Bifurcation plot of cascaded Logistic map.

2.2. *Cascade 2D Hénon Map*

The Hénon map is a 2D chaotic discrete-time map that demonstrates superior chaotic behavior compared to 1D discrete maps [33, 34]. The mathematical formulation of the Hénon map [35] is explained as follows:

$$x_n = 1 - a(x_{n-1})^2 + by_{n-1}, \tag{3}$$

$$y_n = x_{n-1}, \tag{4}$$

These two equations can be written as a two-step recurrence relation as follows:

$$C_1(x_n) = 1 - a(x_{n-1})^2 + bx_{n-2}, \tag{5}$$

$$C_2(C_1(x_n)) = x_{n+1}. \tag{6}$$

Figure 3 shows bifurcation and LE diagram for X dimensions of seed and cascaded Hénon Map. In our work, we choose $a = 1.4$ and $b = 0.3$.

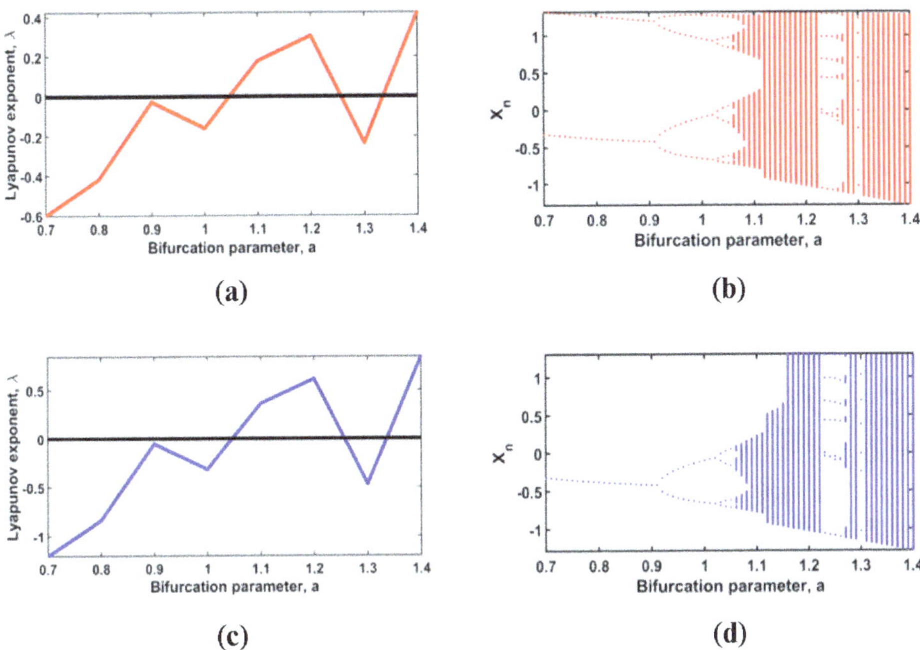

Fig. 3. Seed and Cascaded Hénon map characteristics analysis. (a) LE plot of Hénon map (seed map). (b) Bifurcation plot of Hénon map (seed map). (c) LE plot of cascaded Hénon map. (d) Bifurcation plot of cascaded Hénon map.

3. Conventional RC Architecture

In a traditional RC system (also referred to as a shallow RC system) [36], there are essentially three distinct layers: the input layer, the reservoir layer, and the output layer [8, 37]. Since our focus is on predicting 1D and 2D chaotic time series, our architecture requires increased memory and nonlinearity to capture information. To enhance the reservoir states within the network, we have employed the masking technique in our system. A comprehensive schematic of the mask-based RC system is depicted in Fig. 4.

3.1. *Input Layer*

In the pursuit of our investigation, we embarked on the generation of time series sequences encompassing 1,000 steps each. These sequences were derived from chaotic map equations, serving as the input for our analysis. We employed an additional

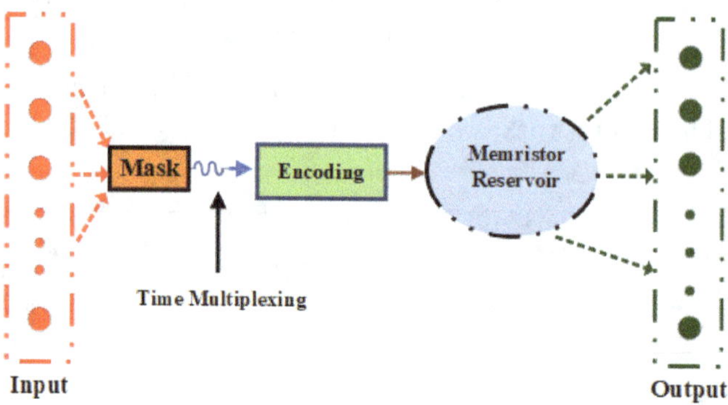

Fig. 4. Schematic of masked memristor-based RC system.

1,000-step target signal for training purposes, incorporating a one-step delay. This setup is intended to guide each data point in the time series to strive toward accurately predicting the next data point. The objective underpinning this methodology is developing a predictive model that could reliably anticipate future data points based on the current state of information.

In the subsequent testing phase, we adhered to the use of an equivalent amount of data; however, the dataset was distinct, again extracted from the chaotic map equations. By adopting this approach, we could examine our model's predictive capability maintaining the essence of the chaotic system in the input while providing new, unseen data.

3.2. *Mask*

In the pre-processing stage of input data, we utilized binary masks to introduce variability in the reservoir states. The masking process within the network serves two purposes: it renders input data sequential and maximizes dimensionality in the network. The mask functions as a time-division multiplexing system in the architecture [38]. In the masking technique, the input signal first passes through a nonlinear node, where nonlinear conversion occurs, before being directed to the delay line of the virtual nodes [3]. The resulting masked output consists of piecewise input data with a constant interval [38]. In a previous study [22], the authors employed a binary mask with a length of 4, resulting in a total of 16 fixed possible combinations. Conversely, we utilized a random mask, generating infinite combinations for the same mask length and introducing greater nonlinearity to the system. The equation for a random mask is provided below:

$$Mask = 2 * rand(N, M) - 1, \tag{7}$$

where random values are between 0 and 1, N is the number of memristors and M is for mask length.

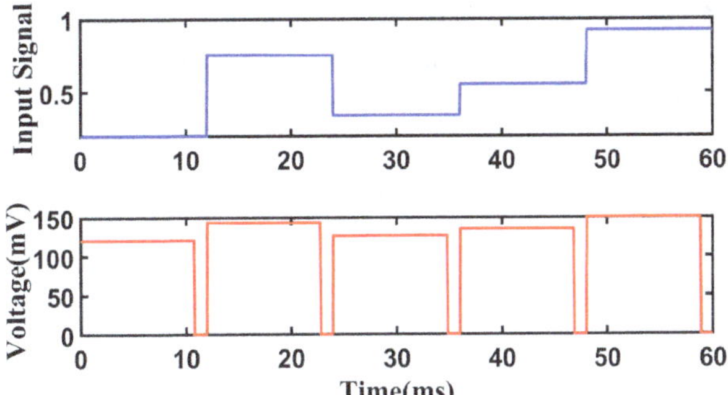

Fig. 5. Schematic of voltage encoding with respect to incoming data for biomolecular memristor. Here A portion (5 steps) of the original signal has been taken to show the encoding.

3.3. *Encoding Technique*

In this study, we implemented a voltage encoding technique where the voltage is encoded with changes in the input. Figure 5 illustrates the voltage encoding method for biomolecular memristors. With this approach, the voltage varies from 120 mV to 150 mV in relation to the data received from the mask.

3.4. *Memristive Reservoir*

A key feature distinguishing memristive elements from other elements is that their resistance changes over time, depending on the charge flowing through them [9, 39]. In our analysis, both types of memristors are voltage functions, meaning their responses vary with voltage changes. The solid-state memristors [22] operate at 2.5V, while the biomolecular memristors [40, 41] function between 100 mV and 170 mV. To construct the RC system, we utilized 32 memristors for logistic map prediction and 50 memristors for the Hénon map.

3.5. *Output Layer*

Linear Regression is a supervised machine learning algorithm employed for predicting unknown inputs [42]. In the RC network, linear regression is carried out in the output or readout layer. The objective of the readout layer is to manage the non-temporal task of mapping input data to target data. The subsequent linear equation is utilized to obtain the output weights [43]

$$W_{out} = Y_{Target} * X^{-1},\qquad(8)$$

where W_{out} is the trained weights used to predict the signal in the testing phase, Y_{Target} is the actual target signal, and X^T is the transposed output from the reservoir.

4. Hierarchical RC Architecture

In an effort to enhance prediction performance, we investigated hierarchical reservoir architecture, drawing inspiration from a relevant article [36]. Numerous studies have proposed methods to augment the reservoir's richness [44–47]. In this study, we employed wide reservoirs (parallel-connected independent sub-reservoirs) and deep reservoirs (sub-reservoirs stacked in series). A schematic of the wide RC architecture is presented in Fig. 6.

In this architecture, distinct voltage encoding is performed for the two parallel connections while maintaining the same mask parameters. In the readout layer, weights from both reservoirs are combined before executing linear regression.

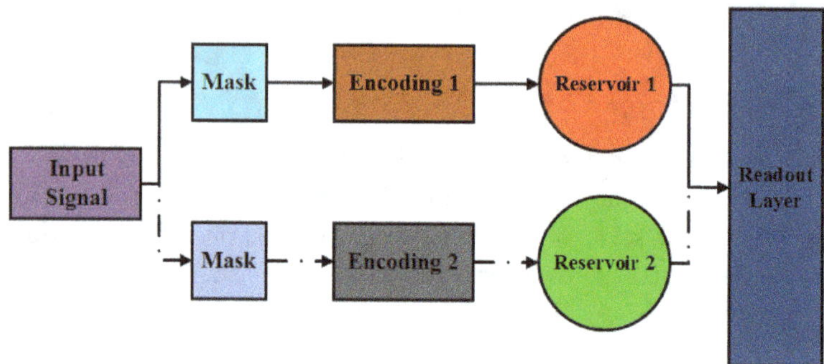

Fig. 6. The schematic of the wide RC system, where two reservoirs are parallelly connected to the readout layer.

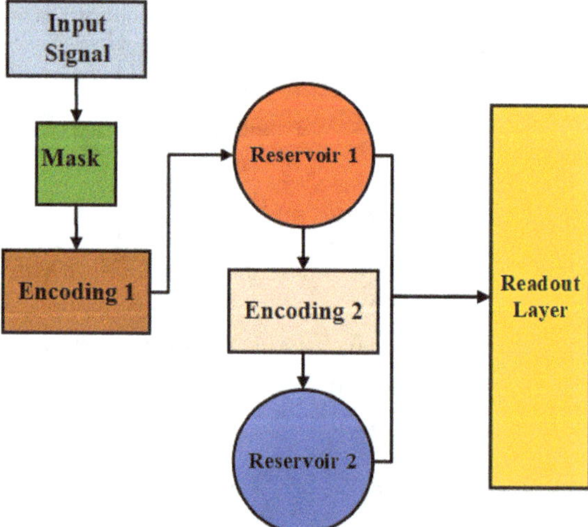

Fig. 7. The schematic of the Deep RC system, where two reservoirs are series connected and reservoir 2 depends on the output of reservoir 1.

A schematic of the deep RC architecture is displayed in Fig. 7. In this setup, data passes through reservoir1 after the initial voltage encoding. The second voltage encoding is carried out based on the output of the first reservoir, and both reservoirs are connected to the readout layer. It is important to note that we used half the number of memristors in each reservoir within the hierarchical reservoir to ensure a fair comparison across all architectures.

5. Simulation Results and Analysis

We performed training and testing on the 1D seed and cascaded logistic map as well as the 2D seed and cascaded Hénon map, utilizing three RC architectures with two types of memristors.

To compare the results, we employed NRMSE (Normalized Root Mean Square Error) in our analysis to assess the goodness of fit between the actual target signal and the predicted output signal [48, 49]. NRMSE is expressed as follows:

$$NRMSE = \sqrt{\frac{mean\left\{(y(t)_{output} - y(t)_{target})^2\right\}}{variance\left\{y(t)_{target}\right\}}} \tag{9}$$

Here $y(t)_{target}$ and $y(t)_{output}$ are the original target and predicted output signals of the chaotic map, respectively.

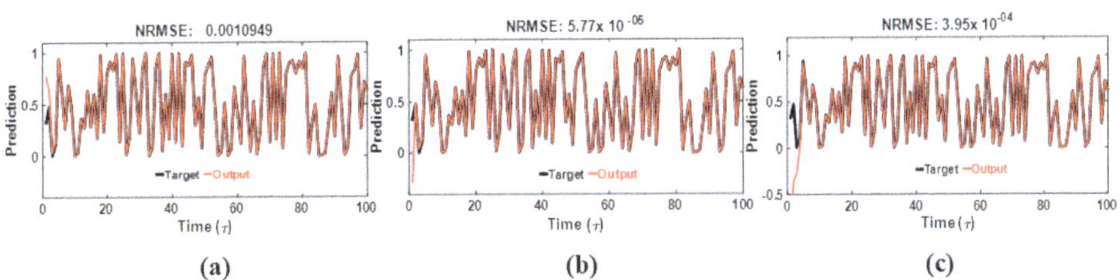

Fig. 8. The target vs. output signal results obtained by biomolecular memristive RC network for cascaded 1D Logistic map. (a) Shallow Reservoir (b) Wide Reservoir (c) Deep Reservoir

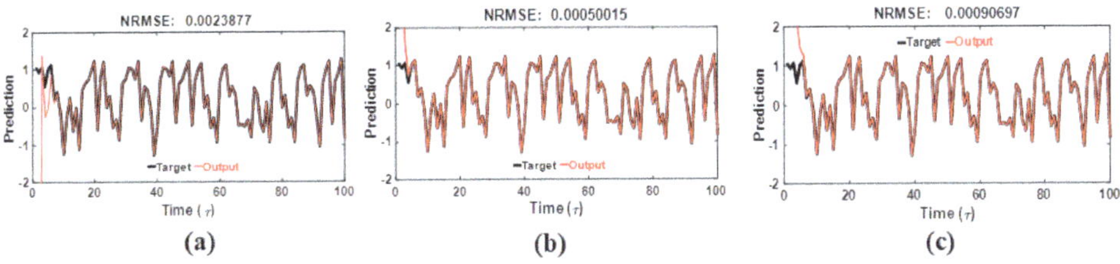

Fig. 9. The target vs. output signal results obtained by biomolecular memristive RC network for cascaded 2D Hénon map. (a) Shallow Reservoir (b) Wide Reservoir (c) Deep Reservoir

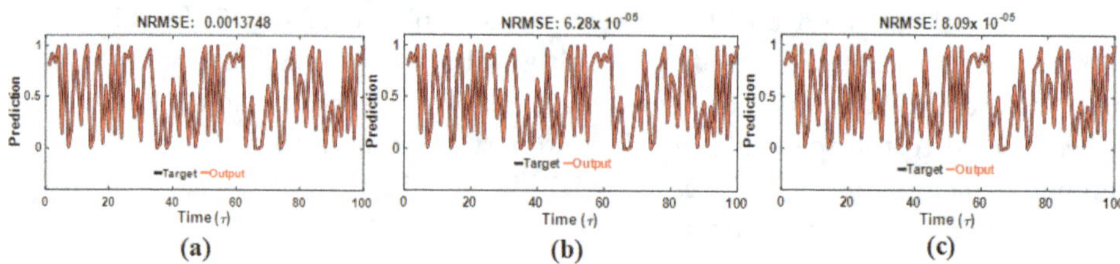

Fig. 10. The target vs. output signal results obtained by solid-state memristive RC network for cascaded 1D Logistic map. (a) Shallow Reservoir (b) Wide Reservoir (c) Deep Reservoir

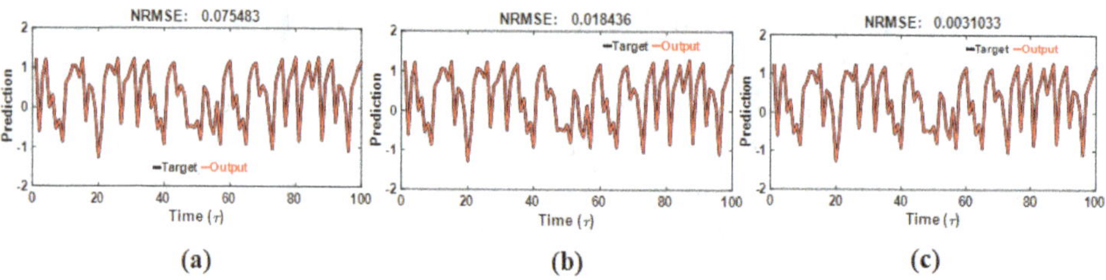

Fig. 11. The target vs. output signal results obtained by solid-state memristive RC network for 2D cascaded Hénon map. (a) Shallow Reservoir (b) Wide Reservoir (c) Deep Reservoir

Figures 8–11 display the predicted results acquired from three RC architectures using two distinct memristor types. It is evident from these figures that the hierarchical reservoir demonstrates superior performance compared to the shallow reservoir architecture. This is attributed to the hierarchical reservoir's ability to capture a broader range of temporal dynamics than the conventional reservoir [36].

To examine the nonlinear transformation necessary for the task, we explored our work with a linear network to assess the network's capacity to predict chaotic time series. In this case, we multiplied our original inputs with the following random scaling factors:

$$X_{linear} = rand(0, 1) \tag{10}$$

In the case of a linear network, there is no nonlinear transformation provided by the reservoir. Figure 12 illustrates the predictions for the 1D cascaded logistic map and the 2D cascaded Hénon map using a linear network. From these figures, it is evident that nonlinear transformation and memory properties, which are supplied by the memristive RC network, are necessary for addressing the chaotic time series problem. Additionally, we have conducted chaotic time series prediction using the Echo State Network (ESN) [48,50]. Here, we utilized the same number of reservoir nodes as in the conventional physical reservoir. Figure 13 displays the predictions for the 1D cascaded logistic map and the 2D cascaded Hénon map using a linear network. A summary of the performances of different RC networks is provided in Table 1.

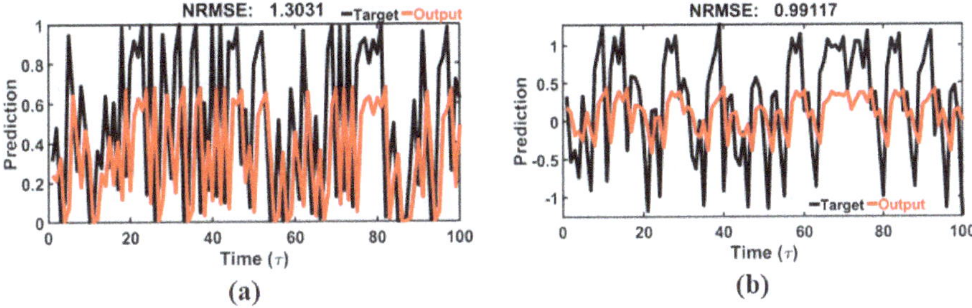

Fig. 12. The target vs. output signal results obtained by the linear network for cascaded 1D Logistic map and cascaded 2D Hénon map.

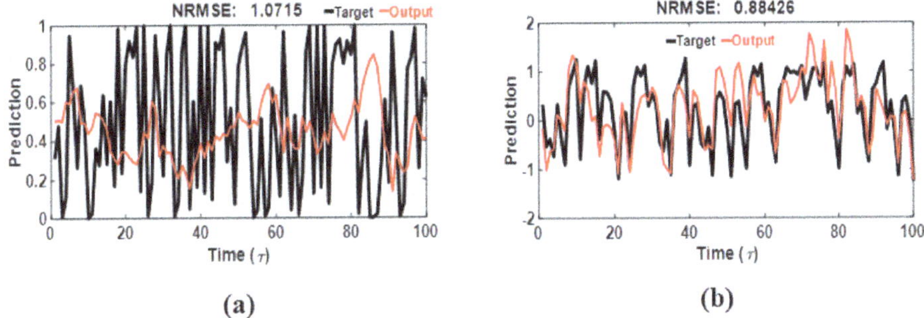

Fig. 13. The target vs. output signal results obtained by Echo State Network for cascaded 1D Logistic map and cascaded 2D Hénon map.

Table 1. Performance comparison on the different architecture of networks for 1D and 2D chaotic time series prediction.

Architecture	Logistic Map (NRMSE)	Hénon Map (NRMSE)	Cascaded Logistic Map (NRMSE)	Cascaded Hénon Map (NRMSE)
ESN	0.01192	0.01301	1.07151	0.88426
Linear Network	1.27071	1.04781	1.30311	0.99117
Biomolecular Memristor:				
Conventional RC	7.42×10^{-5}	2.33×10^{-5}	0.00109	0.00238
Wide RC	4.87×10^{-6}	1.19×10^{-6}	5.77×10^{-5}	0.00055
Deep RC	4.96×10^{-6}	9.66×10^{-6}	3.95×10^{-4}	0.00092
Solid State Memristor:				
Conventional RC	2.096×10^{-6}	0.00192	0.00137	0.07548
Wide RC	1.076×10^{-6}	0.00012	6.278×10^{-5}	0.01843
Deep RC	6.891×10^{-7}	3.421×10^{-5}	8.087×10^{-5}	0.00313

6. Conclusion and Future Work

Memristor-based RC networks have garnered significant attention recently due to their promising potential for chaotic time series prediction. In the present study, we focused on predicting both 1D seed and cascaded logistic maps as well as 2D seed and cascaded Hénon maps, utilizing a memristor-based RC network. These tasks were selected to showcase the versatility and capability of the proposed system to handle complex and nonlinear time series data.

Our results indicate that the memristor-based RC network offers a robust and efficient platform for predicting chaotic time series. The intrinsic properties of memristors, such as short-term memory and nonlinearity, contribute to the network's improved performance. The incorporation of memristors in the RC network allows for more efficient processing and storage of information, ultimately leading to better prediction results.

Moreover, we have demonstrated that employing a hierarchical RC system can further enhance prediction performance. This hierarchical approach takes advantage of the inherent properties of the memristor, enabling more accurate and reliable predictions of chaotic time series. This finding underscores the importance of nonlinear transformation in chaotic time series prediction, as it allows the system to adapt to the complex dynamics of the input data.

In addition to these findings, we have identified several areas for future work. One such area is implementing the software-based approach in hardware, enabling real-time applications of the memristor-based RC network. By transitioning from a software-based model to a hardware implementation, we expect to see significant improvements in processing speed and power consumption, thereby increasing the practicality of the system for a wide range of applications.

Another area for future research is the enhancement of prediction performance, particularly for 2D maps. While our current results are promising, there is still room for improvement. Potential strategies for enhancing prediction performance may include the development of novel memristor-based architectures, optimizing memristor properties, and incorporating advanced machine learning techniques.

In conclusion, our study demonstrates the promising potential of memristor-based RC networks for chaotic time series prediction. By predicting 1D seed and cascaded logistic maps and 2D seed and cascaded Hénon maps, we have shown the versatility and capability of this approach. Furthermore, we have highlighted the importance of nonlinear transformation and the benefits of a hierarchical RC system in enhancing prediction performance. As we continue exploring the potential applications and improvements for this technology, we are confident that memristor-based RC networks will become an essential tool in chaotic time series prediction.

References

1. Mead C., Neuromorphic electronic systems, *Proc. IEEE.* **78**(10):1629–1636, 1990.
2. Han S. H., Kim K. W., Kim S., Youn Y. C., Artificial neural network: understanding the basic concepts without mathematics, *Dement. Neurocogn. Disord.* **17**(3):83–89, 2018.
3. Appeltant L., Soriano M.C. *et al.*, Information processing using a single dynamical node as complex system, *Nat. Comm.* **2**(1):1–6, 2011.

4. Schmidhuber J., Deep learning in neural networks: An overview, *Neural Netw.* **61**:85–117, 2015.

5. Schrauwen B., Verstraeten D., Van Campenhout, J., An overview of reservoir computing: theory, applications and implementations, *Proceedings of the 15th European Symposium on Artificial Neural Networks*, pp. 471–482, 1990.

6. Verstraeten D., Schrauwen B. *et al.*, An experimental unification of reservoir computing methods, *Neural Netw.* **20**(3):391–403, 2007.

7. Maass W., Natschläger T., Markram H., A model for real-time computation in generic neural microcircuits, *Adv. Neu. Info. Process. Syst.* **15**, 2002.

8. Hossain M. R. *et al.*, Reservoir Computing System using Biomolecular Memristor, *IEEE 21st Int. Conf. on Nanotechnology* pp. 116–119, 2021.

9. Chua L., Memristor-the missing circuit element, *IEEE Trans. Circuit Theory* **18**(5):507–519, 1971.

10. Strukov D. B., Snider G. S. *et al.*, The missing memristor found, *Nature* **453**(7191):80–83, 2008.

11. Pershin Y. V., Di Ventra M., Neuromorphic, digital, and quantum computation with memory circuit elements, *Proc. IEEE* **100**(6):2071–2080, 2011.

12. Waser R., Aono M., Nanoscience and technology: a collection of reviews from nature journals, *World Scientific*, pp. 158–165, 2010.

13. Yang J. J., Strukov D. B., Stewart D. R., Memristive devices for computing, *Nat. Nanotech.* **8**(1):13–24, 2013.

14. Prezioso M. *et al.*, Training and operation of an integrated neuromorphic network based on metal-oxide memristors, *Nature* **521**(7550):61–64, 2015.

15. Sheridan P. M. *et al.*, Sparse coding with memristor networks, *Nat. Nanotech.* **12**(8):784–789, 2017.

16. Indiveri, G. *et al.*, Integration of nanoscale memristor synapses in neuromorphic computing architectures, *Nanotechnology* **24**(38):384010, 2013.

17. Thomas A., Memristor-based neural networks, *J. Phys. D: Appl. Phys.* **46**(9):093001, 2013.

18. Najem J.S., Memristive ion channel-doped biomembranes as synaptic mimics, *ACS Nano* **12**(5):4702–4711, 2018.

19. Zhong Y. *et al.*, Dynamic Memristor-based reservoir computing for high-efficiency spatiotemporal signal processing, *Resear. Squ.* 1–17, 2020.

20. Moon J. *et al.*, Temporal data classification and forecasting using a memristor-based reservoir computing system, *Nature Electron.* **2**(10):480–487, 2019.

21. Zhuo Y. *et al.*, A dynamical compact model of diffusive and drift memristors for neuromorphic computing, *Adv. Elec. Mat.* **8**(8):2100696, 2022.

22. Zhong Y., Tang J. *et al.*, Dynamic memristor-based reservoir computing for high-efficiency temporal signal processing, *Nat. Commun.* **12**(1):1–9, 2021.

23. Strogatz S. H., *Nonlinear dynamics and chaos: with applications to physics, biology, chemistry, and engineering*, CRC Press, 2018.

24. Zhou Y., Hua Z. *et al.*, Cascade chaotic system with applications, *IEEE Trans. Cyber.* **45**(9):2001–2012, 2014.

25. Tian Y., Lu Z., NS-box: LL cascade chaotic map and line map, *Ima. Graph., Springer, Cham* pp. 297–309, 2015.

26. Paul P. S. *et al.*, Design of a low-overhead random number generator using CMOS-based cascaded chaotic maps, *Proceedings of Great Lakes Symposium on VLSI*, pp. 109–114, 2021.

27. Hasan M. S. *et al.*, Design of a weighted average chaotic system for robust chaotic operation, *IEEE International Midwest Symposium on Circuits and Systems*, pp. 954–957, 2021.

28. Sadia M. *et al.*, Design and application of a novel 4-transistor chaotic map with robust performance, *International Conference on Electronic Circuits and System,* pp. 1–5, 2021.

29. Sadia M. *et al.*, Design and analysis of a multi-parameter discrete chaotic map using only three SOI four-gate transistors, *SoutheastCon.* pp. 1–7, 2021.

30. Hasan M. S. *et al.*, Integrated circuit design of an improved discrete chaotic map by averaging multiple seed maps, *SoutheastCon.* pp. 1–6, 2021.

31. Zhou Y., Bao L., Chen, C. P., Image encryption using a new parametric switching chaotic system, *Signal Process.* **93**(11):3039–3052, 2013.

32. Paul P. S. *et al.*, Cascading CMOS-based chaotic maps for improved performance and its application in efficient RNG design, *IEEE Access* **10**:33758–33770, 2022.

33. d'Alessandro G. *et al.*, On the topology of the Hénon map, *J. Phys. A: Math. Gen.* **23**(22):5285, 1990.

34. Falcolini C., Tedeschini-Lalli L., Backbones in the parameter plane of the Hénon map, *Chaos* **26**(1):013104, 2016.

35. Wen, H., A review of the Hénon map and its physical interpretations, *School of Phy. Georgia Ins. of Tech.* pp. 30332–0430, Atlanta, GA, 2014.

36. Moon J., Wu Y., Lu W. D., Hierarchical architectures in reservoir computing systems, *Neuro. Comput. Eng.* **1**(1):014006, 2021.

37. Gauthier D. J. *et al.*, Next generation reservoir computing, *Nat. Comm.* **12**(1):1–8, 2021.

38. Appeltant L. *et al.*, Constructing optimized binary masks for reservoir computing with delay systems, *Sci. Rep.* **4**(1):1–5, 2014.

39. Williams R. S., How we found the missing memristor, *IEEE Spec.* **45**(12):28–35, 2008.

40. Hasan, M. S. *et al.*, Biomimetic, soft-material synapse for neuromorphic computing: from device to network, *DCAS Conf.*, pp. 1–6, 2018.

41. Hasan, M. S. *et al.*, Response of a memristive biomembrane and demonstration of potential use in online learning, *Nanotechnology Materials and Device Conference,* pp. 1–4, 2018.

42. Roli A., Melandri L., Introduction to Reservoir Computing Methods, 2013-2014.

43. Lukoševičius M., Jaeger H., Reservoir computing approaches to recurrent neural network training, *Comput. Sci. Rev.* **3**(3):127–149, 2009.

44. Gallicchio C., Micheli A., Pedrelli L., Deep reservoir computing: A critical experimental analysis., *Neurocomp.* **268**:87–99, 2017.

45. Malik Z. K. *et al.*, Multilayered echo state machine: A novel architecture and algorithm, *IEEE Trans. Cybern.* **47**(4):946–959, 2016.

46. Gallicchio, C. *et al.*, Design of deep echo state networks, *Neural Netw.* **108**:33–47, 2018.

47. Ma Q. *et al.*, DeePr-ESN: A deep projection-encoding echo-state network, *Info. Sci.* **511**:152–171, 2020.

48. Rodan A., Tino P., Minimum complexity echo state network, *IEEE Trans. Neur. Netw.* **22**(1):131–144, 2010.

49. Herzfeld D. J., Beardsley, S. A., Neuromorphic electronic systems, *J. Neu. Eng.* **7**(4):046012, 2010.

50. Cui H., Liu X., Li L., The architecture of dynamic reservoir in the echo state network, *Chaos* **22**(3):033127, 2012.

https://doi.org/10.1142/9789811283765_0012

PCB Security Modules for Reverse-Engineering Resistant Design

Shuai Chen[*] and Lei Wang

*Department of Electrical and Computer Engineering,
University of Connecticut, Storrs, CT 06269, USA*
[*]*shuai.chen@uconn.edu*

As the crisis of confidence and trust in overseas foundries arises, the industry and academic community are paying increasing attention to Printed Circuit Board (PCB) security. PCB, the backbone of any electronic system hardware, always draws attackers' attention as it carries system and design information. Numerous ways of PCB tampering (e.g., adding/replacing a component, eavesdropping on a trace and bypassing a connection) can lead to more severe problems, such as Intellectual Property (IP) violation, password leaking, the Internet of Things (IoT) attacks or even more. This paper proposes a technique of active self-defense PCB modules with zero performance overhead. Those protection modules will only be activated when the boards are exposed to the attacks. A set of PCBs with proposed protection modules is fabricated and tested to prove the effectiveness and efficiency of the techniques.

1. Introduction

An article from Bloomberg Businessweek in 2018 stupendously claimed that overseas foundries had developed back doors to servers built for Amazon, Apple and others by inserting millimeter-size chips into circuit boards, as shown in Fig. 1. The companies involved and the U.S. Department of Homeland Security have run deep examinations and refuted the claims in the article.

Although those claims are proven to be inaccurate, the anxiety about hardware supply chain security swept the industry and academic community. People started to focus on the uncertainty of overseas foundries and tried to introduce authorization mechanisms or protections to secure the board and chips fabricated overseas [3, 4, 5].

Among all the electronic components fabricated overseas, PCBs are the most vulnerable because their large feature size makes them easy to be probed, brute force copied, or even revised for Trojan implantations and back door insertions. As a result, PCB design has been a place of no law for a long time. Competitors usually brute force copy a PCB for a shorter turn-around time (TAT) and lower design cost. In industry, the turn-around time of copying a six layers board can be as low as 24 hours; and the cost of the

---

[*]Corresponding author.

Fig. 1. Illustration of PCBs Trojan.

reverse engineering is usually marked as less than one cent per soldering point. Attackers can apply reverse engineering to PCBs to acquire the internal structure for further Trojan implantations or hacking actions. Also, as the metal traces in PCBs sometimes carry important design information, sensitive data, or critical control signals on a running device, attackers can eavesdrop, control, or disable the device.

Due to the lack and inefficiency of IP protection mechanisms and techniques, designers implicitly agreed that the design logic in PCBs is unprotected from attackers and rivals for design copying and revisions, although the design always carries important design and system information. Those PCB-based attacks and tampering are discussed in Section 2.

To deal with PCB hardware IP infringements, some countermeasures have been proposed. In [6], Paley and his co-authors introduced an active monitoring and prevention design for PCBs to defend against physical tampering. Piliposyan, Khursheed and Rossi also proposed a new power analysis method to detect alien components on a PCB, which can be regarded as a potential Hardware Trojan [7]. Yu designed an authentication system for PCBs to prevent counterfeits created by cloning or recycling [9]. In [10], the authors proposed a security module to protect circuit components from unauthorized access. Zhang introduced an authentication methodology to form a unique signature for each PCB, which can reflect the process variations in PCB traces and overall impedance using a PCBs Trace-Based Ring Oscillator [11]. The ROPA can provide both IC and PCBs authentication independently of external equipment and allows remote authentication for the user.

To the best of our knowledge, little effort has been made to counter PCB-level reverse engineering because the large feature size of PCBs' metal traces makes it extremely easy for attackers to apply reverse engineering.

Thus, this paper proposes a technique of active self-defense PCB modules based on transformable vias against reverse engineering. Those modules are realized by adding vias material pairs (magnesium (Mg) and magnesium oxide (MgO)) to the fabrication process of PCBs. Magnesium defines conducting vias, and magnesium oxide is regarded as a part of insulator vias for different metal layers. The mechanism will act from delayering through imaging, during which Mg will be oxidized into MgO, and all the vias material pairs will be identified as MgO [17, 18, 14]. Thus, attackers are unable to distinguish MgO from Mg, which will lead to a routing pattern with extra metal traces. A specific application is to place extra metal traces near a high-frequency bus line. When attackers reverse-engineer the board and mis-identify the MgO vias as conductive vias, the extra metal traces will act as the receptor of the high-frequency metal trace to generate noise in the new routing pattern (see details in Sections 3 and 4). Note that since these extra metal traces are connected to the real circuit nodes (by MgO vias), attackers will not be able to detect them as non-functional.

The remainder of this paper is briefly organized as follows: Section 2 identifies the attack models. Sections 3 and 4 explain the design of transformable-based PCBs design and its feasibility. Section 5 presents the noise-generating model. Section 7 shows the experimental result of a fabricated PCB. Section 8 draws the conclusion.

2. Attack Models

2.1. *PCBs Brute Force Copying*

Generally, the layout drawings or Gerber files, the design files for reproducing, are extracted via PCBs-level reverse engineering. The framework is shown in Fig. 4 as follows:

(1) Bare PCBs can provide some physical information, e.g., physical sizes, number of layers and accurate test points.
(2) Different methods will be conducted to acquire the layers' images. Attackers sometimes destructively remove the material of each layer to image the routing patterns underneath. Figure 2 illustrates an example of a destructive method of delayering [12]. Other approaches for material removal include wet/plasma etching, grinding and polishing. Also, X-ray scanning can serve as a non-destructive method for imaging. Figure 3 is an illustration of X-ray images of PCBs.
(3) Identify metal traces, vias and dielectric materials in the images.
(4) Translate the information identified from the images into a CAD file.
(5) Run DRC (Design Rule Check) to cancel any violation in the design file.
(6) Output the Gerber design file for PCBs reproduction.

Note that attackers can repeat each step listed above for a desirable result before the next steps.

2.2. *PCBs Hacking*

Altered component replacement: The schematic usually reflects the designer's intent and logic most accurately. However, the circuit on the PCBs is much more complicated. A minor component variation on the board can cause serious problems, such as performance drop, overheating, or even power failure.

Attackers can use a maliciously altered version of the component in production to expect damage. This attack is hard to detect as the counterfeit components look similar to the real ones. Here's an example in Fig. 5, this pair of FT232RL USB to serial UARTs seems quite similar. Still, the one on the right is a counterfeit based on a mask-programmable microcontroller and only works with older drivers [15] — a desirable result before the next steps.

Additional components/Trojan Insertion: Hardware hacks might need the inclusion of an extra, surreptitious component. The framework of Trojan insertion using non-destructive Imaging method is shown in Fig. 7. In that case, a spot on the board with many small components is the place to hide it. Modern passive components can be mere

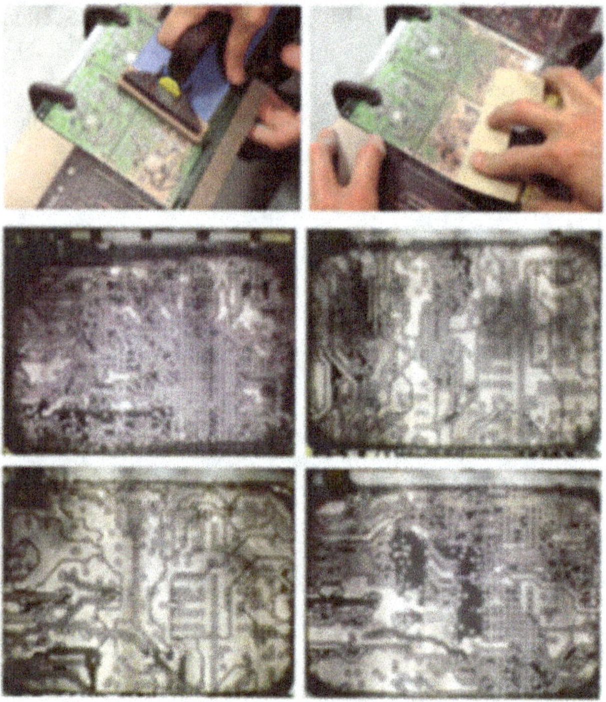

Fig. 2. Example of Destructive Imaging method.

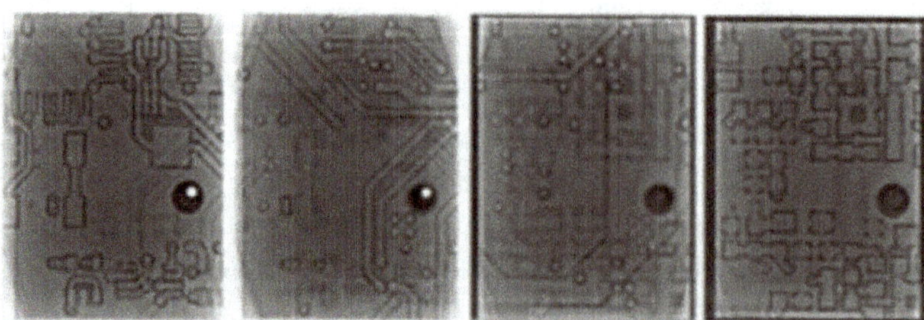

Fig. 3. Example of Non-Destructive Imaging method.

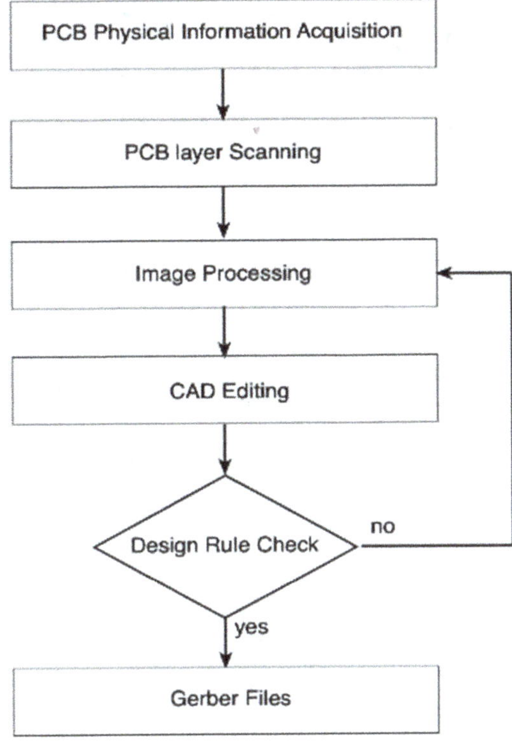

Fig. 4. PCB-Level Reverse Engineering Framework.

Fig. 5. Normal Chip and Counterfeit Chip.

millimeters in size and invisible to the unaided eyes. Here, the motherboard in Fig. 6 [13] is used as an example. The massive passive components area in Fig. 8(a) [13] can be the camouflage for Trojan or additional components. The report that triggered the community's anxiety, as shown in the figure, is an excellent example of Additional components/ Trojan Insertion.

Fig. 6. Illustration of a Motherboard.

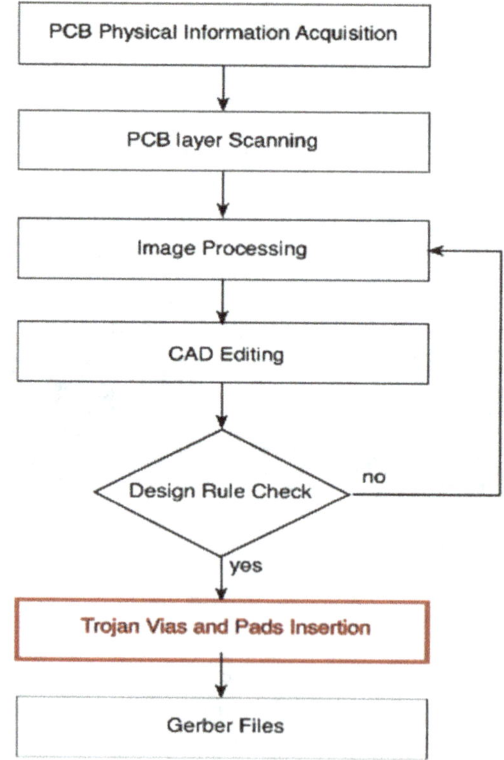

Fig. 7. Example of Trojan Insertion Using Non-destructive Imaging Method.

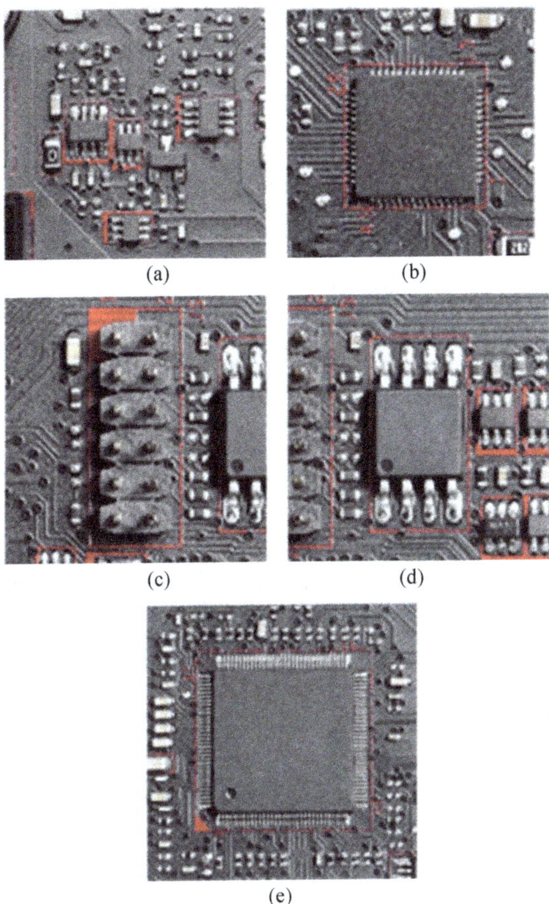

Fig. 8. Tempting Targets of the Board for Hacking: (a) Massive Passive Components Area. (b) Power Controller of the Board. (c) Low Pin Count Bus of the Board. (d) BIOS Flash Memory of the Board. (e) Super I/O Chip of the Board.

Taking control of and eavesdropping on certain data buses: Data buses usually carry important design and system information at runtime. Taking control of or eavesdropping on certain data buses means taking control of the whole system. Here, we still use Fig. 6(a) [13] as the example:

(1) The Power Controller in Fig. 8(b) [13] is a particularly fruitful target because it controls all of the DC voltages that power the CPU, the graphics card and more. It is under the control of the System Management Bus. So, if a hack enables people to seize control of the SMBus, they could reset voltages to damage a computer or limit its operation.

(2) The connector is attached to the LPC bus in Fig. 8(c) [13], which can link the CPU to specific legacy devices and the fans and physical switches on the chassis. Perhaps just

as important to hackers, the LPC bus can connect to a secure microcontroller called a Trusted Platform Module (TPM), which deals with encryption keys and various other security functions.

(3) The Basic Input/Output System (BIOS) flash memory in Fig. 8(d) [13] holds the data needed to initialize hardware during boot-up. It sits on the Serial Peripheral Interface (SPI) bus. Seizing control of the SPI bus would enable a hacker to alter hardware configurations so that a path would be open to inserting malicious code into the computer.

(4) The Super I/O chip in Fig. 8(e) [13] controls the inputs to various low-bandwidth devices, sometimes including keyboards, the mouse, specific sensors, fans and floppy disks. The chip sits on the Low Pin Count (LPC) bus. Seizing control of the LPC bus could let hackers reduce the fan speed so that a computer will overheat.

3. PCB Attacks and Reverse Engineering

To our best knowledge, reverse engineering serves as the footstone of all the PCB attack models discussed in Section 2.

In PCBs brute force copying, reverse engineering provides the attacker with exact physical information of the board to retrieve the Gerber files. In PCBs hacking, reverse engineering gives attackers insight into the schematic's design logic when attackers apply altered component replacement attacks and additional components/Trojan Insertion attacks. Furthermore, attackers can gain physical information on the metal traces using reverse engineering for Additional components/Trojan Insertion and Taking control of and eavesdropping on certain data buses; those spatial position relations are critical for busline probes setups [16].

Especially for those boards with more than four layers, the layouts of each layer are critical for the attacker. However, the top layer and bottom layers are the natural protections for the layers in between. Reverse engineering is the only way for the attackers to revive the contents of those sandwiched layers.

Our proposed technique is to create protection mechanisms for PCBs against reverse engineering, which will protect PCBs from all attacks. Note that there's no way to stop the attackers from applying reverse engineering to the board as the boards are to be distributed. However, our protection mechanisms (see Section 3) can significantly introduce problems/errors and create cost overheads for the attackers.

4. Transformable-Vias Structure in PCBs

Figure 9 shows that PCBs with more than two layers typically have three kinds of vias: through vias, buried vias and blind vias. The proposed technique exploits the "transformable" property of the Mg/MgO pair as a countermeasure to PCBs' reverse engineering. Buried vias material is replaced by Mg. MgO vias are deliberately placed in some locations with Cu metal traces.

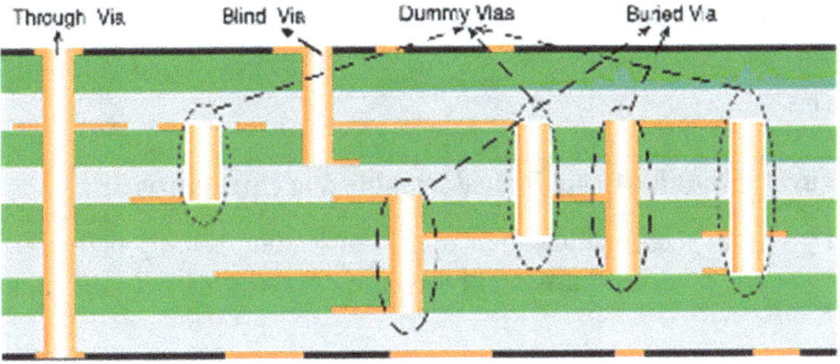

Fig. 9. Cross-Section of a PCB.

Note that the resistivity of Mg is 44.7 nΩ·m, which displays excellent electrical conductivity. And MgO is a dielectric material with a resistivity larger than 1000 Ω·cm. [18]. Thus, Mg Via can serve as normal via material, and Cu traces connected by MgO will be disconnected from the circuit regularly.

The layer imaging process will trigger the defense mechanism itself.

Suppose attackers apply a destructive imaging method to the board. Mg buried vias will oxidize into MgO and blend with the deliberately placed MgO vias. In the following imaging process, the oxide film formed on Mg has a dense morphology at the nanoscale resolution. This will conceal the original morphology of Mg material.

If attackers scan the PCBs with an X-ray, the best way to identify the presence and absence of material is to distinguish the brightness difference of the dielectrics and via material [1]. As shown in Fig. 10, the vias surrounded by bright shadows are those connected to theinterconnects, and those without the shadows are unconnected to the circuit. However, according to [19], little brightness difference between Mg and MgO is expected when exposed to X-ray.

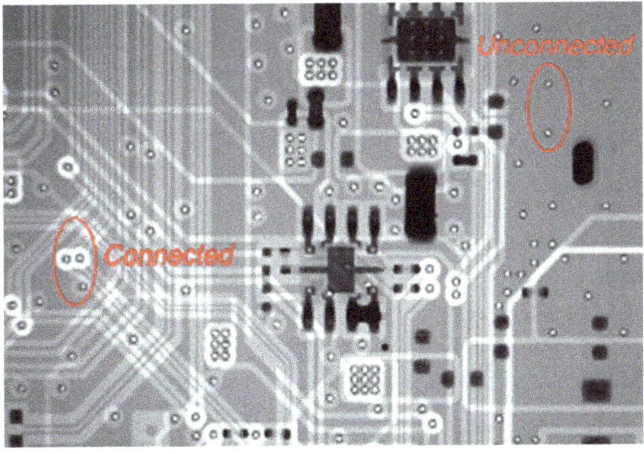

Fig. 10. Vias Identification of a PCB.

This means the original non-conductive MgO vias will blend with Mg buried Vias and mislead the attacks to another routing pattern [20, 8] as those MgO will be identified as conductive vias.

5. PCB Security Modules Using Transformable-Vias Structure

This section gives a specific application of the misleading routing pattern after reverse engineering: the "extra metal traces" will increase the crosstalk between high-frequency signals.

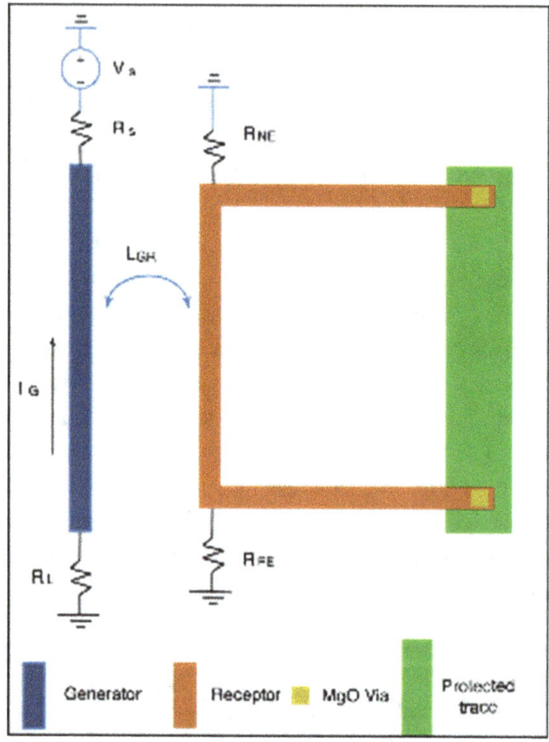

Fig. 11. Diagram of Crosstalk between Metal Traces Created by Transformable-vias.

In Fig. 11, the blue and green metal traces are in the original design. The orange trace is the green trace protection module connected with MgO vias, which disconnects it from the working circuit. The orange trace is deliberately placed near the blue high-frequency signal trace. Once exposed to reverse engineering, the circuit will include the orange trace in the design. The blue metal trace(generator) will generate noise in the orange trace(receptor); the near-end and far-end voltage can be expressed as Equations (1) and (2):

$$V_{NE}(t) = \frac{R_{NE}}{R_{NE}+R_{FE}} L_{GR} \frac{DI_G}{dt} + \frac{R_{NE}R_{FE}}{R_{NE}+R_{FE}} C_{GR} \frac{DV_G}{dt} , \tag{1}$$

$$V_{FE}(t) = \frac{R_{FE}}{R_{NE}+R_{FE}} L_{GR} \frac{DI_G}{dt} + \frac{R_{NE}R_{FE}}{R_{NE}+R_{FE}} C_{GR} \frac{DV_G}{dt} \ . \tag{2}$$

The generator is driven by Vs with the impedance R_S and connected with a load resistor R_l; the far-end and near-end load resistors are R_{FE} and R_{NE}, respectively; I_G is the current in the generator; mutual inductance and capacitance are modeled as L_{GR} and C_{GR}, respectively.

As generator is the high-frequency signal trace, $V_G(t)$ and $I_G(t)$ can be expressed in Equations (3) and (4):

$$V_G(t) = \frac{R_L}{R_S+R_L} V_s(t), \tag{3}$$

$$I_G(t) = \frac{1}{R_S+R_L} V_s(t). \tag{4}$$

Therefore, V_{NE} (t) and V_{FE} (t) are given as in Equations (5) and (6):

$$V_{NE}(t) = \left(\frac{R_{NE}}{R_{NE}+R_{FE}} L_{GR} \frac{1}{R_S+R_L} + \frac{R_{NE}R_{FE}}{R_{NE}+R_{FE}} C_{GR} \frac{R_L}{R_S+R_L} \right) \frac{dV_s(t)}{dt}, \tag{5}$$

$$V_{FE}(t) = \left(-\frac{R_{FE}}{R_{NE}+R_{FE}} L_{GR} \frac{1}{R_S+R_L} + \frac{R_{NE}R_{FE}}{R_{NE}+R_{FE}} C_{GR} \frac{R_L}{R_S+R_L} \right) \frac{dV_s(t)}{dt}. \tag{6}$$

Thus, intuitively, the higher the frequency signal in the generator, the higher the inductive and capacitive couplings will perform better protection modules.

6. Eye Diagram Analysis and Q Factor

We use Eye Diagram Analysis and Q Factor to evaluate the noise disturbance over the system.

The eye diagram provides a visual indication of how noise might impact system performance, as shown in Fig. 12, where $\mu 1$ and $\mu 1$ mean values of the signal levels for a "0" and a "1", and $\sigma 0$ and $\sigma 1$ represent the sum of the noise values at those two signal levels assuming Gaussian noise and the probability of a "0" and "1" transmission being equal.

The Q factor can be expressed as Equation (7), which measures the quality of a transmission signal in terms of its signal-to-noise ratio (SNR). It considers physical impairments to the signal, which can degrade it and cause bit errors.

$$Q = \frac{|\mu_1 - \mu_0|}{\sigma_1 + \sigma_0} \ . \tag{7}$$

Q-Factor represents the quality of the SNR in the "eye" of a digital signal, the "eye" being the human eye-shaped pattern on an oscilloscope that indicates transmission system performance. The best place for determining whether a given bit is a "1" or a "0" is the sampling phase with the most significant "eye-opening." The larger the eye-opening is, the more significant the difference between the mean values of the signal levels for a "1" and a "0" is. The more significant that difference is, the higher the Q factor and the better the BER performance.

In the industry or practical circuit design, a system's maximum Q factor must be smaller than 6% (raw BER of 10^{-9}) [2]. We will use the Q factor to evaluate the effectiveness of implanted protection modules in Section 4.

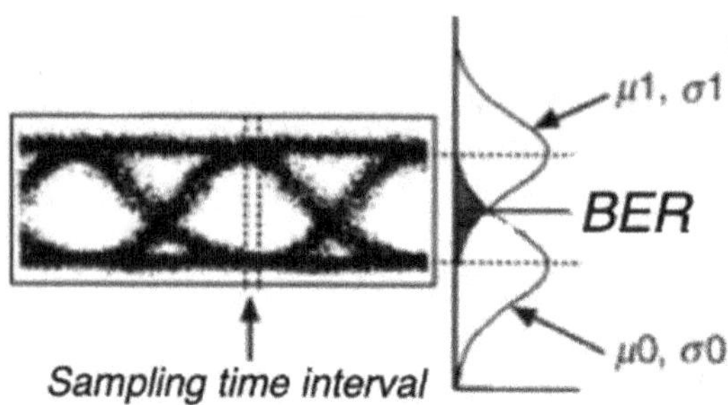

Fig. 12. Eye Diagram Example.

7. Experimental Results

We fabricated and tested the PCBs with the protection modules to prove the effectiveness and efficiency of the proposed protection modules. The fabricated board is shown in Fig. 15.

The board's parameters are as follows: the core thickness is 1.5 mm, the metal trace width is 0.1778 mm, the minimum PCB trace spacing is 0.1778 mm, and the metal layer thickness is 0.075 mm.

Here, we define the unprotected metal trace as the control group and the protected metal trace in the same physical parameters with protection modules as the protection group. Thus, the noise introduced by the protection modules can be measured as the difference between the protection and control groups.

As discussed in Section 5, to maximize the noise introduced by the protection modules, we used switchback routings to maximize the inductive and capacitive couplings for the test, illustrated in Fig. 15. Note that switchback routing is usually used for signal integrity and signal delay adjustment in PCBs, which makes protections legitimate from the attackers' perspective.

During the test, the frequency of the protected signal and generator signals in the test is set to 10MHz and 3MHz–15MHz, respectively. Testing results are shown in Figs. 14, 17 and 19. Significant noise can be observed in the figure. To better evaluate the noise introduced, we generated square waves using the sine waves in the protection groups and compared them with those generated from the control groups. The threshold voltage is set as 0.73VDD.

Figure 13 shows the schematic of Group 1 in Fig. 15. The protection metal traces are wired parallel to the protected trace. The square waves are shown in Fig. 14, and the Q factors are listed in Table 1. As mentioned in Section 6, the design with dummy metal length of 5.08 mm will fail with a Q factor larger than 6%. The protections with lengths of 11.43 mm and 60.96 mm can significantly protect the information carried in the metal trace.

Figure 16 shows the schematic of Group 2 in Fig. 15. The protection metal traces are wired vertically to the protected trace. The square waves are shown in Fig. 17, and the Q factors are listed in Table 2. All four lengths of protection will fail the system if attackers copy the board design.

Figure 18 shows the schematic of Group 3 in Fig. 15. The protection metal traces are spirally wired in nearby layers to the protected trace. The square waves are shown in Fig. 19 and the Q factors are listed in Table 3. All three lengths of protection will fail the system if attackers copy the board design.

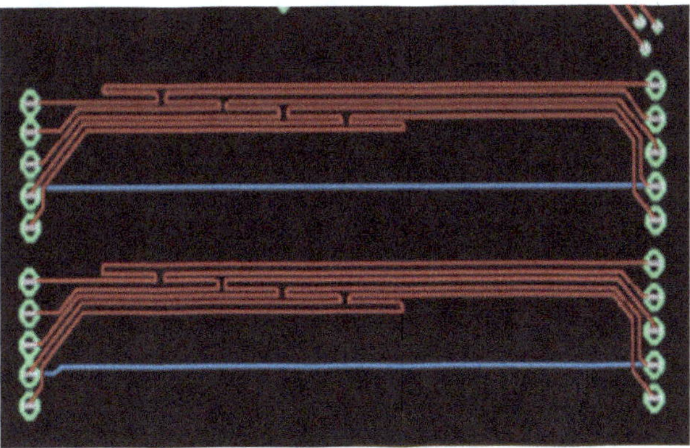

Fig. 13. Schematic of Group 1.

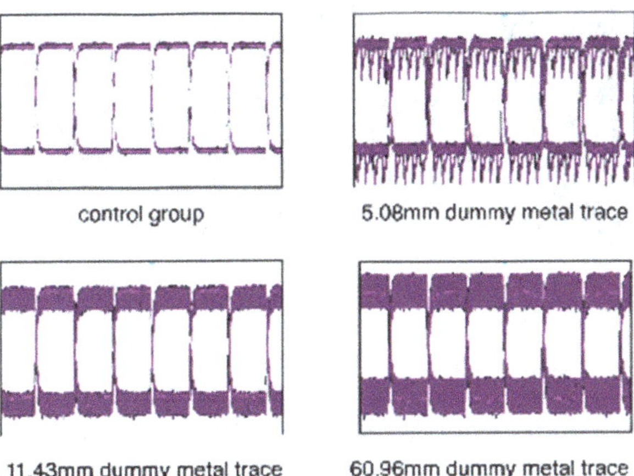

Fig. 14. Results of Module 1 with Different Length.

Table 1. Q factors of Module 1

Dummy Metal length	0	5.08 mm	11.43 mm	60.96 mm
Q factor		40.32%	8.3%	4.2%	3.8%

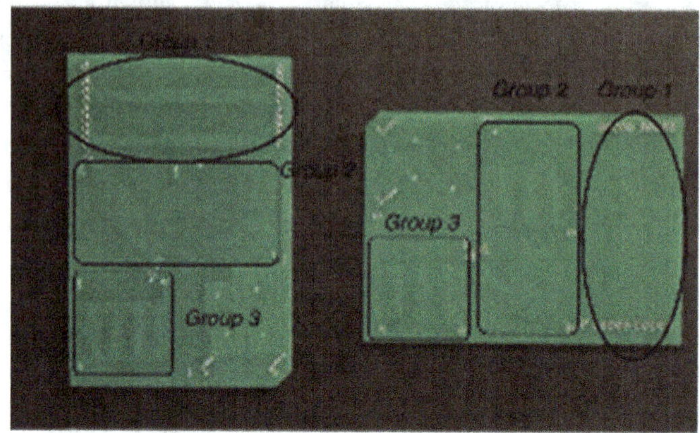

Fig. 15. The PCBs Fabricated for Protection Module Testing.

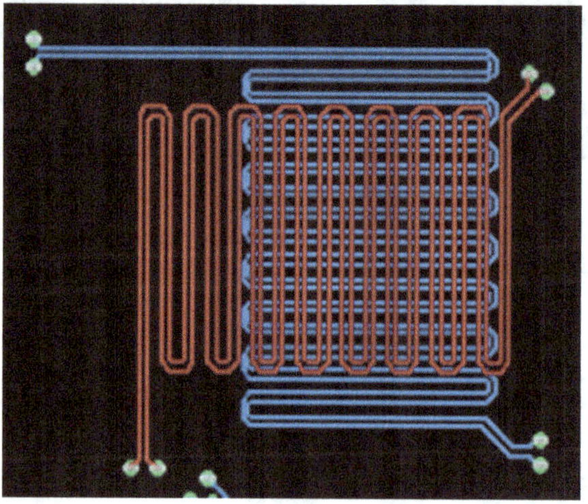

Fig. 16. Schematic of Group 2.

Table 2. Q factors of Module 2

Dummy Metal length	0	10.16 mm	17.78 mm	25.40 mm	40.46 mm
Q factor	40.32%	4.48%	3.80%	1.70%	1.45%

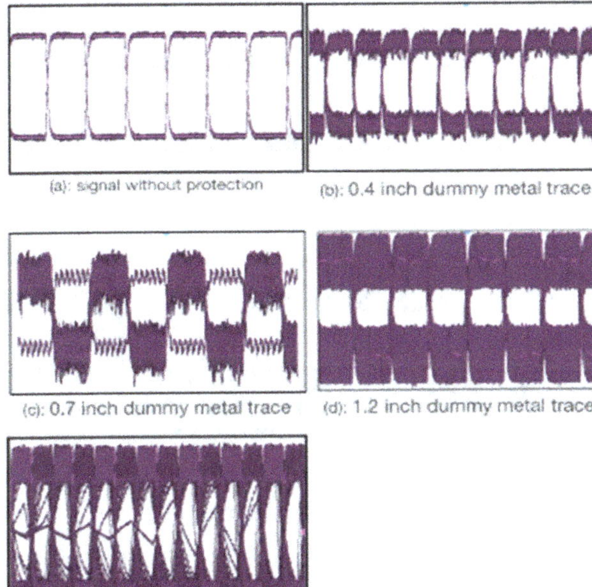

Fig. 17. Results of Module 2 with Different Length.

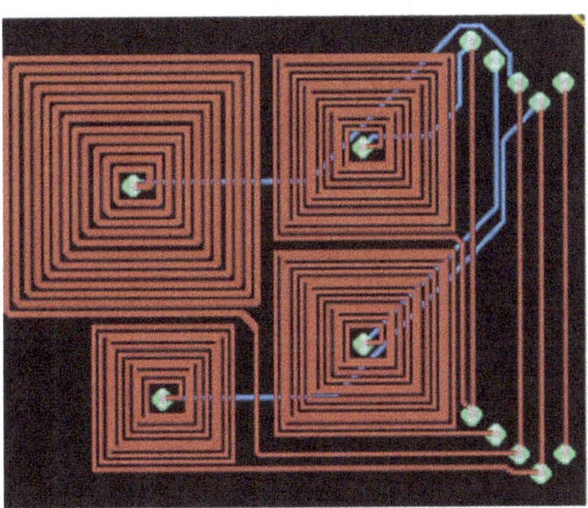

Fig. 18. Schematic of Group 3.

Table 3. Q factors of Module 3

Dummy Metal length	0	20.32 mm	30.48 mm	45.72 mm
Q factor	40.32%	3.92%	3.20%	1.35%

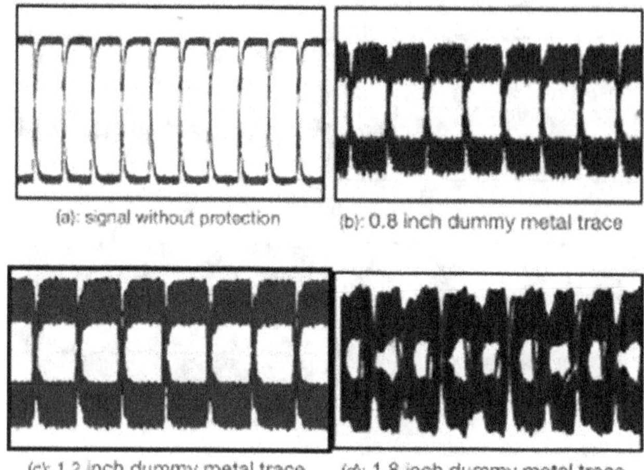

Fig. 19. Results of Module 3 with Different Length.

8. Conclusion

Transformable vias (MgO vias and Mg vias) based self-defense modules for PCBs design and their feasibility has been elucidated. Protection modules using the crosstalk model to protect sensitive data are proposed and analyzed. The experimental result of the fabricated PCBs is presented. Future work will focus on more protection module development to cause different failures for attackers, such as power failure, Electro Magnetic Interference (EMI) failure, and overheating.

References

1. N. Asadizanjani, M. Tehranipoor and D. Forte, "PCB Reverse Engineering Using Nondestructive X-ray Tomography and Advanced Image Processing," in IEEE Transactions on Components, Packaging and Manufacturing Technology, vol. 7, no. 2, pp. 292–299, Feb. 2017, doi:10.1109/TCPMT.2016.2642824.
2. https://www.itu.int/rec/T-REC-O.201/en
3. N. Vashistha, M. L. Rahman, M. S. U. Haque, A. Uddin, M. S. U. I. Sami, A. M. Shuo, P. Calzada, F. Farahmandi, N. Asadizanjani, F. Rahman and M. Tehranipoor, ToSHI-Towards Secure Heterogeneous Integration: Security Risks, Threat Assessment, and Assurance. Cryptology ePrint Archive, 2017.
4. F. Ganji, S. Tajik, J. P. Seifert and D. Forte, Blockchain- enabled cryptographically-secure hardware obfuscation. Cryptology ePrint Archive, 2019.
5. V. Gohil, H. Guo and S. Patnaik, ATTRITION: Attacking Static Hardware Trojan Detection Techniques using Reinforcement Learning. arXiv:2208.12897, 2022.
6. S. Paley, T. Hoque and S. Bhunia, Active protection against PCB physical tampering. 2016 17th International Symposium on Quality Electronic Design (ISQED), Santa Clara, CA, 2016, pp. 356–361, doi:10.1109/ISQED.2016.7479227.

7. G. Piliposyan, S. Khursheed and D. Rossi, "Hardware Trojan Detection on a PCB Through Differential Power Monitoring," IEEE Transactions on Emerging Topics in Computing, doi:10.1109/TETC.2020.3035521.

8. S. Chen and L. Wang, Reverse engineering resistant ROM design using transformable via-programming structure. 2016 IEEE International Symposium on Circuits and Systems (ISCAS), IEEE, 2016, pp. 2627–2630.

9. A. Hennessy, Y. Zheng and S. Bhunia, JTAG-based robust PCB authentication for protection against counterfeiting attacks. 2016 21st Asia and South Pacific Design Automation Conference (ASP-DAC), Macau, 2016, pp. 56–61, doi:10.1109/ASPDAC.2016.7427989.

10. M. T. Enevoldsen et al., Security module for protection circuit components from unauthorized access. U.S. Patent No. 10,009,995. 26 June 2018.

11. D. Zhang, Q. Ren and D. Su, "A Novel Authentication Methodology to Detect Counterfeit PCB Using PCB Trace-Based Ring Oscillator," in IEEE Access, vol. 9, pp. 28525–28539, 2021, doi:10.1109/AC- CESS.2021.3059100.

12. J. Grand, Printed circuit board deconstruction techniques. 8th USENIX Workshop on Offensive Technologies (WOOT 14). 2014.

13. S. H. Russ and J. Gatlin, "Ways to hack a printed circuit board: PCB production is an underappreciated vulnerability in the global supply chain," in IEEE Spectrum, vol. 57, no. 9, pp. 38–43, 2020, doi:10.1109/MSPEC.2020.9173902.

14. S. Chen and L. Wang, "Transformable electronics implantation in ROM for anti-reverse engineering," in International Journal of High Speed Electronics and Systems, vol. 28, no. 03n04, p. 1940021, 2019.

15. https://arstechnica.com/gadgets/2021/06/chip-shortages-lead-to-more-counterfeit-chips-and-devices/

16. J. S. Götte and B. Scheuermann, "Can't touch this: Inertial HSMs thwart advanced physical attacks," in IACR Transactions on Cryptographic Hardware and Embedded Systems, pp. 69–93, 2022.

17. S. Chen, J. Chen and L. Wang, "A chip-level anti-reverse engineering technique," in ACM Journal on Emerging Technologies in Computing Systems (JETC), vol. 14, no. 2, pp. 1–20, 2018.

18. S. Chen *et al.*, "Chip-level anti-reverse engineering using transformable interconnects." 2015 IEEE International Symposium on Defect and Fault Tolerance in VLSI and Nanotechnology Systems (DFTS), IEEE, 2015.

19. S. Durdu, A. Aytac and M. Usta, "Characterization and corrosion behavior of ceramic coating on magnesium by micro-arc oxidation, in Journal of Alloys and Compounds, vol. 509, no. 34, pp. 8601–8606, 2011.

20. W. Stark, S. Chen and L. Wang, "Reverse engineering protection using obfuscation through electromagnetic interference," in International Journal of High Speed Electronics and Systems, vol. 31, no. 01n04, p. 2240003, 2022.

Next Generation RF Modules for 5G, IoT, AR/VR and RFID Applications

Marvin Joshi[*], Kexin Hu, Genaro Soto-Valle, Hani Al Jamal and Manos Tentzeris

*School of Electrical and Computer Engineering,
Georgia Institute of Technology,
85 5th St NW, Atlanta, Georgia, 30308, USA*
[*]mjoshi5@gatech.edu

The rapid development and deployment of 5G/mm-Wave technologies for communication, sensing and energy harvesting applications have been on the rise. Consequently, the need for low-cost, scalable, agile and compact RF modules has become more prominent than ever. This paper presents a review of recent efforts in utilizing additive manufacturing techniques such as inkjet printing to sustainably accelerate the massive deployment of 5G/mm-Wave. First, a novel flexible and massively scalable multiple-input, multiple-output (MIMO) tile-based phased array enabled by additively manufactured microstrip-to-microstrip transitions is presented. Next, a novel Rotman-Based harmonic mmID tag for Ultra-Long-Range localization is presented. Finally, low-power, low-cost mm-Wave backscattering modules for localization and orientation sensing are demonstrated.

Keywords: Additive manufacturing; inkjet printing; packaging; machine learning; RFID; radar.

1. Introduction

Over the past decade, millimeter-wave (mm-Wave) communication and sensing have emerged as key technologies for a wide range of applications, including 5G wireless communication, smart cities, and industrial sensing. The use of mm-Wave frequencies offers several advantages over traditional communication frequencies, including higher bandwidths, lower interference, and more directional communication. However, the adoption of mm-Wave frequencies for communication and sensing has been limited by the high cost and complexity of traditional fabrication techniques for mm-Wave RF modules. Recently, additive manufacturing techniques such as inkjet printing have emerged as promising solutions for scalable, low-cost, highly-customizable and on-demand fabrication of 5G-enabled wireless devices.

This paper provides a review of the latest state-of-the-art efforts in 5G/mm-Wave RF modules. Section 2 discusses a tile-based approach to fabricating phased arrays, enabling the realization of on-demand very large antenna arrays capable of conforming to practically any platform. Section 3 presents the use of a fully-passive dual Rotman-lens-based harmonic mmID tag for ultra-long range localization, being able to maintain a large wide

---

[*]Corresponding author.

angular coverage. Finally, Section 4 discusses low-power backscattering RFID modules operating at mm-Wave frequencies to alleviate the limitations in localization and orientation accuracy due to limited bandwidth and the large system size of UHF tags. The paper is summarized and an outlook on future developments is presented in Section 5.

2. Highly-Scalable Additively Manufactured Tile-Based Phased Array

In recent years, emerging technologies such as 5G and mm-Wave have enabled high data rates, wideband operation, and low latency, making them ideal for future Internet of Things (IoT) and massive MIMO systems. To support these applications, large antenna arrays are necessary to provide adequate link budgets. However, traditional antenna arrays are bulky and heavy, which makes integration with various platforms of different sizes and shapes a challenge. Previous research designs have integrated the entire array in a multilayer rigid board of fixed size, which leads to high complexity and cost. To overcome the limitations in scalability and manufacturability, an additively-manufactured tile-based antenna array structure has been proposed [1]. This structure consists of small and lightweight unit-cell antenna arrays assembled into a large arbitrary-sized array by a tiling layer, as shown in Fig. 1. By inkjet printing planar phased arrays onto flexible substrates, the array shape can be customized and a flexible and scalable array can be built at a low cost.

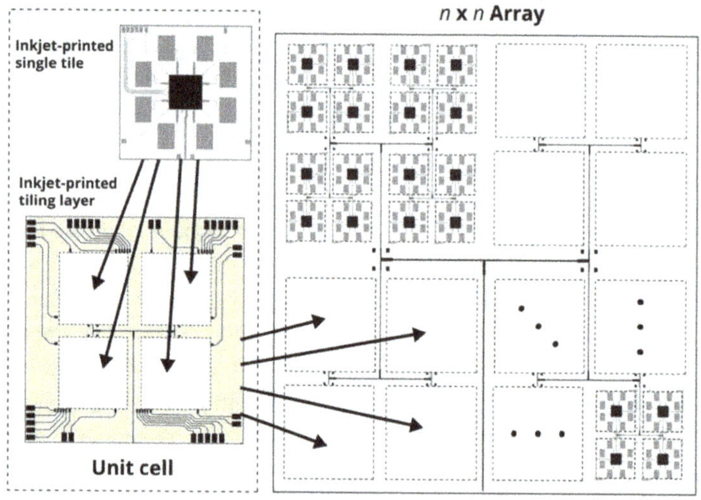

Fig. 1. Scalability demonstration of the tile-based phased array [1].

The single tile phased array comprises eight microstrip patch antenna elements and an Anokiwave Beamformer IC AWS-0102, with a center frequency of 19 GHz. The amplitude and phase of each antenna element are controlled by the serial peripheral interface (SPI) signals from the IC. The 4-tile array is controlled by four ICs simultaneously through a MATLAB program. Each tile has soldering pads for DC signal connection that can be transferred to the tiling layer. The tiling layer has a feedline with corporate feeding

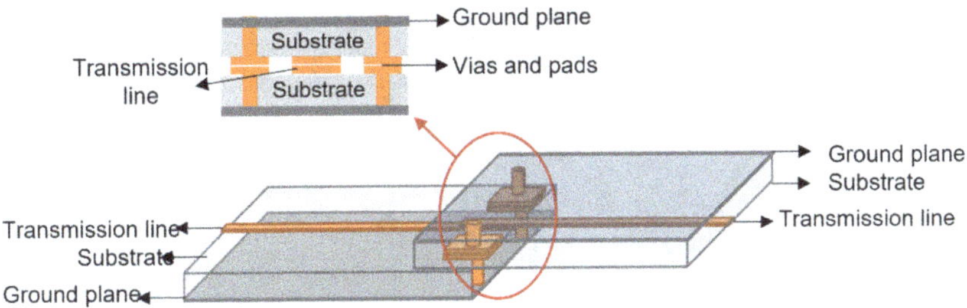

Fig. 2. Microstrip-to-Microstrip transition [1].

structure and a microstrip-to-microstrip transition is used to connect the RF signals from RF input to individual tiles. Figure 2 shows the cross-sectional view of the microstrip-to-microstrip transition, which uses a GSG transition with two vias placed aside the transmission line, connecting ground planes of tiles and tiling layer. Square pads are added at the top of vias for soldering and better reliability. This structure has demonstrated low-loss and wide-band transmission [2]. The tiling layer was inkjet printed on Rogers 3003 substrate with SU8 dielectric ink, which is a photoresist polymer that can be cured using a UV crosslinker. The vias were made through drilling holes with a diameter of 0.2 mm and filling with silver paste. The assembly of the large array was completed by aligning and soldering all the soldering pads on both single tiles and the tiling layer. Header pins were soldered onto the tiling layer to provide DC power and SPI signals. The assembled proto-type shown in Fig. 1 is very stable and can be conformally wrapped over a curved surface without any detachment of the tiles.

The measured radiation pattern and beam steering results of the fabricated prototype are shown in Fig. 3, which were conducted in an anechoic chamber. The measured steering angle for both single tile and 4-tile phased arrays can reach −50 to 50 degrees with a −10 dB side lobe level. The maximum realized gain is 16 dBi when 0 degree phased shift is applied. This tiled phased array can be extended to a larger array by simply adding more single tiles and printing a larger tiling layer. This design of additively-manufactured tiles and tiling layer is ideal for space-limited and power efficient 5G and satellite communi-cation applications and mm-Wave Rx systems for multi-streaming with high throughput that require a large number of antenna arrays.

3. Fully-Passive Rotman-Based Harmonic mmID Tag for Ultra-Long Range Localization and Wide Angular Coverage

As the recent advances in mm-Wave/5G technologies become increasingly available, the expectation for Internet of Things (IoT) devices for ubiquitous sensing is dramatically growing. Billions of IoT devices are expected to be operating by 2025, which implies that an equivalent number of batteries are needed to support their correct functioning. This generates a serious concern about the environmental impact that these technologies could cause, therefore, energy-autonomous solutions are required for a sustainable development

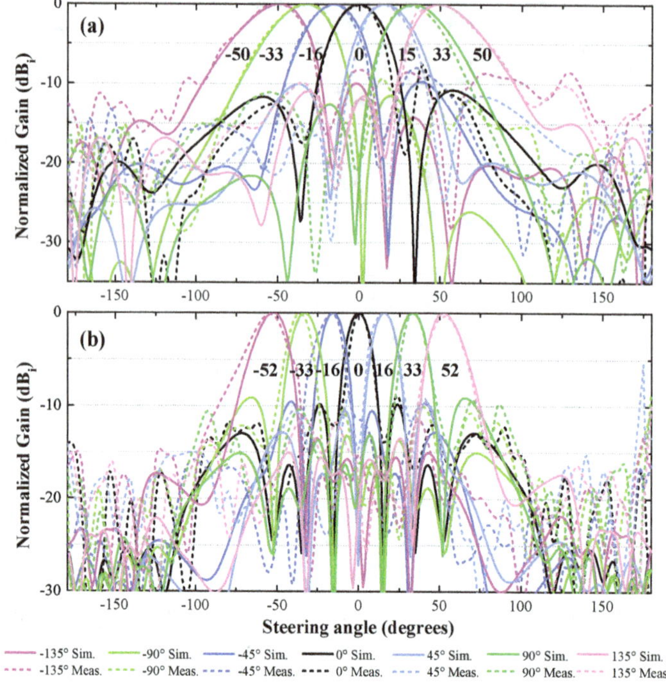

Fig. 3. Simulation and measurement results for a single tile phased array (a) and for the 2×2 tiles ("unit cell") phased array (b) [1].

of the next generation of these wireless sensor networks. RFIDs are a highly favorable solution for energy-autonomous systems as they utilize the received electromagnetic signal to encode its sensing information and reflect it back to the reader, so they do not need to produce their own signal. Multiple types of RFIDs are available, namely, chip-based, chipless and harmonic-based RFIDs; the latter one providing an excellent tradeoff between complexity, reading range and encoding information. Recent works have proposed the use of a harmonic-based millimeter-wave Identification (mmID) tag for ultra-long range localization and wide-angular coverage by combining a Rotman lens design and a frequency-doubler circuit [3–5].

The Rotman lens is a passive beamforming network (BFN) that allows focusing the received power from an interrogating signal into specific beam ports depending on the signal's direction of arrival. In comparison to active beamforming networks, passive BFN's do not rely on any active circuitry. The proposed mmID tag combines two BFNs, one operating at the interrogating signal of 14 GHz (f_0) and the other operating at 28 GHz ($2f_0$) for the backscattered signal. The mmID tag was fabricated on a Rogers 4350B substrate ($\varepsilon = 3.55$, tan $\delta = 0.027$). Each of the Rotman lenses contains eight array ports and six beam ports, as shown in the diagram in Fig. 4(a). Each array port is connected to a linear microstrip patch array that operates either at f_0 or $2f_0$. The angular coverage characterization of each of the Rotman lenses is shown in Figs. 4(b) and 4(c), demonstrating peak gains of 17 dBi and 15 dBi for 14 GHz and 28 GHz, respectively, and a 3 dB bandwidth of ±50 degrees. It can be noted that each of the beaming ports corresponds to a different angle of incidence. The frequency doubling is achieved by using a SMS201, flip-chip, Schottky diode (Macom), which provides high sensitivity and, therefore, higher conversion loss and more efficient passive harmonic generation. Three radial stubs are also utilized for matching network and for suppressing the f_0 components from the doubler output. The conversion loss (CL) characterization of the doubler circuit is presented in Fig. 4(d), which shows that it can be operated at power inputs as low as −45 dBm.

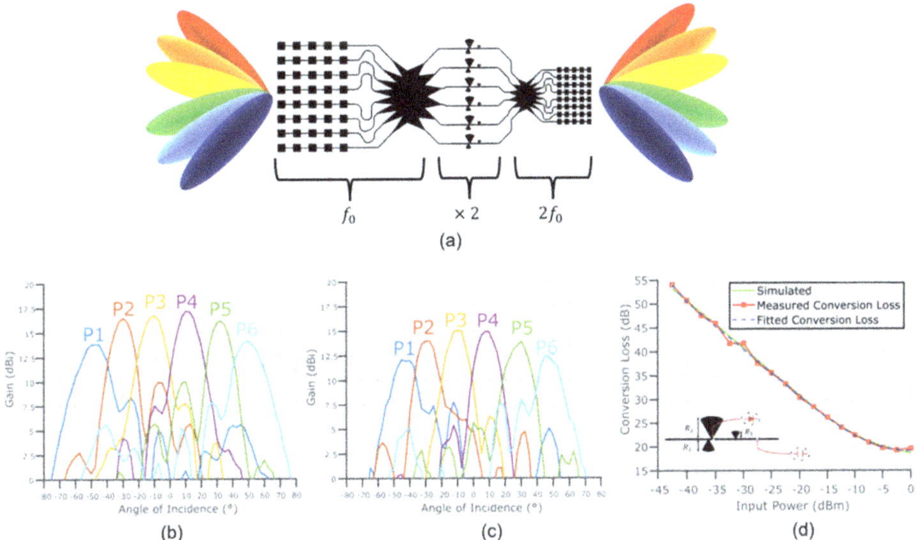

Fig. 4. (a) Diagram of the Rotman-based harmonic mmID. Characterization of gain and angular coverage for the Rotman lens operating at 14 GHz (b) and 28 GHz (c). Conversion loss of the doubler circuit (d) [3, 4].

The performance of the fully-assembled harmonic mmID tag was evaluated by measuring the harmonic radar cross-section (RCS) and the maximum reading range. The harmonic RCS was calculated using the equation $\sigma_{RCS,2f_0} = G_{f_0}G_{2f_0}\lambda^2_{2f_0}CE_{@5m}/(4\pi)$, where $G_{f_0}, G_{nf_0}, \lambda_{nf_0}$ and $CE_{@5m}$, are the fundamental antenna gain, harmonic antenna gain, wavelength of the harmonic frequency, and conversion efficiency assumed to be $1 - CL$ for the input power at 5 m from the reader. As shown in Fig. 5(a), the mmID demonstrates 10 dB beamwidth of ±50 with a maximum harmonic RCS of −35.8 dBsm, which means that the Rotman-based harmonic mmID can provide high detectability even at wide angular ranges. The peaks and nulls can be modified by varying the number of antenna ports and beam ports at the Rotman lens, however, this would reduce the angular coverage at the harmonic frequency. Figure 5(b) shows the received power spectral density from the mmID from 5 m to 65 m, utilizing an equivalent isotropic radiated power (EIRP) level of 51 dBm, demonstrating the ultra-long range capability of the system. Overall, these works have demonstrated the significant benefits of utilizing a dual Rotman

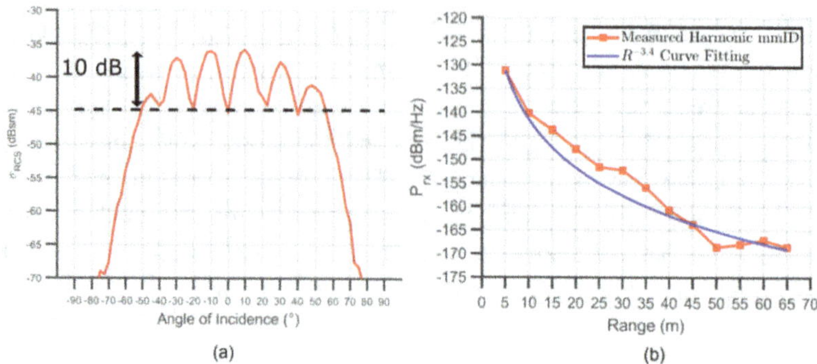

Fig. 5. (a) Estimated angle-dependent harmonic RCS at 28 GHz and range of 5 m. (b) Measured received power spectral density of the fully-passive harmonic mmID at EIRP = 51 dBm [3, 4].

lens-based harmonic mmID tag, which are fully-passive operation, high sensitivity due to the use of a harmonic frequency reducing clutter and self-interference on the reader — and being able to provide high gain and wide angular coverage at very long range. Theoretically, by utilizing the maximum allowed EIRP of 75 dBm for 5G communications, the presented system would be able to operate at ranges in the order of kilometers.

4. Low-Power mm-Wave Backscattering Modules for Localization and Orientation Sensing

In recent times, the integration of Radio Frequency Identification (RFID)-based technologies in the Internet of Things (IoT) ecosystem has increased considerably. As they allow for a low-cost, highly efficient solution, RFID tags have been used for various IoT applications involving wireless sensing and localization. However, these devices operate primarily operate in the Ultra High Frequency (UHF) band, which naturally has limitations in localization accuracy, due to available operational bandwidth, and large overall system size. Thus, to overcome these limitations, the shift up to millimeter-wave (mm-Wave) frequencies is presented in [6]. The proposed RFID, designed to operate in the 60 GHz band, was additively manufactured on Rogers 3003 (ϵ_r = 3.0, tan(δ) = 0.0010) with thickness of 0.127m and can be viewed in Fig. 6. By utilizing Frequency-Modulated Continuous Wave (FMCW) reader, the compact tag was able to achieve highly accurate ranging up to 0.5 m, as the average ranging error is displayed in Fig. 7.

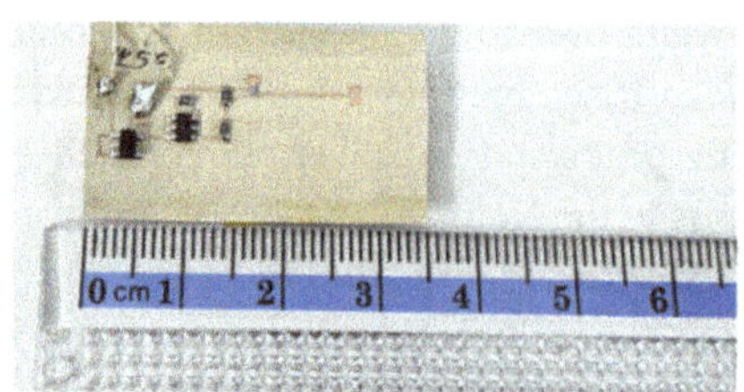

Fig. 6. Proposed 60 GHz RFID tag [6].

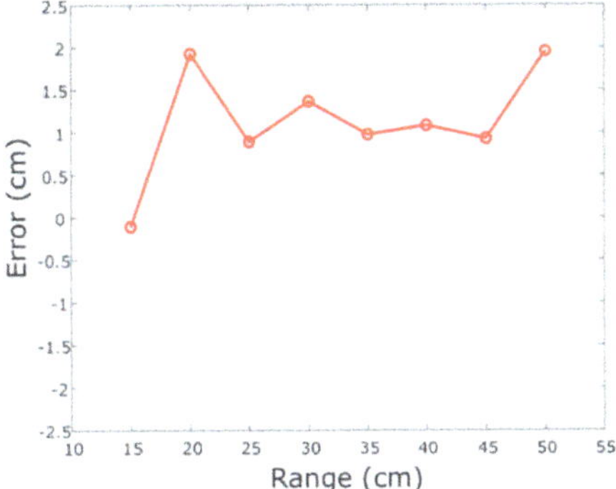

Fig. 7. Ranging accuracy of RFID system [6].

Along with IoT integration, millimeter-wave identification (mmID) tags are a low-cost solution for VR/AR applications. As accurate localization is important for these applications, the ability to track and detect the orientation of an object has equal importance. In [7], a low-power mm-Wave mmID system for rotational sensing is proposed. Operating at 24.125 GHz, the mmID tag, which can be viewed in Fig. 8, has an overall size of 43 mm × 25 mm and was fabricated on RO4350B substrate (ϵ_r = 3.66, tan δ = 0.0037). The RF front end of the tag consists of four antennas, each of which is cross-polarized to improve self-interference rejection, and a low noise FET for modulation. To control the modulation and power the tag, the baseband circuit consists of a 3 V coin cell battery, a low-power oscillator (LTC6906) and a 1.8 V voltage regulator. By integrating a polarization offset between successive antenna elements, as the tag is rotated about the z-axis, each channel will have a different backscattered response which allows for angular ambiguities to be resolved. To avoid harmonic interference, channels A to D had a modulation frequency of 49 kHz, 69 kHz, 85 kHz and 110 kHz, respectively. By extracting the amplitude response from each channel as the tag rotates about the z-axis, a k-Nearest Neighbor (kNN) machine learning algorithm is used to predict the orientation of the tag. Figure 9 shows the ultra-high accuracy of the model when the tag is at a distance of 0.625 m, as a mean error of 0.029° was achieved.

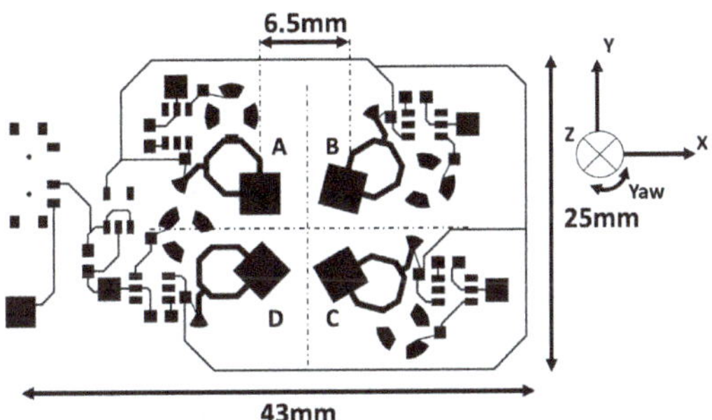

Fig. 8. Schematic of the proposed 24 GHz mmID tag [7].

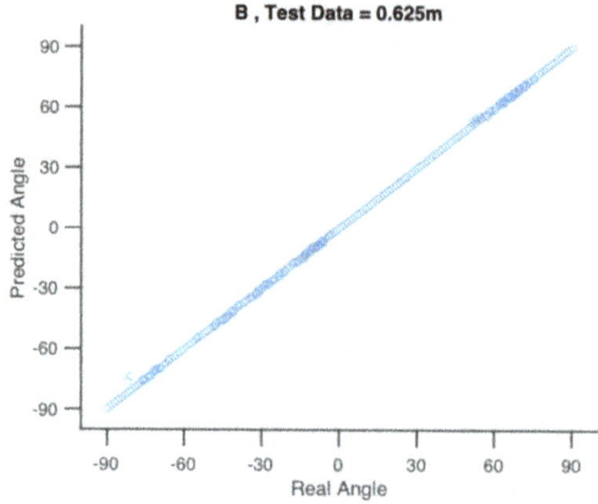

Fig. 9. Performance of orientation detection model at range = 0.625 m [7].

5. Conclusion

This paper has reviewed state-of-the-art 5G/mm-Wave RF modules for 5G, IoT and AR/VR applications. The use of additive manufacturing was shown in the development of a tile-based phased array, as well as mmID tags for localization applications. Additionally, a novel mm-Wave harmonic mmID tag capable of ultra-long-range localization and wide-angular coverage was presented. These advances in mm-Wave RF modules combined with the use of additive manufacturing technologies have allowed for a low-cost, highly scalable solution superior to existing technologies.

References

1. K. Hu, G. Soto-Valle, Y. Cui and M. M. Tentzeris, "Flexible and Scalable Additively Manufactured Tile-Based Phased Arrays for Satellite Communication and 50 mm Wave Applications," *2022 IEEE/MTT-S International Microwave Symposium - IMS 2022, Denver, CO, USA*, 2022, pp. 691-694, doi: 10.1109/IMS37962.2022.9865337.
2. K. Hu, X. He and M. M. Tentzeris, "Flexible and Scalable Additively Manufactured Antenna Array Tiles for Satellite and 5G Applications Using A Novel Rugged Microstrip-to-Microstrip Transition," *2021 IEEE International Symposium on Antennas and Propagation and USNC-URSI Radio Science Meeting (APS/URSI), Singapore, Singapore*, 2021, pp. 37-38, doi: 10.1109/APS/URSI47566.2021.9704748.
3. C. A. Lynch, A. O. Adeyeye, A. Eid, J. Hester and M. M. Tentzeris, "Ultra-Long-Range Dual Rotman Lenses-based Harmonic mmID's for 5G/mm-Wave IoT Applications," *2022 IEEE/MTT-S International Microwave Symposium - IMS 2022, Denver, CO, USA*, 2022, pp. 32-35, doi: 10.1109/IMS37962.2022.9865357.
4. C. Lynch, A. O. Adeyeye, A. Eid, J. G. D. Hester and M. M. Tentzeris, "5G/mm-Wave Fully-Passive Dual Rotman Lens-Based Harmonic mmID for Long Range Microlocalization Over Wide Angular Ranges," in *IEEE Transactions on Microwave Theory and Techniques*, vol. 71, no. 1, pp. 330-338, Jan. 2023, doi: 10.1109/TMTT.2022.3227925.

5. A. Eid, J. G. D. Hester and M. M. Tentzeris, "Rotman Lens-Based Wide Angular Coverage and High-Gain Semipassive Architecture for Ultralong Range mm-Wave RFIDs," in *IEEE Antennas and Wireless Propagation Letters*, vol. 19, no. 11, pp. 1943-1947, Nov. 2020, doi: 10.1109/LAWP.2020.3002924.

6. C. A. Lynch, A. O. Adeyeye, J. G. D. Hester and M. M. Tentzeris, "When a Single Chip becomes the RFID Reader: An Ultra-low-cost 60 GHz Reader and mmID System for Ultra-accurate 2D Microlocalization," *2021 IEEE International Conference on RFID (RFID), Atlanta, GA, USA,* 2021, pp. 1-8, doi: 10.1109/RFID52461.2021.9444319.

7. A. Adeyeye, C. Lynch, J. Hester and M. Tentzeris, "A Machine Learning Enabled mmWave RFID for Rotational Sensing in Human Gesture Recognition and Motion Capture Applications," *2022 IEEE/MTT-S International Microwave Symposium - IMS 2022, Denver, CO, USA,* 2022, pp. 137-140, doi: 10.1109/IMS37962.2022.9865432.

Fabrication of Multi-Bit SRAMs Using Quantum Dot Channel (QDC)-Quantum Dot Gate (QDG) FET

Raja Hari Gudlavalleti, Jacques Goosen, Tao Liu, Hunter Bradley, Elisa Parent,
Abdulmajeed Almalki, Erik Perez and Faquir Jain[*]

*Department of Electrical and Computer Engineering, University of Connecticut,
371 Fairfield Way, Unit 4157, Storrs, CT 06268, USA*
faquir.jain@uconn.edu

This paper presents fabrication of multi-state inverters incorporating SiO_x-cladded Si quantum dot in the channel and gate region of driver, load, and access transistors. Experimental characteristics are presented exhibiting 3-state behavior in Quantum-dot Channel (QDC)-Quantum-dot Gate (QDG) FETs having Si quantum dots. It is shown that QDC-QDG-FETs-based enhancement mode inverter configurations are the building blocks of a multi-bit static random access memory (SRAM). QDC-QDG-FETs exhibiting four states can also be used to implement compact 4-state logic and nonvolatile memories or random access nonvolatile memories.

Keywords: Quantum dots; QDG; QDG-inverter; QDC-QDG FET fabrication.

1. Introduction

Quantum dots in the channel region, such as Quantum Dot Channel (QDC) FET, and on the gate dielectric, such as Quantum Dot Gate (QDG) FET, exhibited intermediate states, which have been reported in the literature [1, 2]. Unlike conventional MOSFET which exhibits two states (logic '0' and logic '1') in digital logic circuits, the QDG- and QDC-FETs due to more intermediate states can increase the processing capability and allow us to create new logic architectures. The number of intermediate states in the device can be increased by having QDs in both channels and on the gate dielectric within the same device, called QDC-QDG FET. This paper, therefore, discusses the fabrication and further demonstrates the experimental results.

2. QDC-QDG-FET Schematic and Theory

2.1. *Device Cross-section*

The schematic cross-section of QDC-QDG-FET is shown in Fig. 1. In comparison to conventional n-MOS devices, the channel region of the QDC-QDG-FET is recessed and SiOx-Cladded Si Quantum dots (QDs) are self-assembled. The upper layer of SiOx-Si

---

[*]Corresponding author.

Quantum-dots (QDs) is oxidized at 825°C which acts as a tunnel oxide layer. Then, the top two layers of quantum dots are self-assembled in the gate region. The metal layer is deposited that acts as the gate contact layer. Figure 2 shows the energy band diagram of the QDC-QDG-FET across the gate region of the device. E_C, E_V and E_F represent the conduction, valency and fermi levels of the p-type silicon (from the left side of the figure). The SiOx-cladded Si quantum dot has cladded and core regions with 4 nm and 0.5 nm thickness, respectively. The SiOx-cladded region has energy band separation of 8.9 eV

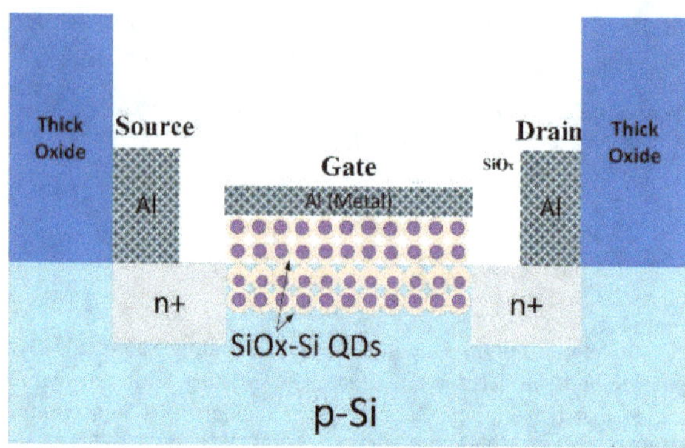

Fig. 1. Schematic cross-section of SiOx-Cladded Si Quantum-dot Channel (QDC)-SiOx-Cladded Si Quantum-dot Gate (QDG) field-effect transistor (FET).

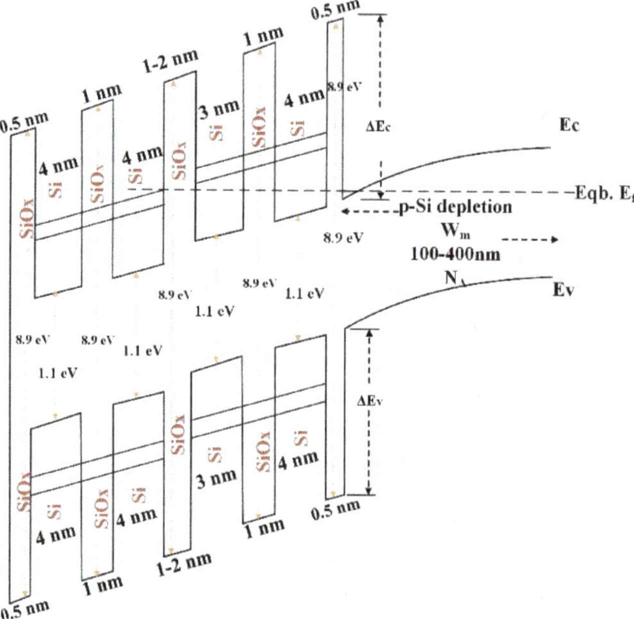

Fig. 2. The energy band diagrams for SiOx-Cladded Si QDC-QDG-FET.

and core region has an energy band separation of 1.1 eV, as shown represented in Fig. 2. However, the top of the two layers of the SiOx-Si quantum dots in the channel region is oxidized, hence a thickness of ~1–2 nm is shown in Fig. 2.

2.2. *Theory*

The array of quantum dots in a QD layer, in the channel and gate regions, forms discrete mini-bands as the wavefunction in one quantum dot overlaps with another due to the thin cladding layer. The intermediate state in the QDC-QDG-FET is manifested due to tunneling of electrons from the inversion layer into the mini-energy sub-bands formed by the quantum dot layers, which forms a quantum dot superlattice (QDSL). The tunneling probability has been reported before [3].

$$P_{w \to d} = \frac{4\pi}{\hbar} \sum_{w,d} |\langle \psi_d | H_t | \psi_w \rangle|^2 (f_w - f_d)\delta(E_d - E_w) \, . \tag{1}$$

The electron transfer in the gate region increases the threshold voltage and results in an intermediate state. Finally, when the gate voltage is increased, electrons occupying the mini-energy sub-bands of QDSL are transferred to the gate and FET returns to the ON state. This formation of the intermediate state can also be observed via the characteristic equations of the drain current for the QDG-FET given in (3).

$$V_{T,QD} = V_T + \Delta V_T$$

$$V_{T,QD} = V_T - \frac{q}{C_{ox}}\left[\sum \frac{x_{QD1} n_1 N_{QD1}}{x_g} + \sum \frac{x_{QD2} n_2 N_{QD2}}{x_g}\right] \, , \tag{2}$$

$$I_{D,lin} = \mu C_{ox} \frac{W}{L}\left((V_{GS} - V_{T,QD})V_{DS} - \frac{V_{DS}^2}{2}\right) . \tag{3}$$

where μ_n is the electron mobility, C_{ox} is the gate capacitance, W is the width, L is the length of the device, respectively, V_{GS} is the gate voltage, V_{DS} is the drain voltage and V_{Teff} is the effective threshold voltage. For a particular gate voltage range, charges begin to tunnel from the channel into the quantum dots via Fowler–Nordheim tunneling, which causes the drain current to become essentially invariant to the change in gate. The effective voltage that describes this occurrence is comprised of two parts, namely[1]

$$V_{Teff} = V_{TH} + \Delta V_{TH} \tag{4}$$

where V_{TH} is the threshold voltage of a conventional n-MOS FET and ΔV_{TH} is the modified threshold voltage as described below[1]:

$$\Delta V_{TH} = -\frac{q}{C_{ox}}\left[\sum \frac{x_{QD1} n_1 N_{QD1}}{x_g} + \sum \frac{x_{QD2} n_2 N_{QD2}}{x_g}\right] \tag{5}$$

where X_{QD} refers to the distance between the core of the quantum dots and the gate contact, n refers to the number of quantum dots in the layer, x_g is the distance between the interface of the gate dielectric and the substrate to the gate contact, and N_{QD} is the charge in each dot, which involves the probabilistic tunneling of charges through the QDSL.

3. Fabrication

The fabricated FET array is shown in Fig. 3(a) and an inset in Fig. 3(b) shows the individual device. The device was fabricated on p-Si substrate. The p-Si samples had a field oxide of ~1200Å grown on them utilizing a wet oxidation process. The process steps from growing the field oxide proceeded as follows: (1) diffusing n-type impurities to create the source and drain wells, (2) exposing the gate region and growing a ~40Å SiO_2 layer via a dry thermal oxidation process, (3) creating a recess in the channel region, (4) depositing and annealing two layers of Si quantum dots in the channel region via the bottom-up self-assembly method, (5) oxidizing the top layer of the Si quantum dots at 825°C, (6) depositing and annealing two layers of Si quantum dots on the gate via the bottom-up self-assembly method, (7) opening the contact regions and depositing ~1000Å of gold and (8) etching the excess metal and annealing. The annealing of the Si quantum dots after the self-assembly is done at 750°C.

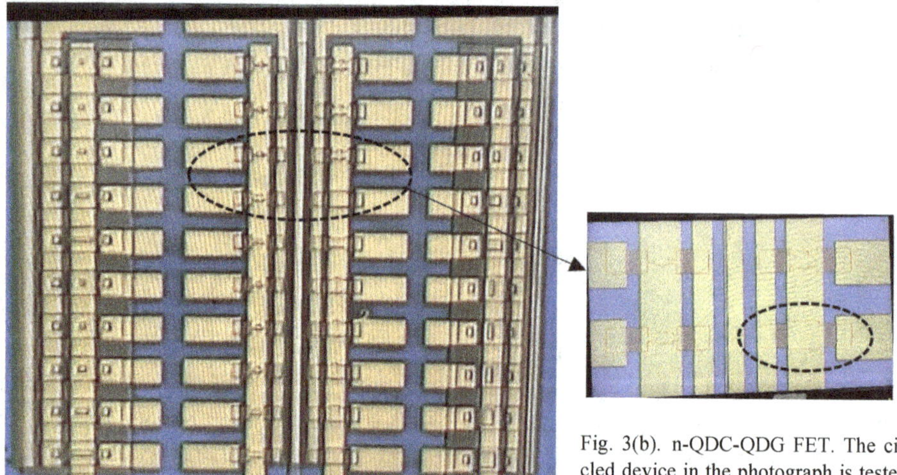

Fig. 3(b). n-QDC-QDG FET. The circled device in the photograph is tested. The aspect ratio of the device is (26μm/ 25μm).

Fig. 3(a). Fabricated n-QDC-QDG-FET device array.

4. Results and Discussion

4.1. *Experimental Results*

The experimental results, I_D-V_G for the fabricated device shown in Fig. 3(b), are shown in Fig. 4. The step-like characteristics in the figure show that intermediate states appear due to multiple thresholds in the device, which are formed due to mini-energy bands in SiOx-Cladded Si QDs in the channel and gate region of the QDC-QDG-FET.

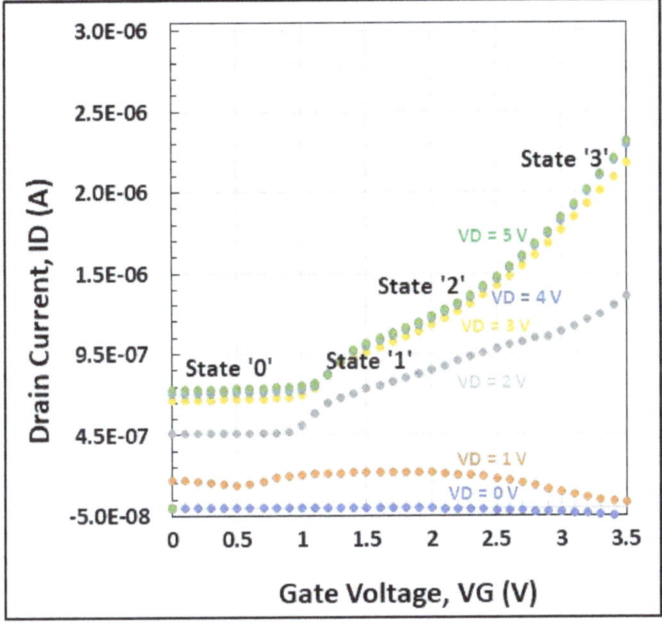

Fig. 4. The output characteristics of the SiOx-Cladded Si QDC-QDG-FET. The threshold values for the FET are ~1V, 1.2V and 1.8V, with an intermediate state holding at ~1.2V, and ~2.3V.

4.2. *Discussion*

The QDC-QDG FET shown in Fig. 1 can be used to implement an nMOS enhancement type inverter. M1 and M2 shown in Fig. 5 form a multistate QDC-QDG-FET-based inverter. The inverter can further be utilized to an SRAM when inverters are connected in back-to-back configuration with access transistors M5 and M6, as shown in Fig. 5. A multibit SRAM would increase the storage density.

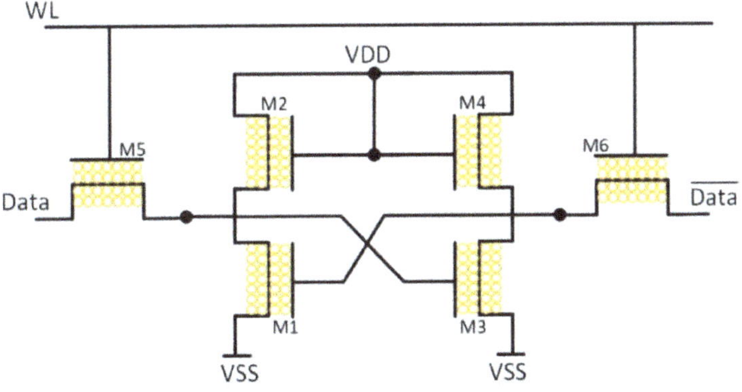

Fig. 5. QDC-QDG-FET-based multibit SRAM.

5. Conclusion

In this work, the 4-state capabilities of the QDC-QDG-FET were demonstrated using SiOx-cladded Si QD layers with different core diameters in the recessed channel and on the gate oxide of the transistor. Unlike the conventional approach of gate dielectric deposition or device, the gate oxide layer is formed by oxidizing the top layer of the Si QD layers in the recessed channel region. The presence of four states demonstrated in the experimental characteristics can further be utilized in the development of logic and memory circuits with multistate processing and storage capability, respectively. The logic circuits with n-QDC-QDG-FET only would have static current during operation and hence, research towards the development of p-QDC-QDG-FET is under progress.

In conclusion, these devices show potential in the development of advanced architectures for high-speed processing and multibit storage in volatile/non-volatile memory devices [4].

Acknowledgments

The authors would like to thank the Department of Electrical and Computer Engineering for the lab support. We would like to thank Dr. Ron LaComb for metal deposition on the device.

References

1. F. C. Jain, E. Heller, S. Karmakar, and J. Chandy, Device and circuit modeling using novel 3-state quantum dot gate FETs, *2007 International Semiconductor Device Research Symposium*, 2007, pp. 1-2.
2. M. Lingalugari, K. Baskar, P.-Y. Chan, P. Dufilie, E. Suarez, J. Chandy, E. Heller, and F.C. Jain, Novel multi-state quantum dot gate FETs using SiO2 and lattice-matched ZnS-ZnMgSZnS as gate insulators, Journal of Electronic Materials, **42**(11), 3156-3163 (2013).
3. P.-Y. Chan, M. Lingalugari, E. Heller, and F. Jain, An investigation on quantum dot superlattice (QDSL) diode, International Journal of High Speed Electronics and Systems, **23**(01n02), 1420004 (2014).
4. F. Jain, R. Gudlavalleti, R. Mays, B. Saman, P.-Y. Chan, J. Chandy, M. Lingalugari, and E. Heller, Quantum dot channel FETs harnessing mini-energy band transitions in GeOx-Ge and Si QDSL for multi-bit computing, International Journal of High Speed Electronics and Systems, **31**(01n04), 2240012 (2022).

Enhancing Number of Bits Via Mini-Energy Band Transitions Using Si Quantum Dot Channel (QDC) and Ge Quantum Dot Gate (QDG) FETs and NVRAMs

F. Jain[*,§], R. H. Gudlavalleti[*], A. Almalki[*], B. Saman[†], P-Y. Chan[*], J. Chandy[*],
F. Papadimitrakopoulos[*] and E. Heller[‡]

[*]*University of Connecticut, CT, USA*
[†]*Taif University, Saudi Arabia*
[‡]*Synopysis Inc., Ossining, NY, USA*
[§]*faquir.jain@uconn.edu*

This paper presents multi-state QDC-QDG FET structures that has the potential to introduce additional states (8 or 16) by utilizing additional mini-energy sub-bands. Mini-energy bands are formed in Si quantum dot channel (QDC) comprising two silicon oxide cladded Si quantum dots (QDs). Quantum simulations are presented to show more states when additional two germanium oxide cladded Ge dots are added on top of two Si QD layers in the gate region. With the addition of a control gate oxide layer, we transform the QDC-QDG-FET into a quantum dot (QD) nonvolatile random access memory (NVRAM). Quantum simulations are presented.

Keywords: Quantum dot FETs; multi-state FETs; QD-NVRAMs; 16-state FETs.

1. Introduction

Cladded Si Quantum dot channel (QDC) FETs incorporating Si Quantum dot gate layers have been reported to experimentally exhibit distinct 4-state ID-VD and ID-VG characteristics [1]. In addition to ON and OFF, two intermediate states are attributed to carrier transfer in the inversion layer mini-energy sub-bands manifested in QDC due to the formation of quantum dot superlattice (QDSL) shown in Figs. 1(a) and 1(b).

Figure 2(a) shows schematic of a spatial wavefunction switched (SWS) FET and Fig. 2(b) shows the conduction in lower and upper channels for a fabricated device exhibiting 4-states [2].

Unlike, SiOx-cladded Si quantum dots, the GeOx-cladded Ge dot form mini-energy bands are separated by smaller energies and hence have the potential of manifesting additional logic states in SWS-QDC-QDG-FETs and QDC-QDAC-NVRAMs.

Figure 3(a) shows the computed mini-energy bands in quantum dot superlattice formed by 4nm Ge quantum dots with 0.5nm GeOx cladding. Figure 3(b) depicts the density of states (DOS). Figure 3(c) shows the carrier density as a function of Fermi level location for

---

[§]Corresponding author.

direct gap mini-energy sub-bands for 4nm, 5nm, and 6nm GeOx-cladded Ge quantum dots. Carrier density for indirect and direct mini-energy gap sub-bands has also been reported [3].

Recently, we reported [3] an 8-state SWS-QDC-based sub-9nm FET using two Ge quantum dot channels (QD2 and QD1) shown in Fig. 4. The occupation of mini-energy bands is listed in terms of wavefunction location in lower or upper channel, and various sub-bands fully or partially occupied and enabling current conduction.

This paper presents multi-state QDC-QDG and QD-NVRAM structures that have the potential to introduce more than 8 states by utilizing additional mini-energy sub-bands in the channel and gate regions. This is presented next in Section 2.

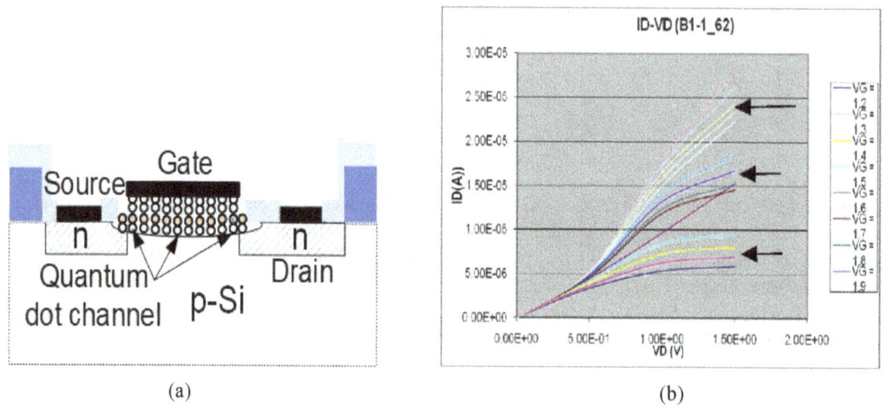

(a)   (b)

Fig. 1. (a) Schematic of QDC-QDG FET. (b) ID-VD characteristics showing distinct characteristics as a function of gate voltage.

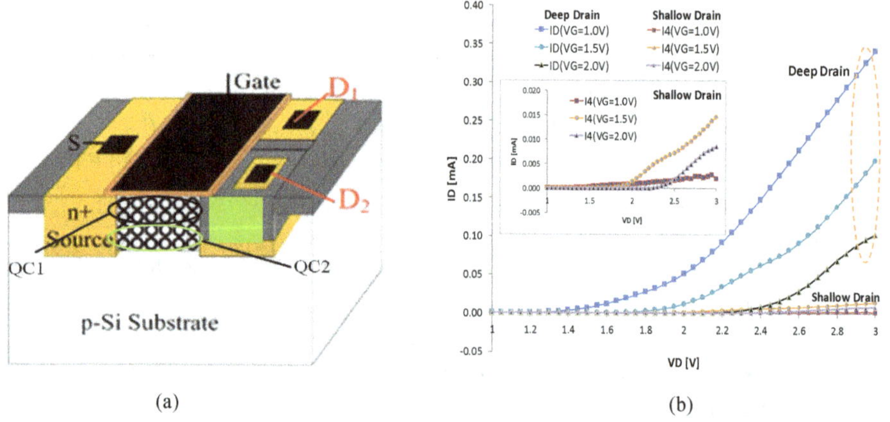

(a)   (b)

Fig. 2. (a) Schematic of 2-Si quantum dot channel SWS-NMOS. Deep drain D2 is connected to QC2 comprising of lower most two QD layers, and shallow drain D1 is connected to QC1, upper QD layers. (b) I_D-V_D characteristics of 2-QD channel SWS-NMOS.

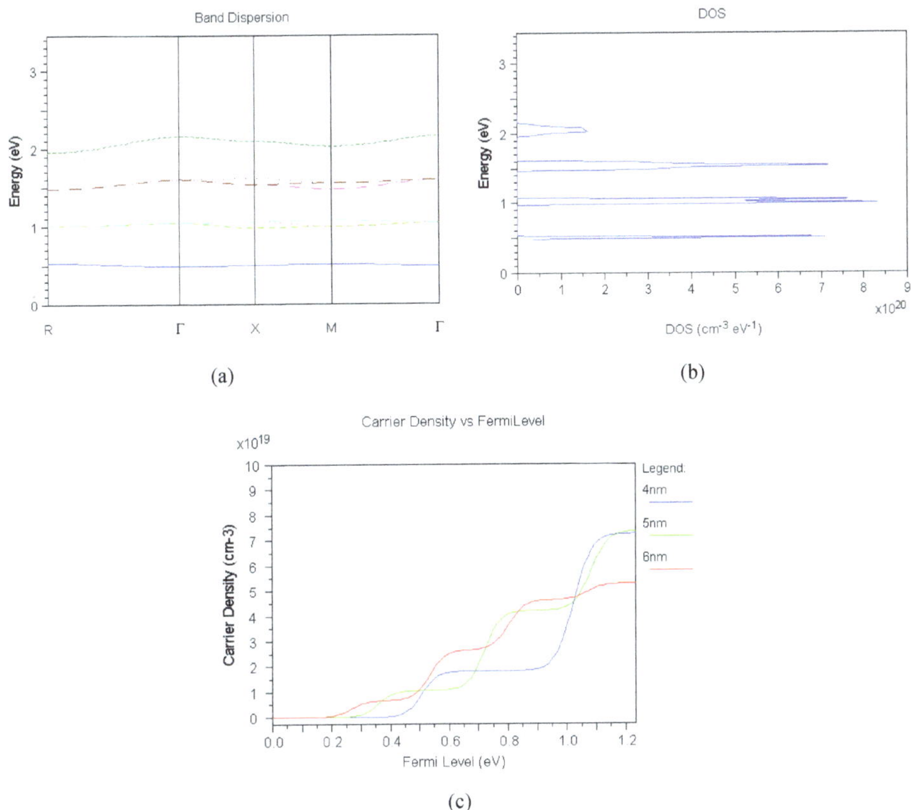

Fig. 3. (a) Energy band diagram of GeOx (barrier/cladding 0.5nm) – Ge quantum dot (4nm) QDSL. (b) Density of states (DOS) in sub-bands in Ge QDSL. (c) Simulated carrier density in direct mini-energy sub bands.

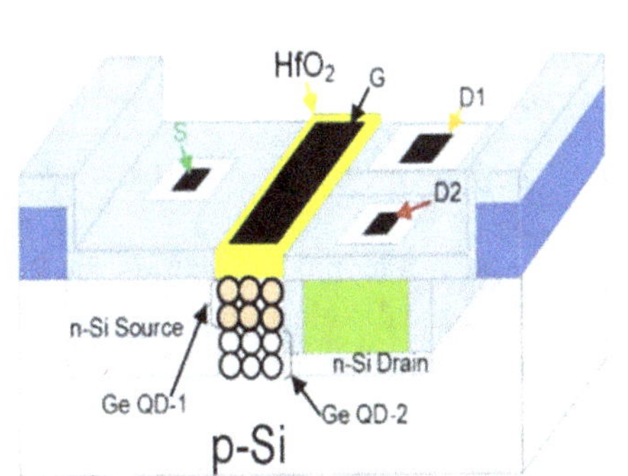

The eight states are:
(i)   no electrons in any channel (00),
(ii) electron in first sub-band of QC2 (0_B1),
(iii) electrons in first and second sub-bands of QC2 (0_B2B1),
(iv) electrons in three sub-bands of QC2 (0_B3B2B1),
(v)   electrons in first sub-band of QC1 and two sub-bands of QC2 (B1_B2B1),
(vi) electrons in first and second sub-bands of QC1 and first sub-bands of QC2 (B2B1_B1),
(vii) electrons in three subbands of QC1 and first sub-band of QC2 (B3B2B1_B1),
(viii) electrons in three subbands of QC1 and none in QC2 (B3B2B1_0).

Fig. 4. GeO$_x$-Ge QDC-FET with two-quantum dot channels QD2 and QD1. The listing shows the mini-energy sub-band transport states enabling current flow and wavefunction switching from QD2 to QD1.

2. Integrating Additional Oxide Cladded Ge Quantum Dot Layers in the Gate Region

In this section, schematic of a QDC-QDG-FET structure is shown in Fig. 5(a). QDC-QDG-FET builds on experimentally reported QDC-QDG FET [Fig. 1(a)] with two Ge QD layers in the gate region. Figure 5(b) presents the quantum simulations. We envision 4 Ge QD layers on Si QD layers in the gate region, which will permit additional states (e.g. 16).

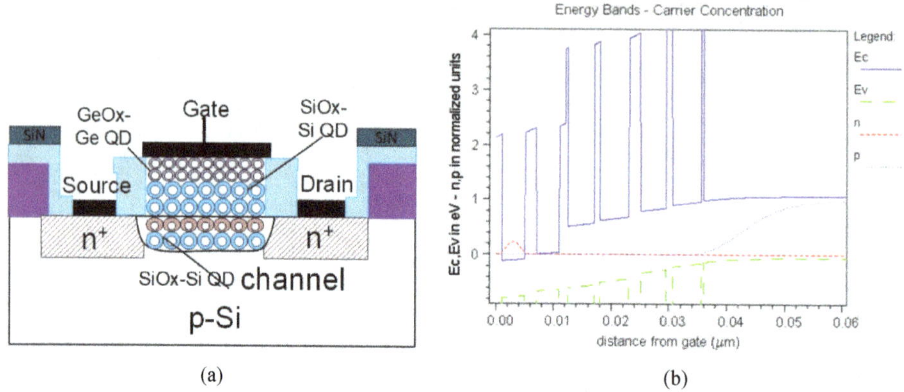

Fig. 5. (a) Asymmetric SiOx-cladded Si quantum dot channel with Si and Ge QD layers in the gate region. (b) Simulation results indicate the wavefunction is bound to the top Ge QD layers. (RIGHT)

Figure 5(a) QDC-QDG structure could be modified by incorporating two quantum dot channels as shown in Fig. 2(a). Here, we have two symmetric and two asymmetric Si QD layers forming SWS-QDC-FET. In addition, it can be further expanded by having two symmetric Si QD layers, and two Ge QD layers forming quantum dot gate (QDG) is shown in Fig. 5(a). Variation of Ge QD and Si QD diameters as well as cladding oxide thickness will influence the location of wavefunction in QDG.

Simulation of the energy bands, such as that in Fig. 5, was performed by solving the Poisson equation (1) and the Schrödinger equation (3), self-consistently [4, 5].

$$\nabla \cdot (\varepsilon \nabla \varphi) = q(n_{QM} + n - N_D^+ + N_A^- - p) \tag{1}$$

In equation (1), ε is the permittivity, φ is the electrostatic potential, and n_{QM} is the 2D electron gas in the quantum well layers. Also, n and p are the 3D electron and hole concentrations, respectively. Lastly, N_D^+ and N_A^- are the ionized donor and acceptor concentrations, respectively. The 2D carrier concentration, n_{QM}, is expressed by equation (2), where E_F is the Fermi level, E_n denotes the eigen energies of the bound states, and ψ_n denotes the corresponding wavefunctions. Finally, Θ, is the Heaviside Step function.

$$n_{QM} = \sum_n \frac{m^*}{\pi \hbar^2} k_B T \Theta(E_F - E_n) \ln\left[1 + \exp\left(\frac{E_f - E_n}{kT}\right)\right] |\psi_n|^2 \tag{2}$$

The Schrödinger equation provides the E_n and ψ_n for a given potential energy profile V, which is a function of position along the heterostructure. V, is defined by the individual material properties of each layer, such as energy gap and electron affinity, plus the electrostatic potential, φ, from the Poisson equation. These equations are solved iteratively until a self-consistent solution is determined.

$$\frac{\hbar^2}{2}\nabla\cdot\left(\frac{1}{m*}\nabla\psi_n\right)+(E_n-V)\psi_n=0 \tag{3}$$

The QDSL band structure and density of states (DOS), such as those in Fig. 3, were determined by approximating the quantum dots as 3D cuboids, and solving the Schrödinger equation (3), by applying Kronig Penny (K-P) model. In this case, V (eqn 4), is the potential profile due to differences between the QD core and cladding properties, and each, E(q) (eqn 5), is the energy solution of the K-P characteristic equation, for each direction, at wavevector q [6].

$$V(r)=V_X(x)+V_Y(y)+V_Z(z) \tag{4}$$

$$E(q)=E_X(q_X)+E_Y(q_Y)+E_Z(q_Z) \tag{5}$$

Finally, the DOS was determined by integrating the band structure over the 1st Brillouin Zone.

The simulation parameters are listed in Table 1. The yellow highlighted HfO_2 and GeO_x cladding and Ge QD core layers are for QD-NVRAM structure shown in Fig. 6(a).

3. QD-NVRAM with Asymmetric SiOx-Si QDC for Additional States

Figure 6(a) QD-NVRAM with asymmetric SiOx-cladded Si quantum dot channel and Si QD floating gate and Ge quantum dot access channel (QDAC).

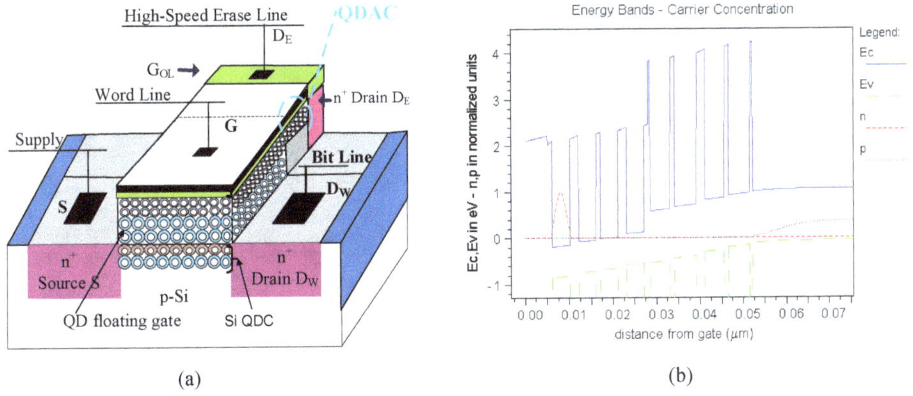

Fig. 6. (a) QD-NVRAM with asymmetric SiOx-cladded Si quantum dot channel and Si QD floating gate and Ge quantum dot access channel (QDAC). (b) Wavefunction transfer to Ge QD floating gate layer adjacent to HfO_2 gate.

The parameters used in quantum simulations of QD-NVRAM of Fig. 6(a) are shown in Fig. 6(b). Here, the gate region over the floating gate comprises germanium oxide cladding layer and HfO_2 layers serving as control gate oxide. The QDAC part does not have HfO_2 layer and thus permits fast erase via dedicated drain D2 similar to prior reporting [7]. Two-dimensional SiOx-cladded Si and GeOx-cladded Ge Quantum Dot Arrays vertically stacked for CMOS-X Logic, SRAMs, NVRAMs and IR imaging and multi-bit Computing have been recently reported [8].

Table 1. Parameters used in quantum simulations.

Layer	Thick (um)	Chi (eV)	Eg (eV)	me	mh	er	Nd (cm^{-3})	Na (cm^{-3})
HfO$_2$	0.0050	2.05	5.7	0.16	0.50	13.7	0.0e00	0.0e00
GeOx clad	0.0010	2.25	5.70	0.16	0.16	4.4	0.0e00	0.0e00
GeQD core	0.0040	4.55	0.67	0.08	0.28	16.0	0.0e00	0.0e00
GeOx clad	0.0020	2.25	5.70	0.16	0.16	4.4	0.0e00	0.0e00
GeQD core	0.0040	4.55	0.67	0.08	0.28	16.0	0.0e00	0.0e00
GeOx clad	0.0010	2.25	5.70	0.16	0.16	4.4	0.0e00	0.0e00
GeQD core	0.0040	4.55	0.67	0.08	0.28	16.0	0.0e00	0.0e00
GeOx clad	0.0020	2.25	5.70	0.16	0.16	4.4	0.0e00	0.0e00
GeQD core	0.0040	4.55	0.67	0.08	0.28	16.0	0.0e00	0.0e00
GeOx clad	0.0010	2.25	5.70	0.16	0.16	4.4	0.0e00	0.0e00
SiOx clad	0.0005	0.9	9.0	0.5	0.5	3.9	0.0e00	0.0e00
Si QD	0.0045	4.15						
SiOx clad	0.001	0.9	9.0	0.5	0.5	3.9	0.0e00	0.0e00
Si QD	0.0050	4.15						
SiOx clad	0.0005	0.9	9.0	0.5	0.5	3.9	0.0e00	0.0e00
SiOx clad	0.0015	0.9	9.0	0.5	0.5	3.9	0.0e00	0.0e00
Si QD	0.0045	4.15						
SiOx clad	0.0010	0.9	9.0	0.5	0.5	3.9	0.0e00	0.0e00
Si QD	0.0050	4.15						
.SiOx clad	0.0005	0.9	9.0	0.5	0.5	3.9	0.0e00	0.0e00
Si	0.5000	4.15	1.12	0.19	0.49	11.9	0.0e00	1.0e16

4. QDC-FETs on Poly-Si Thin Film and Potential to Implement 3D FETs and NVMs

We reported [1] the experimental data of QDC-FET fabricated on poly-Si layer. The QDC-FET cross-sectional schematic is shown in Fig. 7(a). The ID-VD and ID-VG characteristics are shown in Figs. 7(b) and 7(c). The computed electron mobility is about 80 cm^2/V-sec.

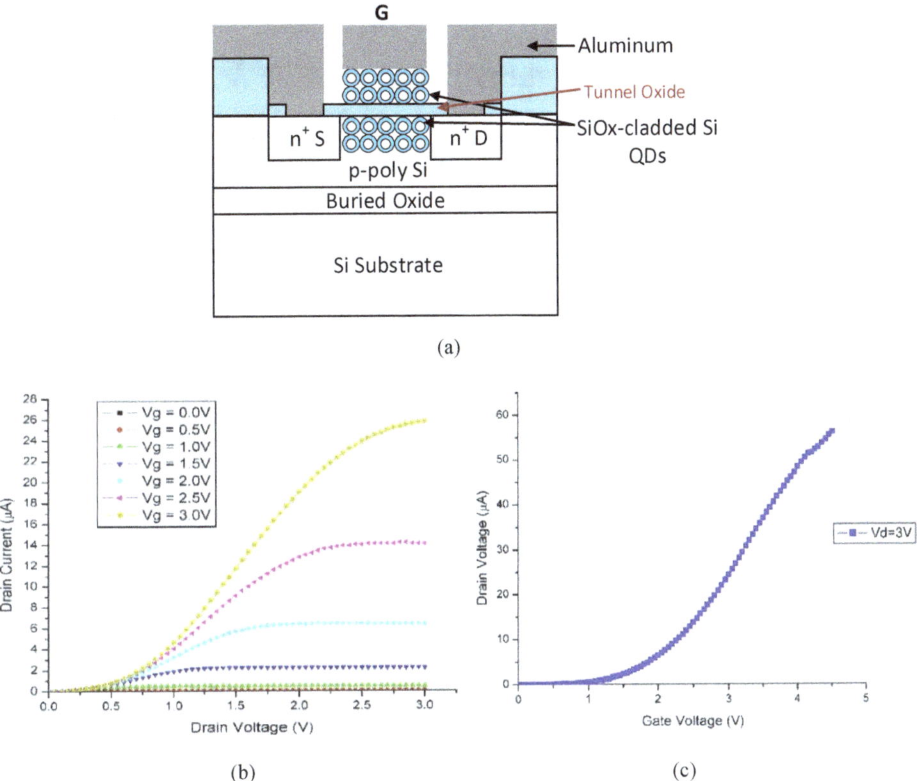

(a)

(b)   (c)

Fig. 7. (a) Cross-sectional schematic. (b) ID-VD characteristics of a QDC-FET on p-poly-Si, (c) ID-VG plot.

The electron mobility is two orders of magnitude higher than commonly exhibited in p-poly-Si FETs.

The high mobility in quantum dot channel QDC realized in p-poly Si thin films can be utilized to improve the performance of 3D QDC-NAND flash cells and QDC-QDG-FETs, which will be dealt with in Section 5.

5. 3D Stack of QDC-FETs and QDC-NAND with Gate All Around (GAA)

Figure 8(a) shows schematically a poly-Si device having a high-mobility Si Quantum dot Channel (QDC), which hosts the inversion channel in 3D NAND, and yields higher Write speeds. Figure 8(b) shows floating gate also comprising quantum dot layers, which will enable multi-bit storage due to quantum dot superlattice formation. We also envision, as shown in Fig. 8(c), a 3D stack of QDC-QDG FET in gate all around (GAA) configuration.

The novel feature is hosting the inversion layer in the Si quantum dot channel (QDC) on poly-Si core in gate all around (GAA) configuration. GAA configuration has been reported for V-NANDs [9].

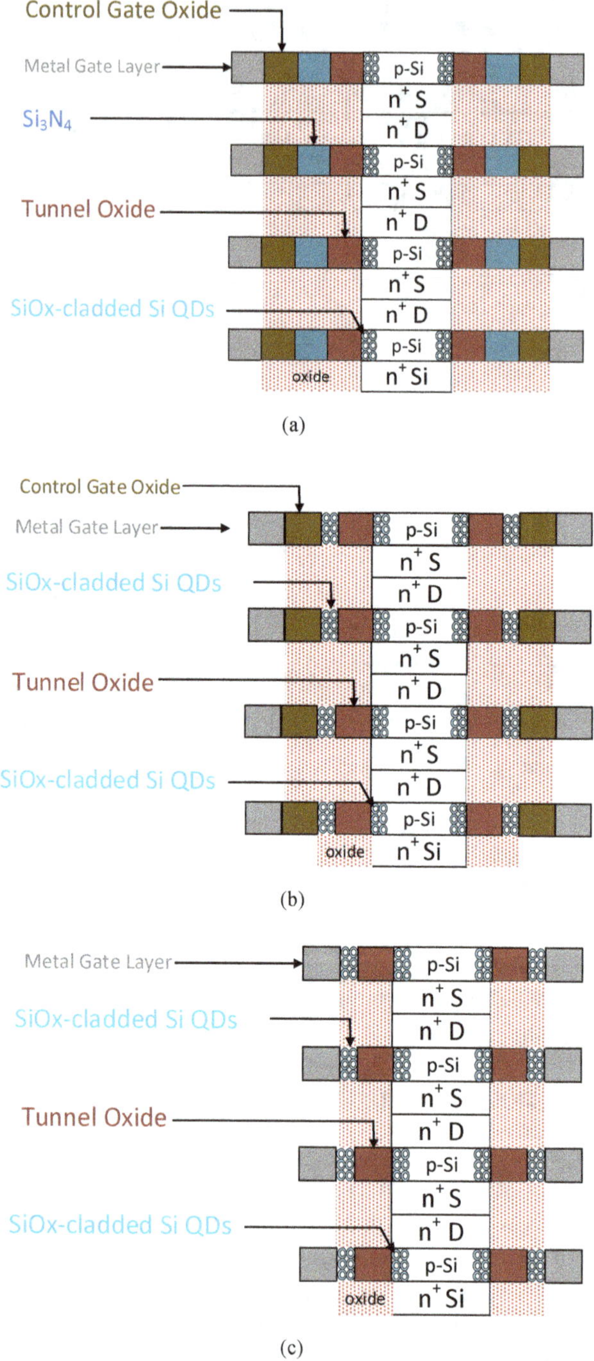

Fig. 8. (a) 3D high-mobility QDC-NVM cells in gate-all-around (GAA) configuration. (b) 3D quantum dot floating gate QDC-QD-NVM cells in GAA configuration. (c) 3D QDC-QDG FETs in gate all around (GAA) configuration.

6. Conclusion

Novel QD structures and quantum simulations show the potential of achieving 8-16 states in NanoFETs and QD-NVRAMs. Four-state and 8-states (3-bit) inverters and SRAMs have been reported [8, 10]. This work opens the way for 4-bit/16-state QD-based logic and memory circuits. 3-bit/4-bit QD-SRAM increases the memory density.

Increasing the states is obtained in QDC, SWS and QDG configuration by methods including: (1) changing the thickness of upper cladding of upper QD of QD channel, (ii) changing the QDC SiOx-cladded Si QD layers, (iii) making two QD layers asymmetrical by controlled oxidation of upper SiO2 layer (see Fig. 1a), (iv) changing the Ge core diameter of two uppermost Ge QD layers.

Additionally, by utilizing multistate QD-SRAM with spatial wavefunction switched (SWS)-QDC-FETs-based dot-product multiplier, increases data transfer efficiency in AI/ML data computation systems. Raja *et al.* [10] reported using QD-SRAMs for compute in memory (CIM) applications. The CMOS compatible QD structure forms a new paradigm to implement architecture based on multi-state/multi-bit logic, SRAMs, QD-NVRAMs, and CIM / In-Memory Computing [11].

The integration of high-mobility quantum dot channels in stacked p-poly Si cores with gate all around (GAA) configuration sets the stage for a new paradigm. Additionally, we have shown in Fig. 8, 3D stacks of multistate QDC-NVM, QDC-QD-NVM, and QDC-QDG-FETs. This methodology will be an interim room temperature alternative to qubit realized on Si [12] and opens the pathway to QDSL-based quantum computing using more than 1-2 electrons per QD. In addition, multi-state FETs and CMOS-compatible QD-NVRAMs and SRAMs have the potential for applications in-memory computing, compute-in-memory (CIM), and artificial intelligence (AI) hardware platform with higher speeds and much smaller silicon footprint and power dissipation.

References

1. F. Jain, S. Karmakar, P.-Y. Chan, E. Suarez, M. Gogna, J. Chandy and E. Heller, Quantum dot channel (QDC) field-effect transistors (FETs) using II-VI barrier layers, *J. Electron. Mater.* **41**, 2775 (2012).
2. F. Jain, M. Lingalugari, B. Saman, P.-Y. Chan, P. Gogna, E.-S. Hasaneen1, J. Chandy and E. Heller, Multi-state sub-9 nm QDC-SWS FETs for compact memory circuits, *46th IEEE Semiconductor Interface Specialists Conference (SISC)*, 2–5 December, 2015.
3. F. Jain, R. Gudlavalleti, R. Mays, B. Saman, P-Y. Chan, J. Chandy, M. Lingalugari, and E. Heller, Multi-State Quantum Dot Channel (QDC) FETs for Multi-Bit Computing, *52nd IEEE Semiconductor Interface Specialists Conference (SISC)*, 8–11 December, 2021.
4. F. Jain, M. Lingalugari, J. Kondo, P. Mirdha, E. Suarez, J. Chandy and E. Heller, Quantum dot channel (QDC) FETs with Wraparound II-VI gate insulators: Numerical simulations. *J. Electron. Mater.* **45**, 5663 (2016).
5. E. K. Heller, S. K. Islam, G. Zhao and F. C. Jain, *Solid-State Electron.* **42**, 901–914 (1999).
6. C.-W. Jiang and M. A. Green, *J. Appl. Phys.* **99**, 114902 (2006).

7. M. Lingalugari, P.-Y. Chan, E. K. Heller, J. Chandy, and F. C. Jain, Quantum dot floating gate nonvolatile random access memory using quantum dot channel for faster erasing, *Electron. Lett.* **54**, 36 (2018).

8. F. Jain, R. Gudlavalleti, R. Mays, B. Saman, P-Y. Chan, J. Chandy, M. Lingalugari and E. Heller, Two-dimensional SiOx-cladded Si and GeOx-cladded Ge quantum dot arrays vertically stacked for CMOS-X Logic, SRAMs, NVRAMs and IR imaging and multi-bit computing, SISC, December 2022.

9. R. Meyer, Y. Fukuzumi and Y. Dong, 3D NAND scaling in next decade, *Proc. IEDM*, 26.1.1-26.1.4, pp. 599–602, December 2022.

10. R. H. Gudlavalleti, E. Heller, J. Chandy and F. Jain, Computing-In-Memory SRAM Cell using multistate spatial wavefunction switching (SWS)-quantum dot channel (QDC) FET, *Int. J. of High Speed Electron. Syst.*, **32**, 2350012, 2023.

11. D. Lelmini and H-S. P. Wong, In-memory computing with resistive switching devices, *Nat. Electron.* **1**, 333–337 (2018).

12. A. J. Sigillito, J. C. Loy, D. M. Zajac, M. J. Gullans, L. F. Edge and J. R. Petta, Site-selective quantum control in an isotopically enriched 28Si/Si0.7Ge0.3 quadruple quantum dot, *Phys. Rev. Appl.* **11**, 061006 (2019).

https://doi.org/10.1142/9789811283765_0016

Filtration Methods for Microplastic Removal in Wastewater Streams — A Review

U. Salahuddin[*], J. Sun[*], C. Zhu[*,†] and P. Gao[*,†,‡]

[*]*Institute of Materials Science, University of Connecticut, Storrs, CT 06269-3136, USA*
[†]*Department of Materials Science and Engineering, University of Connecticut, Storrs, CT 06269-3136, USA*
[‡]*puxian.gao@uconn.edu*

Microplastics are commonly recognized as environmental and biotic contaminants. The prevalent presence of microplastics in aquatic settings raises concerns about plastic pollution. Therefore, it is critical to develop methods that can eliminate these microplastics with low cost and high effectiveness. This review concisely provides an overview of various methods and technologies for removing microplastics from wastewater and marine environments. Dynamic membranes and membrane bioreactors are effective in removing microplastics from wastewater. Chemical methods such as coagulation and sedimentation, electrocoagulation, and sol-gel reactions can also be used for microplastic removal. Biological methods such as the use of microorganisms and fungi are also effective for microplastic degradation. Advanced filtration technologies like a combination of membrane bioreactor and activated sludge method show high microplastic removal efficiency.

Keywords: Microplastics; filtration; quantification; water treatment.

1. Introduction

Microplastics are plastics with a size of less than 5 mm, usually formed by degradation and exfoliation of large quantities of plastics released into the ecosystem [1]. A plethora of reports suggest accumulation of these microplastics in the marine environment and their harmful effects on the aquatic life [2, 3]. Microplastics in the ocean can be attributed from different resources, as shown in Fig. 1(a). In 2018, 35% of microplastics in the ocean come from synthetic textiles; 28% comes from car tires, and 24% is from city dust. It is estimated that microplastics account for 94% of the Great Pacific Garbage Patch in the ocean [4, 5]. Besides the challenge of a large quantity of microplastics accumulation, microplastics have a broad size distribution and are hard for accurate quantification which makes the microplastics study even more complicated. As indicated in Fig. 1(b), microplastics size distribution can range from 10 μm to 32000 μm. The particle size distribution is highly dependent on the quantification method and locations, as shown in a comparison of data between Cózar *et al.* (2014) [6] and Enders *et al.* (2015) [7]. What is more, waste plastics vary significantly in chemical composition, physical form, size, texture, and shape. These characteristics can even evolve while in use and after discard. Hence, the removal of these

---

[‡]Corresponding author.

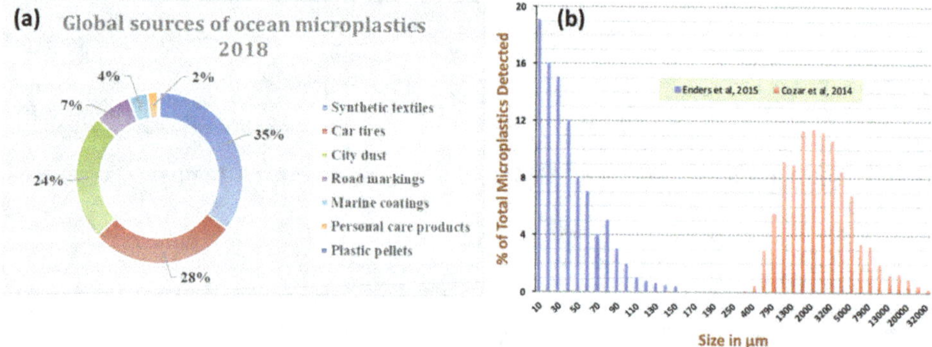

Fig. 1. Global sources of ocean microplastics in 2018 and (b) Microplastic size distributions depend on sampling location (geographically and vertically in the water column) and analytical methods applied [4].

microplastics is a critical yet challenging topic that needs immediate attention of the research community. Currently, many novel technologies and approaches have been applied to achieve efficient microplastics removal from fresh and saltwater environments. This document entails the methods and technologies most widely reported in the literature for microplastics' removal from wastewater or marine environments.

Based on the source of plastics, the waste plastics can be characterized into primary and secondary microplastics. Primary microplastics are plastics directly released into the environment in their original form. They can be a voluntary addition to products such as scrubbing agents in toiletries and cosmetics or they can come from the abrasion of large plastic objects during manufacturing, use or maintenance, such as the erosion of tires when driving or of the abrasion of synthetic textiles during washing. Secondary microplastics are microplastics originating from the degradation of larger plastic items into smaller plastic fragments once exposed to marine environment. This happens through photodegradation and other weathering processes of mismanaged waste such as discarded plastic bags or from unintentional losses such as fishing nets [8, 9].

Microplastics can be toxic to marine life and the ecosystem. The potential toxicity of microplastics can be from unreacted monomers, oligomers and chemical additives that tend to leak out from the plastic in the long run or physical effect when absorbed and transferred to living bodies. The toxicity depends also physically on the size and shape of micro-plastics. Biota and humans are affected by toxic effects of microplastics via mechanisms including sorption and aggregation in different organs, ingestion, and exertion of physical damages. Although a lot of research studies have been conducted on the microplastics toxicity, effort is still needed to clarify how microplastics induce tissue changes and pathological disorders [10–14].

As mentioned, accumulation of waste microplastics in the water system has been growing, resulting in significant environmental issues [15]. The use of plastics will only increase in the foreseeable future. Thus, it is vital to find an effective solution to the microplastic pollution and eventually solve the problem from the source. Recently, many efforts have been made and much progress has been achieved for waste microplastics removal in the water system. Projects organizations like the Ocean cleanup project and Planet Care and actively working to reduce and mitigate plastic waste from the environment [16, 17]. Moreover, research groups such as Hale *et al.* [18], Kvale *et al.* [19] and Wang *et al.* [20] have been actively working on understanding and developing novel solutions for microplastic removal from the oceans. Thus, in-time summary and understanding of the technologies involved are necessary to guide the future study.

There are reviews that provide us with a wholesome view on microplastic concentration levels in the ocean, health issues related to microplastic contamination as well as future of microplastic pollution [21–24], however, reviews focusing on summarizing all the available technologies for microplastic filtration are lacking. Here, in this short review, we summarize the recent methods and technologies that are used for efficient microplastics removal from waste and marine water systems.

2. Filtration Methods

There are some conventional methods such as rapid sand filters, dissolved air flotation, disk filters as well as membrane filters. Some of these methods are already being used in the industry. Other emerging technologies are less prevalent and are only proven at a smaller scale in research laboratories. For ease of understanding, we have divided the methods into four categories, conventional, which includes the most commonly used methods at present, chemical, biological, and advanced technologies, as schemed in Fig. 2.

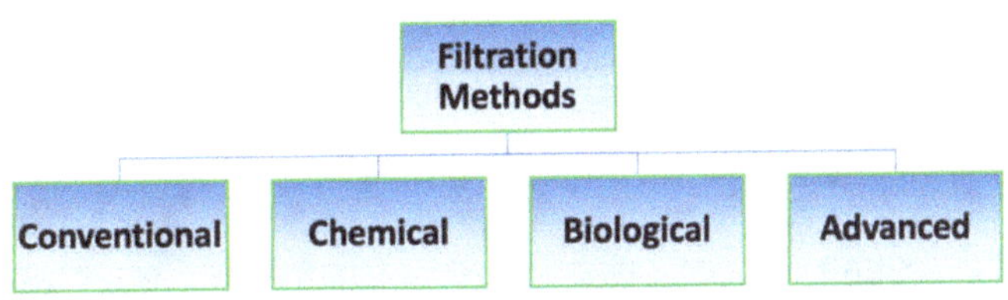

Fig. 2. Classification of filtration method based on category.

2.1. *Conventional Methods*

Microplastics can be separated from wastewater using dynamic membranes. As shown in Fig. 3, Li *et al.* reported the effects of variables like particle concentration and influent flux on the removal efficiency of membranes formed on a diatomite platform with a 90 μm supporting mesh [25]. The membrane reduced the turbidity from 195 NTU to less than 1 NTU in just under 20 minutes [26]. Higher influxes of influents and higher concentrations of microplastics facilitate the formation of dynamic membranes.

Membrane bioreactor is another technique that offers higher capacities relative to dynamic membranes [27]. This technique has the advantage of handling a large variety of wastewater and high-strength contaminants. Talvitie *et al.* compared the removal efficiencies of various technologies such as disk filters, rapid sand filters, dissolved air

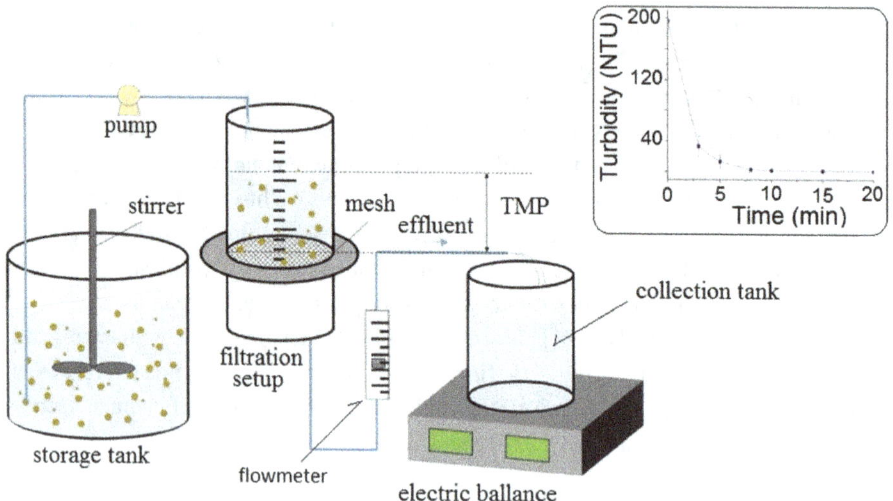

Fig. 3. Dynamic membrane experimental setup and graph showing the decrease in turbidity with time when microplastics are removed. TMP transmembrane pressure, NTU nephelometric turbidity unit [10].

floating and membrane bioreactors [28]. They showed that membrane bioreactor out-performed all other technologies by eliminating 99.9% of microplastics from the influent stream. The study also showed that membrane bioreactors can remove microplastic particles as small as 20 μm. The shape of the particles also did not seem to affect the efficacy of this method, as FTIR analysis on the effluent stream showed extremely low concentrations of microplastics. This fact hints towards good sorption capabilities of membrane reactors towards microplastics possessing various chemical structures.

2.2. *Chemical Methods*

Since the size range for microplastics is very small ranging from a couple of microns up to several hundred microns, their separation is a big challenge. Many commercial-scale wastewater treatment plants use technologies like coagulation and sedimentation to circumvent this problem [29, 30]. Large contaminant particles that are formed after coagulation and sedimentation processes are easier to separate. These processes usually involve Al- or Fe-based salts that bind tiny microplastic particles using a ligand exchange mechanism [31]. A recent study investigated the coagulation of polyvinyl chloride particles < 50 μm, using ferric and aluminum sulphate [32]. The use of ferric and aluminum sulphate as coagulants resulted in the removal of about 80% of MPs. Specifically, polyvinyl chloride (PVC) MPs with a size less than 50 μm were successfully coagulated by ferric and aluminum sulfate, and MPs with a size of at least 15 μm were eliminated under optimized coagulation conditions (ferric sulphate at 20 mg L^{-1} and pH 7 or aluminum sulphate at 40 mg L^{-1} and pH 7).

Electrocoagulation is another technique used for microplastic removal that shows higher efficiency, incurs lower cost, allows sludge minimization, etc. [33]. Zeng *et al.*

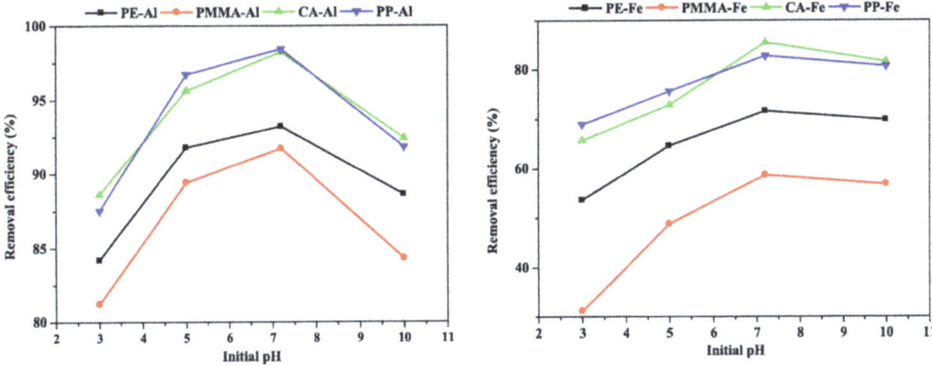

Fig. 4. Effect of initial pH on removal of microplastics during electrocoagulation [34]. (Copyright © 2021 Elsevier B.V.)

reported the findings that aluminum anode was better than iron anode in the removal of microplastics, and the removal rate was above 80% in all experiments [34]. As shown in Fig. 4, the removal rate of four microplastics by electrocoagulation can reach more than 82% in the range of pH 3–10, and the best removal rate was 93.2% for PE, 91.7% for PMMA, 98.2% for CA and 98.4% for PP at pH 7.2.

Herbort *et al.* demonstrated an agglomeration method based on alkoxy-silyl bond formation via a sol-gel reaction [35]. The sol-gel formed in this method resembles a hybrid organic-inorganic silica gel that can be used in a wide variety of applications [36]. Photocatalytic degradation is a well-established method for the destruction of polymeric chains [37]. Li *et al.* suggested an aqueous phase photochemical method for the degradation of microplastics, in which the degradation reaction is promoted by hydroxyl radicals [38].

Moreover, heterogenous photocatalytic degradation of low-density polyethylene over rod-like ZnO nanoparticles in aquatic media has been recently investigated by Tofa *et al.* [39]. Results showed that the nanorod surface area greatly affected the photocatalytic performance of the reaction. Another study showed the efficacy of titania-based nano-devices for microplastic photocatalytic treatment [40]. A similar study shows the ability of protein-based TiO_2 photocatalyst to degrade polyethylene microplastics in solid and aqueous media. The results showed a mass loss of 1.1% and 6.4% respectively for solid and aqueous phases.

2.3. *Biological Methods*

Microorganisms including eukaryotes, bacteria and archaea have excellent potential in degradation of microplastics in coastal sediments and marine environments. Dawson *et al.* studied polyethylene fragmentation and size alteration by Antartic Krill, a planktonic crustacean [41]. The stock suspension contained beads with a mean diameter of 31.5 µm. When particles were isolated from krill, their mean size was 78% smaller than the original beads, with some fragments reduced by 94% of their original diameter, resulting in an average size of 7.1 μm. Fecal material also contained smaller particles with a mean size of 6.0 μm. Moreover, the size distribution of particles within the krill and excreted particles was significantly different from the beads in the exposure stock. The smaller size of plastic particles found in krill and their fecal pellets suggests that the Antarctic krill are physically breaking down the beads after ingestion. Cocca *et al.* reported a four-month long study on the removal of high-density polyethylene microplastics in seawater using two types of indigenous marine communities, namely, Agios consortium and Souda consortium [42]. The study showed that microplastics acted as a rich carbon source for the microorganisms. Paco *et al.* demonstrated the use of fungus Zalerion maritimum for the biodegradation of polyethylene in a batch reactor [43]. The results showed that the fungus used microplastics as a nutrient source. Sangale *et al.* discovered the elite polythene deteriorating fungi (isolated from the rhizosphere soil of Avicennia marina) [44]. The degradation potential of polythene by fungi was evaluated at different pH levels (3.5, 7, and 9.5). The study was carried out by subjecting the polythene samples to continuous shaking at ambient temperature for 60 days and monitoring changes in weight and tensile strength. The BAYF5 isolate (pH 7) exhibited the highest reduction in weight (58.51 ± 8.14), while PNPF15 (pH 3.5) showed the greatest reduction in tensile strength (94.44 ± 2.40). The results of degradation of the polythene strips by the fungal isolates were confirmed by the formation of the cracks/holes/scions and were visualized in Scanning electron microscopic photographs as shown in Fig. 5(B).

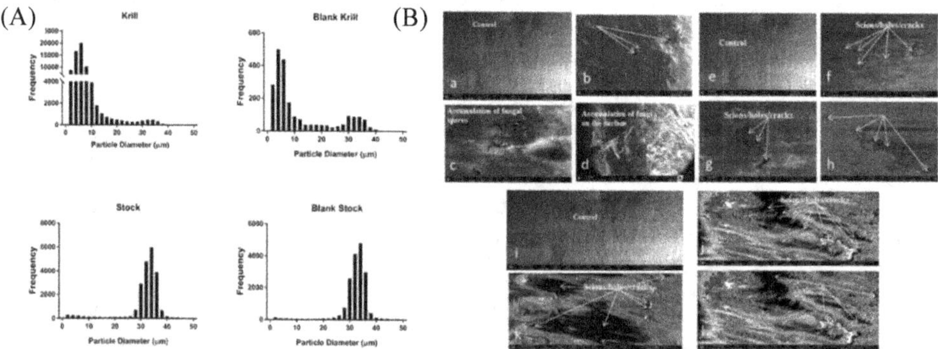

Fig. 5. (A) Size distribution of particles isolated from sample Antarctic krill (Euphausia superba), unhomogenized and enzyme digested Antarctic krill, stock suspension, and stock suspension enzyme digested. Particles between 25 and 50 μm indicate a whole bead; values below this range are fragments. [41] (Copyright ©, Springer Nature 2018), (B) Scanning Electron Microscopic image of the polythene strips: (a–d) SEM of the PE with most %weight gain (24.4%) by JAMNF at pH9.5; (e–h) SEM of PE strip with maximum (94) % loss in TS by PNPF-15 at pH 3.5; (i–l) SEM of PE strips of maximum %WL (41%) by MANGF1 [45]. (Copyright © 2019, Springer Nature Singapore Pte Ltd. 2019)

3. Advanced Filtration Technologies

A combination of membrane bioreactor and activated sludge method was recently studied by Lares *et al.* at a pilot scale [46]. They collected samples every 2 weeks for 3 months.

The pilot scale setup included an aeration tank for microorganism activation, a sedimentation tank for sludge separation and a secondary purifier. A microplastic removal efficiency of 99.4% was reported. Similarly, a study in China showed a comparison of 11 different wastewater treatment plants for the efficiency to remove microplastics [47]. An efficiency of above 90% was witnessed by all the plants that used multiple filtration steps.

Moreover, another study analyzed the removal methodology used by municipal sewage treatment plants for the removal of microplastics [48]. The study showed that the influents were treated initially using a series of processes, such as grit chamber, primary and secondary sedimentation tanks, anerobic, aerobic and anoxic treatments. Advanced treatment processes such as denitrification, ultrafiltration, ozonation, and ultraviolet treatment were later applied to remove the microplastics. The overall efficiency of plastic removal was reported to be 95.16% which was still lower than membrane bioreactors (99.9%). The study also reported that sewage treatment plants release a large quantity of microplastics into the aquatic ecosystem as these plants are not primarily designed for microplastic removal.

4. Challenges and Limitations

All these methods detailed above have their own limitations and challenges. Some of the key challenges are limitations are summarized in Table 1.

Table 1. Comparison of available microplastic filtration methods [27, 28]

Technology	Conventional Filters	Chemical/ Electrochemical Methods	Biological Methods	Other multistage filtration techniques
Filtration size range (µm)	> 90	2000-5000	1-30	> 90
Efficiency (%)	90-99	60-65	< 15	97-99
Regeneration	Highly energy intensive	Not possible	Easy	Complicated
Scalability	Scalable but usually high cost	Scalable but of chemical and or electrical current constraints	Limited due to low plastic consumption rates of microorganisms	Possible but costly due to multistage processes
Limitations	Expensive regeneration & costly membrane filters	Water quality affected by chemical treatment, requiring post-treatments	Slow kinetics of biological digestion	Multistep processes with high cost

5. Conclusions and Outlook

Microplastics are widely acknowledged as contaminants in both the environment and biota. Their omnipresence in aquatic environments has raised concerns about plastic pollution, highlighting the urgent need for effective and affordable removal methods. This paper presents an overview of the various techniques and technologies used to remove microplastics from marine environments and wastewater. Dynamic membranes and membrane bioreactors are effective in removing microplastics from wastewater, while chemical methods such as coagulation and sedimentation, electrocoagulation, and sol-gel reactions can also be used. Biological methods, such as the use of microorganisms and fungi, have also proven effective for microplastic degradation. Advanced filtration technologies, such as a combination of membrane bioreactor and activated sludge method, demonstrate high microplastic removal efficiency.

As the detrimental impacts of microplastics become more apparent, the demand for effective filtration methods continues to grow. In the future, it is likely that research will focus on developing more cost-effective and scalable technologies for removing microplastics from wastewater. Additionally, there may be a shift towards more sustainable and eco-friendly filtration methods, such as biological methods. Advances in nanotechnology may also lead to the development of more efficient and precise filtration methods. This review clearly shows the need for a robust, cheap, and efficient method for microplastic removal from waste streams. Conventional wastewater treatment plants are not typically designed to filter microplastics and hence any supplementary microplastic filtration devices that can be retrofitted to the existing wastewater treatment plants could really solve this problem. However, the retrofitting needs to be modular and scalable in nature to accommodate the variable volumes of wastewater.

As of now, there are no regulations or laws restricting the water treatment plants for assessing and minimizing the microplastics concentrations in their effluent water. If regulated, these treatment plants will be forced to retrofit their treatment plants with microplastic capturing filters. In addition to the methods reported in this paper, there has been an increasing interest in using ceramic-based filters for microplastics filtration. These ceramic filters have proven to be low cost and are an robust alternative to all other filters reported in this review.

Another challenging problem in microplastic filtration is microplastic quantification in water-based samples. Since microplastic particles are neutral in charge, non-reactive and extremely resistant to any solvent or chemical, it is extremely difficult to quantify them. Available methods include microscopic counting and fluorescent tagging, but these methods only quantify in terms of number of particles per unit volume of water. What is needed is a scalable technique that can accurately provide a weight-based quantification of microplastics in water samples. The significance of microplastic quantification is two-fold. First, weight-based quantification is crucial to assessing water quality. Second, it also helps in assessing the efficacy of a given microplastic filtration device. Going forward, the exploration and development of new strategies would be very important toward achieving quantitative differentiation and measurement of microplastics in water stream, such as advanced remote sensing, optical imaging and spectroscopy tools, coupled with various data analytics and machine learning capabilities.

Acknowledgment

The authors acknowledge the financial support from UConn CARIC program and NSF.

References

1. Zhang, Kai *et al.* "Microplastic pollution in China's inland water systems: A review of findings, methods, characteristics, effects, and management." *Science of the Total Environment* **630** (2018) 1641–1653.
2. Van Cauwenberghe, Lisbeth, *et al.*, "Microplastic pollution in deep-sea sediments." *Environmental Pollution* **182** (2013) 495–499.
3. Law, Kara Lavender, *et al.* "Distribution of surface plastic debris in the eastern Pacific Ocean from an 11-year data set." *Environmental Science & Technology* **48**(9) (2014) 4732–4738.
4. Lebreton, Laurent, *et al.*, "Evidence that the Great Pacific Garbage Patch is rapidly accumulating plastic." *Scientific Reports* **8**(1) (2018) 1–15.
5. Chong, Fiona, *et al.*, "High concentrations of floating neustonic life in the plastic-rich North Pacific Garbage Patch." *Plos Biology* **21**(5) (2023) e3001646.
6. González-Fernández, Daniel, *et al.*, "Microplastics in estuarine waters: Surface layer vs. water column." (2022).
7. Čerkasova, Natalja, *et al.*, "A public database for microplastics in the environment." *Microplastics* **2**(1) (2023) 132–146.
8. Boucher, Julien, and Damien Friot, Primary microplastics in the oceans: A global evaluation of sources. Vol. 10. Gland, Switzerland: IUCN, 2017.
9. Kasmuri, Norhafezah, Nur Aliah Ahmad Tarmizi, and Amin Mojiri, "Occurrence, impact, toxicity, and degradation methods of microplastics in environment—a review." *Environmental Science and Pollution Research* **29**(21) (2022) 30820–30836.
10. Padervand, Mohsen, *et al.*, "Removal of microplastics from the environment. A review." *Environmental Chemistry Letters* **18** (2020) 807–828.
11. Parsai, Tanushree, *et al.* "Implication of microplastic toxicity on functioning of microalgae in aquatic system." *Environmental Pollution* (2022) 119626.
12. D'Costa, Avelyno H. "Microplastics in decapod crustaceans: Accumulation, toxicity and impacts, a review." *Science of The Total Environment* (2022) 154963.
13. Salimi, Ahmad *et al.*, "Differences in sensitivity of human lymphocytes and fish lymphocytes to polyvinyl chloride microplastic toxicity." *Toxicology and Industrial Health* **38**(2) (2022) 100–111.
14. da Silva Brito, Walison Augusto *et al.* "Comprehensive in vitro polymer type, concentration, and size correlation analysis to microplastic toxicity and inflammation." *Science of The Total Environment* **854** (2023) 158731.
15. Zhou, Chongyu, *et al.* "The emerging issue of microplastics in marine environment: A bibliometric analysis from 2004 to 2020." *Marine Pollution Bulletin* **179** (2022) 113712.
16. Brodin, Malin, *et al.* "Filters for washing machines: Mitigation of microplastic pollution." (2018).
17. Holst, Rozemarijn Roland, "The Netherlands: The 2018 agreement between the ocean Cleanup and the Netherlands." *The International Journal of Marine and Coastal Law* **34**(2) (2019) 351–371.

18. Hale, Robert C., *et al.* "A global perspective on microplastics." *Journal of Geophysical Research: Oceans* **125**(1) (2020) e2018JC014719.

19. Kvale, Karin F., A. E. Friederike Prowe and Andreas Oschlies, "A critical examination of the role of marine snow and zooplankton fecal pellets in removing ocean surface microplastic." *Frontiers in Marine Science* (2020) 808.

20. Wang, Ziheng, Majid Sedighi and Amanda Lea-Langton, "Filtration of microplastic spheres by biochar: Removal efficiency and immobilisation mechanisms." *Water Research* **184** (2020) 116165.

21. Zhang, Kai, *et al.* "Understanding plastic degradation and microplastic formation in the environment: A review." *Environmental Pollution* **274** (2021) 116554.

22. Sharma, Shivika and Subhankar Chatterjee, "Microplastic pollution, a threat to marine ecosystem and human health: A short review." *Environmental Science and Pollution Research* **24** (2017) 21530–21547.

23. Danopoulos, Evangelos, Maureen Twiddy and Jeanette M. Rotchell, "Microplastic contamination of drinking water: A systematic review." *PloS one* **15**(7) (2020) e0236838.

24. do Sul, Juliana A. Ivar and Monica F. Costa. "The present and future of microplastic pollution in the marine environment." *Environmental Pollution* **185** (2014) 352–364.

25. Li, Lucheng, *et al.*, "Dynamic membrane for micro-particle removal in wastewater treatment: Performance and influencing factors." *Science of the Total Environment* **627** (2018) 332–340.

26. Ersahin, Mustafa Evren, *et al.* "Impact of anaerobic dynamic membrane bioreactor configuration on treatment and filterability performance." *Journal of Membrane Science* **526** (2017) 387–394.

27. Lares, Mirka, *et al.* "Occurrence, identification and removal of microplastic particles and fibers in conventional activated sludge process and advanced MBR technology." *Water Research* **133** (2018) 236–246.

28. Talvitie, Julia, *et al.* "Solutions to microplastic pollution–Removal of microplastics from wastewater effluent with advanced wastewater treatment technologies." *Water Research* **123** (2017) 401–407.

29. Hu, Chengzhi *et al.* "Effect of aluminum speciation on arsenic removal during coagulation process." *Separation and Purification Technology* **86** (2012) 35–40.

30. Shirasaki, N., *et al.*, "Effect of aluminum hydrolyte species on human enterovirus removal from water during the coagulation process." *Chemical Engineering Journal* **284** (2016) 786–793.

31. Chorghe, Darpan, Mutiara Ayu Sari, and Shankararaman Chellam. "Boron removal from hydraulic fracturing wastewater by aluminum and iron coagulation: Mechanisms and limitations." *Water Research* **126** (2017) 481–487.

32. Prokopova, Michaela, *et al.* "Coagulation of polyvinyl chloride microplastics by ferric and aluminium sulphate: optimisation of reaction conditions and removal mechanisms." *Journal of Environmental Chemical Engineering* **9**(6) (2021) 106465.

33. Perren, William, Arkadiusz Wojtasik, and Qiong Cai. "Removal of microbeads from wastewater using electrocoagulation." *ACS Omega* **3**(3) (2018) 3357–3364.

34. Shen, Maocai, *et al.* "Efficient removal of microplastics from wastewater by an electro-coagulation process." *Chemical Engineering Journal* **428** (2022) 131161.

35. Herbort, Adrian Frank, *et al.*, "Alkoxy-silyl induced agglomeration: a new approach for the sustainable removal of microplastic from aquatic systems." *Journal of Polymers and the Environment* **26**(11) (2018) 4258–4270.

36. Nicole, Lionel, *et al.* "Advanced selective optical sensors based on periodically organized mesoporous hybrid silica thin films." *Chemical Communications* **20** (2004) 2312–2313.

37. Yousif, Emad and Raghad Haddad, "Photodegradation and photostabilization of polymers, especially polystyrene." *SpringerPlus* **2**(1) (2013) 1–32.

38. Li, Shengying *et al.*, "Photocatalytic degradation of polyethylene plastic with polypyrrole/TiO2 nanocomposite as photocatalyst." *Polymer-Plastics Technology and Engineering* **49**(4) (2010) 400–406.

39. Tofa, Tajkia Syeed, *et al.* "Visible light photocatalytic degradation of microplastic residues with zinc oxide nanorods." *Environmental Chemistry Letters* **17**(3) (2019) 1341–1346.

40. Sekino, Takahiro, Satoru Takahashi and Kiyoshi Takamasu, "Fundamental study on nano-removal processing method for microplastic structures using photocatalyzed oxidation." *Key Engineering Materials* **523** (2012).

41. Dawson, Amanda L., *et al.*, "Turning microplastics into nanoplastics through digestive fragmentation by Antarctic krill." *Nat. Commun.* **9**(1) (2018) 1–8.

42. Cocca, Mariacristina, *et al.*, eds. Proceedings of the 2nd International Conference on Micro-plastic Pollution in the Mediterranean Sea, Springer Nature (2020).

43. Paço, Ana *et al.* "Biodegradation of polyethylene microplastics by the marine fungus Zalerion maritimum." *Science of the Total Environment* **586** (2017) 10–15.

44. Shahnawaz, Mohd, Manisha K. Sangale, and Avinash B. Ade, "Bioremediation technology for plastic waste." (2019) 978–981.

45. Shahnawaz, Mohd, Manisha K. Sangale, and Avinash B. Ade, "Bioremediation technology for plastic waste." (2019) 978–981.

46. Lares, Mirka, *et al.*, "Occurrence, identification and removal of microplastic particles and fibers in conventional activated sludge process and advanced MBR technology." *Water Research* **133** (2018) 236–246.

47. Ma, Baiwen, *et al.*, "Removal characteristics of microplastics by Fe-based coagulants during drinking water treatment." *Journal of Environmental Sciences* **78** (2019) 267–275.

48. Yang, Libiao, *et al.*, "Removal of microplastics in municipal sewage from China's largest water reclamation plant." *Water Research* **155** (2019) 175–181.

Numerical Investigation of the Electrothermal Properties of SOI FinFET Transistor

Faouzi Nasri[*]

Center for Research in Microelectronics and Nanotechnology (CRMN), Sousse, Tunisia
Laboratory of Thermal Processes, Research and Technology Centre of Energy, Hammam-Lif, Tunisia
[*]Nasrifaouzi90@yahoo.fr

Husien Salama

Computer Systems Institute, Boston, USA
husien.salama@uconn.edu

This paper investigates the non-Fourier transient heat transfer in an SOI FinFET transistor. The calibrated drift-diffusion (D-D) model in conjunction with the ballistic diffusive (BDE) model is used as an electrothermal model to predict phonon and electron transports in the quasi-ballistic regime. The finite element method has been employed to generate the numerical results. The proposed mathematical formulation was found to capture the transfer characteristics and the temporal temperature as given by TCAD simulation and experimental data. On the other hand, we have demonstrated that after 100 ns, the 14 nm Bulk FinFET supports better temperature distribution than the 14 nm SOI FinFET.

Keywords: SOI FinFET; electrothermal modeling; phonon transport; effective mobility.

1. Introduction

The downscaling of Multigate FET structures has led to enormous problems concerning their thermal and electrical performances [1]. Among these types of transistors, the FinFET and the SOI FinFET have immunity against the short-channel effect. Many other problems attack these architectures, including the heat flux and the operating temperature inside the transistor channel region [2]. To overcome these defects, several experimental and numerical investigations have been done [3]. In this paper, to improve the electrical and thermal effects of an SOI FinFET device, we have used a highly nonlinear transient analysis model, which could be used to elucidate the phonon scattering and electron transports in a 20 nm SOI FinFET. We report the drain and gate-source voltages effect and high-k dielectric material role in increasing the operating temperature in the given structure. In the first step, we compared the transfer characteristics with experimental and numerical data. After validation of the electrical part of the proposed model, we compared

---

[*]Corresponding author.

the temporal temperature evolution given by the proposed model with those available in the literature. On the other hand, we have compared the temperature given in the SOI FinFET structure with that obtained in Bulk FinFET. This paper is organized as follows. In Section 2, we provide an overview of the electrical approach as well as theoretical considerations relative to the thermal simulation. In Section 3, we present and discuss the obtained results. Concluding remarks are provided in Section 4.

2. Electrothermal Formulation

To enhance the D-D model and to make it valid in the nanoscale regime, we introduce in this work the effective mobility model given by [4]

$$\mu_{eff} = \frac{\mu_0}{1 + \theta_1(V_G - V_{TH}) + \theta_2(V_G - V_{TH})^2} \tag{1}$$

where θ_1 and θ_2 are represented as fitting parameters, that describe respectively mobility degradation due to phonon scattering and degradation due to surface roughness scattering. Equations mentioned below represent the D-D model which is given by the poison equation coupled with the electron and hole continuity equations [5]:

$$\varepsilon\Delta V = -q(p - n + N_D - N_A) \tag{2}$$

where ε is the permittivity, V is the electrical potential. p, n, N_D, and N_A are, respectively, the hole, electron, donor, and acceptor concentrations, and q is the elementary charge. Electron and hole equations (D-D model):

$$\frac{\partial n}{\partial t} = \nabla \cdot (n\mu_{eff}\, E + D_n\nabla n) + (R - G) \tag{3}$$

$$\frac{\partial p}{\partial t} = -\nabla \cdot (p\mu_p E + D_p\nabla p) + (R - G) \tag{4}$$

where μ_{eff} and μ_p are the electron and hole mobilities used in this work. D_n and D_p are the electron and hole diffusion coefficient.

The key parameter that could give us much information about the heat transfer inside a nanodevice is the volumetric heat generation rate Q due to Joule effect which represents the link between both electrical and thermal mathematical models. The source term Q is given by [6]

$$Q = \vec{J}.\vec{E} + (R - G).(E_g + 3K_BT) \tag{5}$$

The heat transfer equation related to the BDE model is expressed as follows [20]:

$$\tau_R\frac{\partial^2 T(r,t)}{\partial t^2} + \frac{\partial T(r,t)}{\partial t} = \frac{1}{c}\left(k\Delta T(r,t)\right) - \frac{1}{c}\nabla q_b(r,t) + \frac{Q}{c} + \frac{\tau_R}{c}\frac{\partial Q}{\partial t} \tag{6}$$

where τ_R is the relaxation time due to resistive collision, K and C are, respectively, the thermal conductivity and the volumetric heat capacity and q_b is the ballistic heat flux.

In our simulation the ballistic heat flux and the relaxation time are calculated as follows [7]:

$$q_b = k_{eff}\left(\frac{Cw}{2Kn}\right)\frac{\Delta T}{L} \tag{7}$$

In this work, we have analyzed the electrothermal properties of SOI FinFET with channel length of 20 nm, as given in Fig. 1.

All needed parameters are given in Table 1 [8, 9].

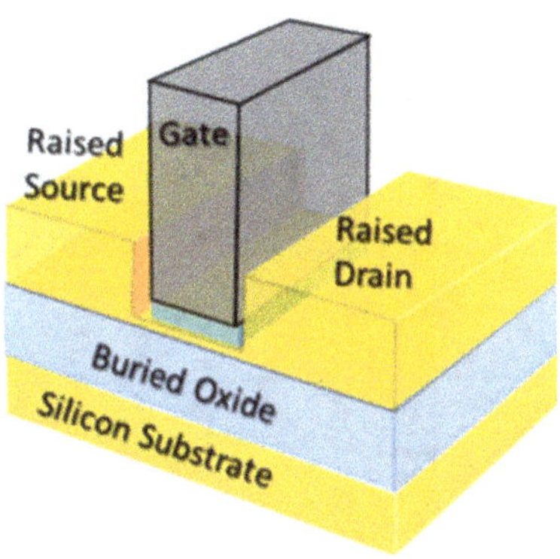

Fig. 1. Schematic geometry of SOI FinFET transistor [10].

Table 1. Physical parameters of the studied device.

Symbol	FinFET transistor
Gate length (nm)	20
EOT (nm)	0.62
Channel p type doping (cm^{-3})	1e15
S/D n type doping (cm^{-3})	1e21

3. Results and Discussions

To validate the proposed electrothermal model, we present in Fig. 2 a comparison of the drain current versus gate bias with TCAD simulation [11] and with experimental data [12]. It is obvious that the proposed model can predict the transfer characteristics as given by TCAD simulation and experimental data.

Figure 3 depicts a comparison of the volumetric heat generation inside a 20 nm FinFET and SOI FinFET transistors at Vds=Vgs=0.7 V, the Joule effect is maximal at the channel region-spacer oxide side interface and decreases at drain region. On the other hand, we can

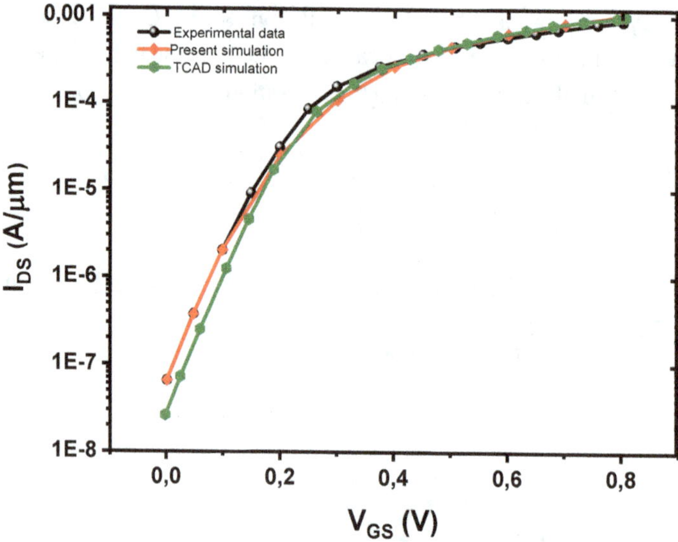

Fig. 2. Comparison of the transfer character of the 20 nm SOI FinFET with experimental data and TCAD simulation at Vds=0.7 V.

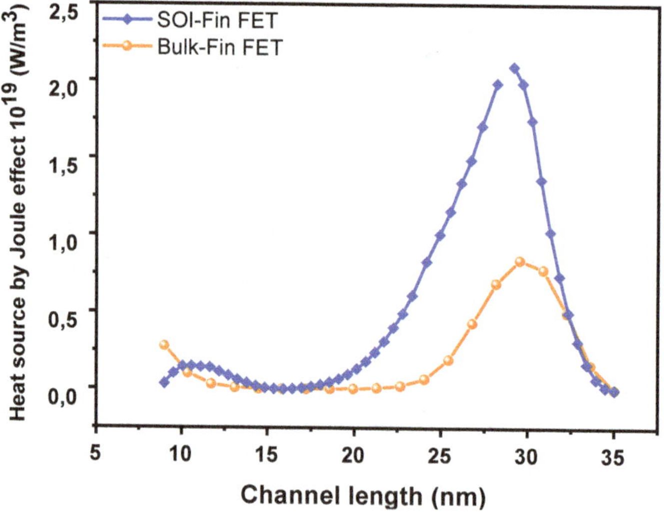

Fig. 3. Comparison of the Joule effect inside FinFET and SOI FinFET transistors at Vgs=Vds=0.7 V.

see that the volumetric heat generation given in SOI FinFET transistor is almost twice as large as that given in FinFET device. The key to the temperature increase is the volumetric heat generation position and its higher value, as given in Fig. 4, which represents a comparison of the temporal temperature evolution in 20 nm FinFET and SOI FinFET devices at Vgs=Vds=0.7 V. From this figure, we show that the FinFET temperature achieved the saturation after 10 ns, however, it achieved the saturation in SOI FinFET case

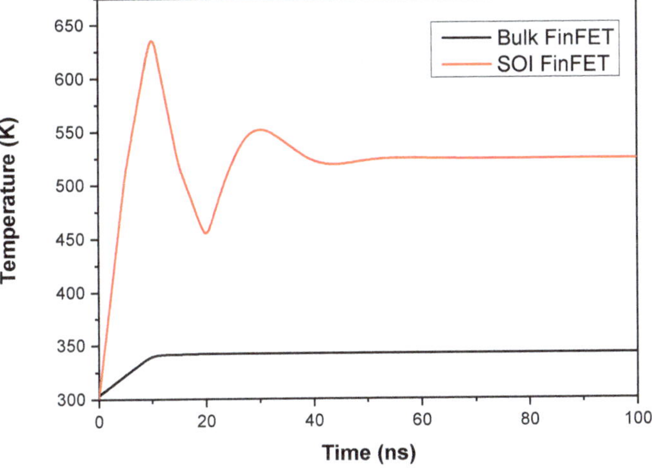

Fig. 4. Comparison of the temporal temperature evolution of FinFET and SOI FinFET devices at Vds=Vgs=0.7 V.

after 45 ns. The temperature variation in SOI FinFET case is due to the presence of the buried oxide which makes the temperature trapped in the channel region.

Figure 5 shows the 2D temperature distribution in the studied SOI FinFET at t=100 ns and VDS=VGS=0.8 V. We show that the temperature is maximal in the channel region of the studied device, and the temperature is concentrated in the channel region drain zone interface. The weak point of this architecture appears when we use the buried oxide to eliminate the leakage currents. On the other hand, the SiO2 buried oxide is responsible for locking up the temperature in the channel region and the source/drain sides.

Fig. 5. 2D temperature card in FinFET at t=100 ns and VDS=VGS=0.8 V.

4. Conclusion

In this work, we have proposed an electrothermal model in conjunction with an effective electron mobility model to investigate the temperature evolution inside a 20 nm SOI FinFET transistor. We have found that our proposed model is able to predict the transfer characteristics as given by experimental data and TCAD simulation. On the other hand, we have found that the buried oxide makes the temperature trapped in the SOI FinFET channel region and the saturated temperature achieved 525 K after 100 ns, however, it achieved 350 K in FinFET case.

Otherwise, the nanotechnology promises significant advances in electronics, materials, biotechnology, alternative energy sources and other applications. Nanocrystals, nanotubes, nanowires, and nanofibers are all next-generation materials [13, 15]. The future direction of this research is to generalize the proposed method to be used in various important applications such as micro-stereolithography.

References

1. F. Nasri and M. Atri, Channel length effect on heat transfer process in nano MOSFETs, in *2017 International Conference on Engineering & MIS (ICEMIS)* (IEEE, 2017), pp. 1–4.
2. F. Nasri and M. F. Ben Aissa, Thermal analysis of phonon temperature in nano transistor based on DPL model coupled with temporal temperature jump, in *2017 International Conference on Engineering & MIS (ICEMIS)* (IEEE, 2017), pp. 1–4.
3. F. Nasri, M. F. Ben Aissa and H. Belmabrouk, Nanoheat conduction performance of black phosphorus field-effect transistor, *IEEE Transactions on Electron Devices* **64** (2017) 2765–2769.
4. H. Rezgui *et al.*, Analysis of the ultrafast transient heat transport in sub 7-nm SOI FinFETs technology nodes using phonon hydrodynamic equation, *IEEE Transactions on Electron Devices* **68**(1) (2021) 10–16.
5. H. Rezgui, F. Nasri, G. Nastasi, M. F. Ben Aissa, S. Rahmouni, V. Romano, H. Belmabrouk and A. A. Guizani, Design optimization of nanoscale electrothermal transport in 10 nm SOI FinFET technology node, *Journal of Physics D: Applied Physics* **53**(49) (2020) 495103.
6. J. W. Lee, Electrical characterization and modeling of low dimensional nanostructure FET, Universite de Grenoble; 239, Korea University, Français, NNT (2011), 2011GRENT070ff, tel00767413f, version 1.
7. F. Nasri, H. Rezgui and M. F. Ben Aissa, Numerical investigation of nano heat transfer process in FET nano devices using ballistic-diffusive model, in *2018 IEEE International Conference on Smart Materials and Spectroscopy (SMS)*, Hammamet, Tunisia (IEEE, 2018), pp. 12–17, https://doi.org/10.1109/SMS44485.2018.9101397.
8. H. Rezgui, F. Nasri, M. F. Ben Aissa, H. Belmabrouk and A. A. Guizani, Modeling thermal performance of nano-GNRFET transistors using ballistic-diffusive equation, *IEEE Transactions on Electron Devices* **65** (2018) 1611–1616, https://doi.org/10.1109/TED.2018.2805343.
9. H. Belmabrouk, H. Rezgui, F. Nasri, M. F. Ben Aissa and A. A. Guizani, Interfacial heat transport across multilayer nanofilms in ballistic–diffusive regime, *The European Physical Journal Plus* **135** (2020) 109, https://doi.org/10.1140/epjp/s13360-020-00180-7.

10. K. Schuegraf, M. C. Abraham, A. Brand, M. Naik and R. Thakur, Semiconductor logic technology innovation to achieve sub-10 nm manufacturing, *IEEE Journal of the Electron Devices Society* **1**(3) (2013) 66–75, doi:10.1109/JEDS.2013.2271582.

11. J. Sun, X. Li, Y. Sun and Y. Shi, Impact of geometry, doping, temperature, and boundary conductivity on thermal characteristics of 14-nm bulk and SOI FinFETs, *IEEE Transactions on Device and Materials Reliability* **20** (2020) 119–127, https://doi.org/10.1109/TDMR.2020.2964734.

12. C.-H. Lin *et al.*, High performance 14nm SOI FinFET CMOS technology with 0.0174μm^2 embedded DRAM and 15 levels of Cu metallization, in *2014 IEEE International Electron Devices Meeting* (IEEE, 2014), pp. 3.8.1–3.8.3, https://doi.org/10.1109/IEDM.2014.7046977.

13. H. Salama, B. Smaani, F. Nasri and A. Tshipamba, Nanotechnology and quantum dot lasers, *Journal of Computer Science and Technology Studies* **5**(1) (2023) 45–51, https://doi.org/10.32996/jcsts.2023.5.1.6.

14. H. Salama, B. Saman, R. Gudlavalleti, R. Mays, E. Heller, J. Chandy and F. Jain, Compact 1-bit full adder and 2-bit SRAMs using n-SWS-FETs, *International Journal of High Speed Electronics and Systems* **29**(1–4) (2020) 2040013, https://doi.org/10.1142/S0129156420400133.

Memristor-Based Material Implication Logic:
Prelude to In-Memory Computing

Anas Mazady[†] and Mehdi Anwar[*,‡]

*Electrical and Computer Engineering Department, University of Connecticut, Storrs, CT 06269, USA
†Northrop Grumman Corporation, Baltimore, MD 212, USA
†aalvee_asad.kausani@uconn.edu
‡a.anwar@uconn.edu

We report experimental demonstration of *Material Implication* (IMP) logic using ZnO nanowire-based memristors. The logic is demonstrated with a high-to-low resistance ratio of only five. This imposes much less stringent requirements on memristor performance that can enable IMP logic operation with lower bit error rates. Process independence on memristor and memristor-based IMP logic performance is demonstrated, and a more practical implementation of logic is made by relaxing the restriction imposed on the ranges of the values of on and off state resistances. IMP logic is validated up to a clock frequency of 100 KHz.

Keywords: IMP logic; nanowire-based memristors; in-memory computing.

1. Introduction

Circuit designers are confronted with two distinct challenges, with one being evolutionary that has to do with scaling power consumption with shrinking device dimensions and the other one being revolutionary that seeks innovative concepts to take technology beyond Moore's Law. At present, the technology of choice refers to the complementary metal oxide semiconductor (CMOS) field effect transistors with their small footprint, low power consumption, and small propagation delay. However, memristors with the promise to deliver a scalable and low power computation platform have the potential to replace the present digital logics as predicted by Di Ventra *et al.* [1]. Designing logics using memristors could allow the use of the same physical unit performing multiple functions, such as memory, logic, and interconnect [2]. This approach has the potential to redefine traditional computer architectures to realize a more advanced architecture that overcomes the "von Neumann bottleneck" of throughput [3]. Traditional von Neumann architectures require sequential processing of fetch, decode, and execution of instructions resulting in reduced data transfer rate. Moreover, this exchange of data between the processor and memory could be prone to side channel attacks leading to security breach. Performing logic

‡Corresponding author.

operations within the memory elements permits the central processing unit (CPU) to simultaneously access both the program memory and data memory utilizing the full potential of the CPU.

Scaling down of power consumption with shrinking device dimensions is an ongoing effort in low power electronics, however, the ever-shrinking technology nodes are not followed by power supply voltage scaling due to restrictions imposed by MOSFET threshold voltage (V_T). Any decrease in V_T results in an exponential increase in leakage current, detrimentally affecting voltage scaling and increasing the static power consumption of the CMOS logic family. For example, with the listed leakage current of 1 μA, the static power dissipation P_S of a 2-input CMOS NAND gate (74HC00) equals 3 μW for a supply voltage V_{DD}=3 V. The dynamic power, for example, for the same 2-input CMOS NAND gate, could be as high as 53.1 μW [$P_D=C_{PD} \times V_{CC}^2 \times f_i \times N + C_L \times V_{CC}^2 \times f_0$], where C_{PD} =22 pF is the dynamic power dissipation capacitance, f_i =100 kHz is the input signal frequency, N =2 is the number of bits switching, C_L=15 pF is the load capacitance including jig and probe capacitances, and f_o =100 kHz is the output signal frequency. For a switching transition time of 19 ns, the switching energy equals 1 pJ that could have been reduced if we were able to scale voltage without increasing leakage current. The switching transition time is defined as the sum of rise and fall times of the output signal, where rise time is the time required for the signal to swing from 10% to 90% of the "logic 1" voltage and fall time is the time required for the signal to go from 90% to 10% of the same voltage amplitude.

The total leakage current in MOSFETs, for technology nodes smaller than 65 nm, is primarily tunneling in nature and becomes more pronounced for thinner oxides [4]. In recent years, high-k dielectrics with the dielectric constants k ranging from 7.5 to 300 have been explored to reduce tunneling current. Ultra high-k materials, such as BST (Ba,Sr)TiO$_3$ with k of 300 may provide the thinnest equivalent oxide thickness (EOT) but suffers from fringing-induced barrier lowering (FIBL) [5, 6], consequently lowering the threshold voltage with a concomitant increase in leakage current. High-k materials with moderately ranging dielectric constants, such as Si$_3$N$_4$ (k=7.5) and Al$_2$O$_3$ (k=10), contain random interface trap (RIT) states resulting in increased Coulomb scattering degrading career mobility. ZrO$_2$ (k=22) and HfO$_2$ (k=25) happen to be the materials of choice for gate oxide. However, incompatibility with poly-silicon gates has limited the use of ZrO$_2$ in MOSFETs while issues such as boron penetration, low crystallization temperature, charge trapping, and low channel mobility still need to be resolved for HfO$_2$ gate oxide.

Improved electrostatic control of multi-gate devices makes them an optimized choice for sub-32 nm technology, with FinFETs being of particular interest due to their compatibility with the traditional planar CMOS fabrication processes [7], reduced short channel effects (SCE), and reduced junction leakage arising from band-to-band tunneling (BTBT) [8]. However, TCAD simulations of bulk FinFET technology suggest leakage current as high as 10 nA/μm for 65 nm technology node resulting in static power dissipation P_S = 12 nW/μm for V_{DD}=1.2 V that for 22 nm node increases to 2 μA/μm (P_S = 1.6 μW/μm) for V_{DD}=0.8 V [9]. In 22 nm node silicon on insulator (SOI) FinFETs, the leakage current is reduced to 0.4 $\mu A/\mu m$ (P_S = 0.32 $\mu W/\mu m$ for V_{DD}=0.8 V) by isolating the device layer from the silicon substrate and similar results may also be obtained for bulk FinFETs by increasing body doping from 10^{16} cm^{-3} to 10^{18} cm^{-3} [9]. Additional reduction in leakage current is obtained by optimizing the isolation oxide height (T_{ins}) between the gate and substrate for bulk FinFETs. As an example, for 32 nm technology node, decreasing T_{ins} from 150 nm to 50 nm decreases the leakage current from 580 nA/μm (P_S = 0.52 $\mu W/\mu m$ for V_{DD}=0.9 V) to 250 nA/μm (P_S = 0.22 $\mu W/\mu m$ for V_{DD}=0.8 V), however, further decrease of T_{ins} increases the leakage current [9].

A decrease in fin widths leads to a reduction in leakage current due to suppression of SCEs [8], while the use of a triangular fin as opposed to a rectangular fin leads to a reduction in leakage current with a concomitant decrease in static power dissipation by 70% [10]. However controlled etching of fins to meet specific shape and dimension requirements is rather challenging. The best performing FinFETs, as has been published in recent years, have typical leakage current of 0.1 $\mu A/\mu m$ [7, 11, 12], while their low power (LP) counterparts have leakage currents as low as 1 nA/μm [13], resulting in static/leakage power dissipation of 80 nW/μm and 0.8 nW/μm, respectively. Assuming a power dissipation capacitance C_{PD} = 2.4×10^{-15} F/μm, the dynamic power dissipation of a FinFET operating at 100 kHz with V_{DD} = 0.8 V and load capacitance of 15 pF/μm is 308 $\mu W/\mu m$ [11, 14].

Memristors, since their experimental demonstration in 2008, provide an alternative platform for low power and scalable electronics [15]. Unlike shorting the power supply and ground buses during switching of CMOS logic circuits, memristors offer a finite resistance during the ON state preventing any short circuit current. The dynamic power consumption of a 50 nm ZrO_x memristor is only 67.5 nW, assuming a SET voltage of 0.15 V, input signal frequency of 100 kHz, and a load capacitance of 15 pF [16]. For a total transition time of 14.3 ps, the switching energy of memristor logic is merely 9.65×10^{-19} J. Assuming a computation cycle of 1ns, energy dissipation during computation is 18.75×10^{-15} W for a memristor ON resistance of 1.2 kΩ. Among other futuristic technologies, carbon nanotube (CNT) FET-based inverter logic has the potential to offer low switching energy. HSPICE simulations of these devices suggest switching energy of 6.08×10^{-17} J which still is two orders of magnitude higher than that of memristor-based logic gates.

The proposed logic circuits employing memristors are based upon "Material Implication" as opposed to the traditional "Logical Implication", however, from an electrical engineering perspective they have identical truth tables and imply similar functionalities. Material Implication (IMP) logic, x → y ("x implies y" or "if x then y") is one of the fundamental building blocks, together with the FALSE logic, of all Boolean logic operations [17, 18]. The importance of IMP logic can be understood by the fact that IMP and FALSE logics together form a computationally complete logic basis and can compute any compound Boolean operation. Logicians use material implication, also known as "material consequence", "implication", "implies", or "conditional", as a binary connective to create new sentences in the form of "if...then" structure. The truth-value of the conditional new sentence entirely depends on the truth-values of the component propositions. As an example, "if you own a dog, then you own an animal" is a material implication in the sense that the conditional statement is false only if someone claims that he owns a dog but does not own an animal.

In its simplest form, material implication logic can be realized by using only a resistor and two memristors and with the physical states of memristors, namely — high resistance state (HRS) or low resistance state (LRS), implemented as the logic inputs for the IMP logic gate is reminiscent of Shannon's formulation [19]. In that sense, memristor-based IMP logic is also a "stateful" logic with the memristors remembering their states and performing the logic operations at the same time. The fact that the same device could serve both as logic and memory in the IMP logic architecture, the demonstration of IMP logic using memristors also means demonstration of *non-von Neumann* architectures.

Use of memristors as logic gates has mostly been reported using phenomenological models or simulations [3]. A simulation model of memristor-based IMP logic is reported by Kvatinsky *et al.* suggesting that the widely used linear ionic drift model is not practical for IMP logic gates [20]. The same group also proposed a possible architecture for 8-bit full adder using only IMP and FASLSE logics [3]. Klimo *et al.* proposed voltage-based fuzzy logic computations using memristors and could be considered as an extension of classical Boolean logic [21]. A hybrid CMOS-memristor logic, known as memristor ratioed logic (MRL), is modeled by performing OR and AND logic operations using memristors and signal sensing and restorations using CMOS inverters [22]. Chang *et al.* proposed a computer architecture implementing memristor-based latch and flip-flops in the arrangement of a 3D-IC [23]. Rajendran *et al.* designed programmable threshold gates, where memristors are used as weights at the inputs [24]. In contrast to memristors, Wang *et al.* reported the use of magnetic tunnel junctions (MTJ) to demonstrate stateful logic operations where magnetic spins are employed as the physical state [25]. Two separate current biases switched the magnetization of the active layer and the operations of six logic gates were demonstrated. Kimura *et al.* combined a ferroelectric memory element in a configurable logic gate by implementing complementary ferro-electric capacitors [26]. It is worth noting that the fabrication of the devices involving Ta/NiFe/MnIr/CoFe/Al layers requires thickness in the range of 7Å that in turn demands very advanced and expensive fabrication processes. Moreover, the use of a Wheatstone bridge for the read circuitry adds to the power requirement of the circuit. However, compared to MTJs, memristors promise faster switching around 120 ps with power consumption in the range of pJs [17].

The only experimental demonstration of IMP logic using memristors to date is reported by Borghetti *et al.* using TiO_2 thin film crossbars [17]. It is important to note that Borghetti *et al.* employed current amplifiers to distinguish between logic outputs that result in an increased power requirement for the demonstration. In addition, fabrication of the crossbar architecture reported by Borghetti *et al.* requires a two-step metallization process that complicates the fabrication process and results in sneak paths [27].

2. Fabrication and Characterization of Memristor

ZnO NW memristors are fabricated using a three-step process. In the first step, ZnO epitaxial layer is grown using metalorganic chemical vapor deposition (MOCVD) that serves as the nucleation layer for subsequent growth. Following the growth of the epitaxial layer, a novel process developed to grow horizontal nanowires using controlled hydrothermal synthesis is performed as reported by Rivera *et al.* [28]. It is to be noted that unlike the standard practice of growing vertical NRs followed by pick-and-place technique to obtain horizontal nanostructures, the present technology allows the direct growth of horizontal nanowires on desired substrates. SEM measurements suggest the NWs to be 2–5 μm long with the diameter varying from 400 nm to 600 nm as shown in the inset of Fig. 1. Microscopic image of the fabricated device is shown in Fig. 1, where electron beam lithography (EBL) is used to pattern the contact electrodes. Al contacts are deposited using physical vapor deposition (PVD) technique followed by lift-off and annealing at 300°C. SEM measurements suggest the NWs to be 2–5 μm long with the diameter varying from 400 nm to 600 nm as shown in the inset of Fig. 1.

Figure 1 shows the fabricated ZnO NW memristive device that has a length of 2 μm. The measured DC I-V characteristic is shown in Fig. 1(b). Depending on the variation of the device diameters, an electroforming voltage between 5V and 8V was usually required to prepare the devices with a predefined conductive path. A positive bias greater than 2 V applied for 10 ms toggles the device from high resistance state (HRS) to low resistance state (LRS). HRS and LRS of the device are defined as logic "0" and logic "1" inputs to

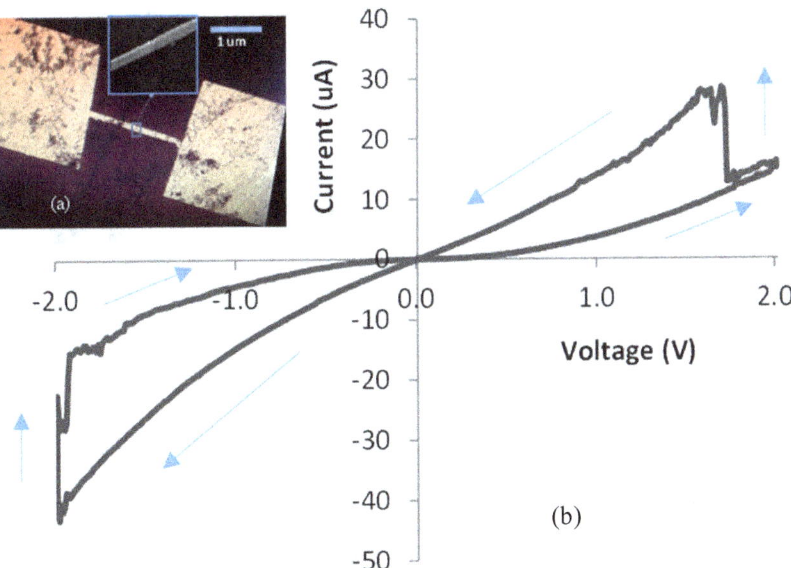

Fig. 1. Measured DC I-V characteristics of an Al/ZnO NW/Al memristive device. Inset: Optical microscopic image of the device fabricated using EBL showing the electrodes.

the device, respectively. A negative voltage, with a magnitude larger than −2 V, resets the device to HRS. For a smaller or zero bias voltage, the device does not switch and remains in the same resistance state, confirming the fundamental property of ReRAM. It is worth mentioning that the measured HRS to LRS resistances in our fabricated memristive devices were 129 kΩ and 189 Ω, respectively. The HRS/LRS ratio of 684 is a significant improvement over the ratio of 3 reported by Chang *et al.* for a Pt/ZnO (100 nm thin film)/Pt memristors and can be used to build large memory banks [29]. As a reference, Amsinck *et al.* reported that a minimum HRS/LRS ratio of 43 must be maintained for a 512×512 array memory to operate [30]. Note that for a 1-bit logic gate, as has been demonstrated in this paper, a HRS/LRS ratio of ~5 was sufficient. This alludes to the fact that devices with HRS/LRS of 684 can support much larger logic arrays.

3. Demonstration of IMP Logic

A schematic of the implemented IMP logic employing memristors is shown in Fig. 2. In this configuration two memristors (M1 and M2) are connected to a common node and to a load resistor R_L=1 kΩ. The input logic voltages (x and y) correspond to the initial states of M1 and M2 (HRS = logic "0" and LRS = logic "1"). Implementation of IMP logic requires stable resistance levels at LRS and that collapses only minimally in subsequent voltage sweeps [31, 32]. LRS of 1.1 kΩ–1.3 kΩ and HRS of 6.25 kΩ–7.15 kΩ at 1V-applied bias are chosen for stable and repeatable switching. Successful demonstration of the IMP logic using memristors of HRS/LRS ratio of only ~5 is a significant improvement over the requirement of HRS/LRS ratio of 100 reported by Borghetti *et al.* [17]. The ability of the logic gates to function with memristors of smaller HRS/LRS ratios indicates less stringent requirements on the memristor performance, resulting in lower bit error rates, improved endurance, and better retention characteristics.

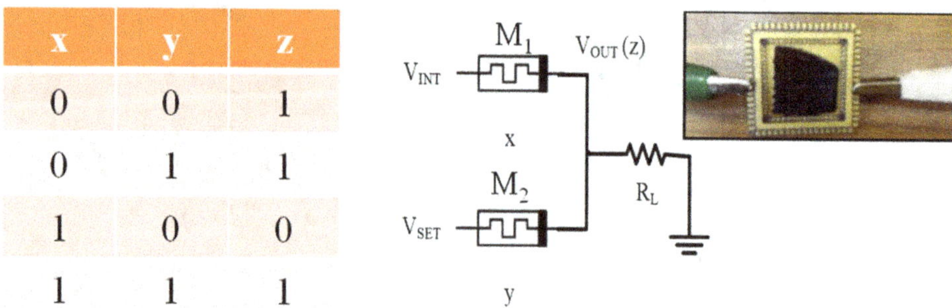

x	y	z
0	0	1
0	1	1
1	0	0
1	1	1

Fig. 2. Truth table and schematic diagram of the IMP logic employing two memristors. The figure also shows the wire-bonded memristors on a chip carrier used in the experimental demonstration.

The memristors are SET to logic "1" by applying a positive bias voltage V_{SET} for 10ms. It is important to note that positive bias is ubiquitously used for digital logic operation than the previously demonstrated IMP operation with negative bias [17]. The magnitude of V_{SET}

is selected as 3 V, higher than the threshold voltage required for switching, to ensure switching in the subsequent voltage pulses. The devices require a negative voltage of –5 V to RESET (logic "0"). Prior to each logic operation of the truth table, a read voltage of 1 V is applied across the device and the resistance is measured. An intermediate voltage V_{INT} = 1.3 V is defined such that $V_{SET} - V_{INT}$ is smaller than the threshold voltage required for switching. V_{INT} is considered as the higher limit of logic "0" throughout this experiment.

The IMP operation is performed by applying a V_{INT} pulse to M1 and a V_{SET} pulse to M2 simultaneously. The circuit is configured as a synchronous logic that triggers at the rising edge of the clock pulse. For initial conditions x = 0 and y = 0, both memristors M1 and M2 are in HRS and V_{OUT} is in logic "0" (z = 0). With the triggering of the clock pulse, V_{INT} and V_{SET} are applied to M1 and M2, respectively. Pulse V_{INT} is not enough for M1 to switch, however, V_{SET} switches M2 to LRS and z becomes a logic "1" as shown in Fig. 3. For initial conditions x = 0 and y = 1, LRS of M2 results in a high V_{OUT}. With the application of the next clock pulse, V_{INT} does not switch the resistance of M1, and V_{SET}

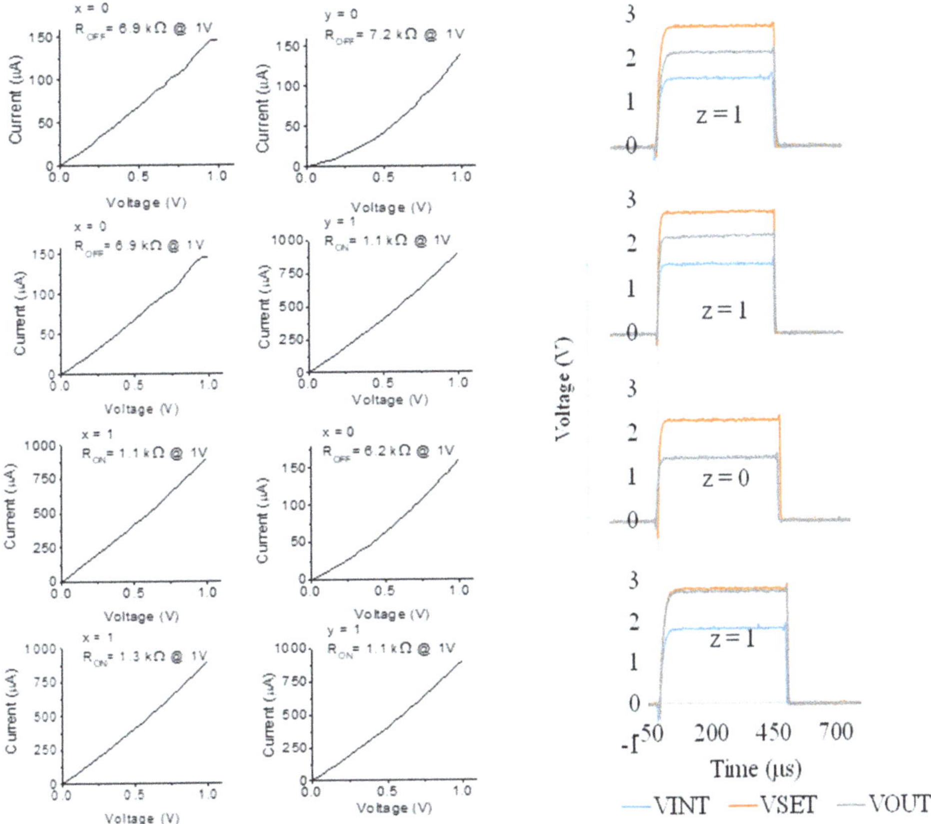

Fig. 3. Experimental demonstration of IMP logic operation using memristors. The four different states as outlined in the truth table are implemented and centered.

does not switch M2 as it already was in LRS. As a result, V_{OUT} remains high and $z = 1$. For initial conditions $x = 1$ and $y = 0$, LRS of M1 brings V_{OUT} close to V_{INT}. In the succeeding clock trigger, the voltage across M2 is $\sim V_{SET} - V_{INT}$, which is not enough to switch M2 to LRS due to such selection of V_{INT} magnitude, and as a result $z = 0$. For initial conditions $x = 1$ and $y = 1$, both the memristors are in LRS and with the application of V_{INT} and V_{SET}, voltage magnitudes $V_{INT} - V_{OUT}$ and $V_{SET} - V_{OUT}$ are not sufficient to switch either M1 or M2.

In order to demonstrate the reproducibility and reliability of the logic switching measurements, a second set of memristors, fabricated using a similar process as mentioned above but carried out at a different time frame to account for the effect of process variability, is implemented. The Boolean logics of the truth table shown in Fig. 2 are demonstrated using the new set of memristors and the results are shown in Fig. 4. The LRS and HRS resistances are not regulated within a tight range, as the experiment mentioned earlier, to emulate a more practical scenario. R_{ON} and R_{OFF} vary in the range between 54 Ω and 237 Ω, and 25 kΩ and 308 kΩ, respectively. During the READ operation of the

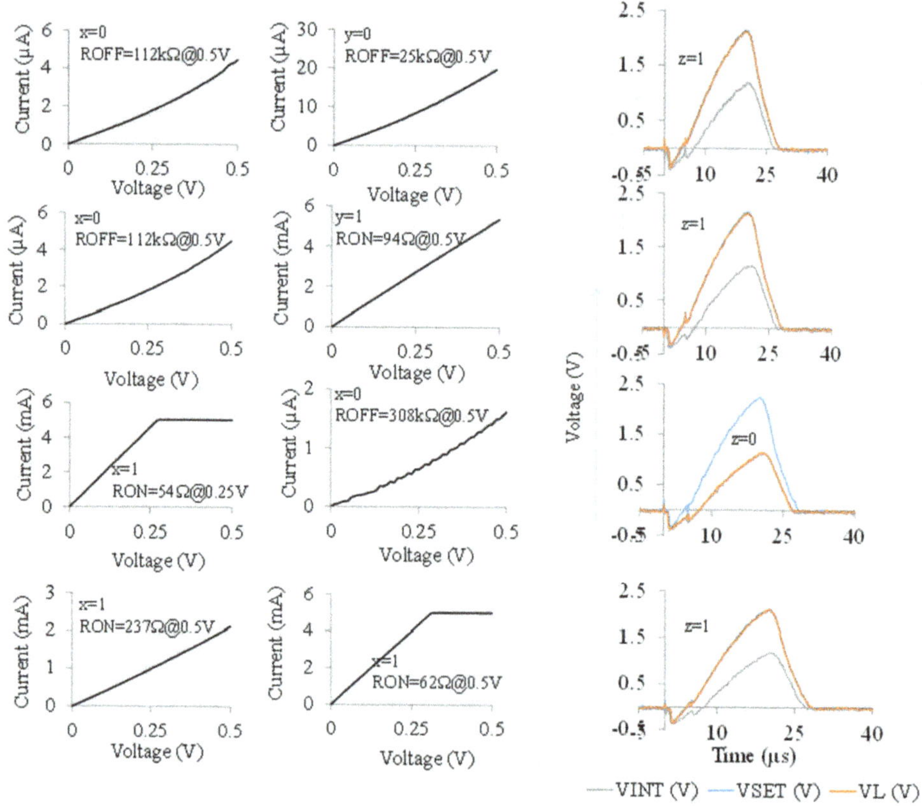

Fig. 4. Experimental demonstration of memristor-based IMP logic at 100 KHz frequency. The four different states outlined in the truth table are demonstrated.

mentioned R_{ON} and R_{OFF} resistances, the supply voltage is swept from 0V to 0.5 V and a current compliance of 5 mA is imposed by the semiconductor parameter analyzer to ensure that the devices do not break down due to excessive current flowing through them. The logic gate is demonstrated to be functional at a clock frequency of 100 kHz. The lone experimental demonstration of IMP logic [17] did not report the frequency of operation and to the best of our knowledge, this is the fastest IMP logic yet to be reported.

It is to be noted that the bias voltage required for the memristor to switch can also be scaled by scaling the device length. It is well known that the role of the bias voltage in memristive systems is to drift the oxygen vacancies from anode to the cathode and thereby creating conductive filaments. This ionic drift process is dictated rather by the applied electric field than the bias voltage. As a result, the required bias voltage decreases linearly as the device scales, described by $E=V/d$, where E is the electric field, V is the applied bias, and d is the length of the memristors. As an example, scaling the ZnO NW memristor from 2 μm to 100 nm results in a V_{SET} of 0.15 V, V_{INT} of 65 mV, V_{RESET} of −0.25 V, and V_{READ} of 50 mV. Such voltage scaling leads to reduced power consumption as shown in Fig. 5 where static power dissipation (PS), dynamic power dissipation (PD), as well as total power dissipation (PTotal) of memristors of different lengths are calculated. The total power consumption of a 2 μm memristor-based IMP logic is only 56.2 μW which is expected to reduce below 0.1 μW for 50 nm ZnO memristors. Shrinking the device lengths from 2 μm to 50 nm also results in 3–4 orders of reduction in the delay of a typical memristor, owing to increased electric field across the electrodes as discussed above. This suggests that replacing the 2 μm memristors in the logic gates with 50 nm memristors will result in approximately 98% reduction in the gate delay. It is worth mentioning that we purposefully limited our analyses here to the core technology of a 1-bit fundamental logic gate implemented using memristors. Integration-related performance metrics such as fan out, drive capability, and so forth are left to be addressed in a future work.

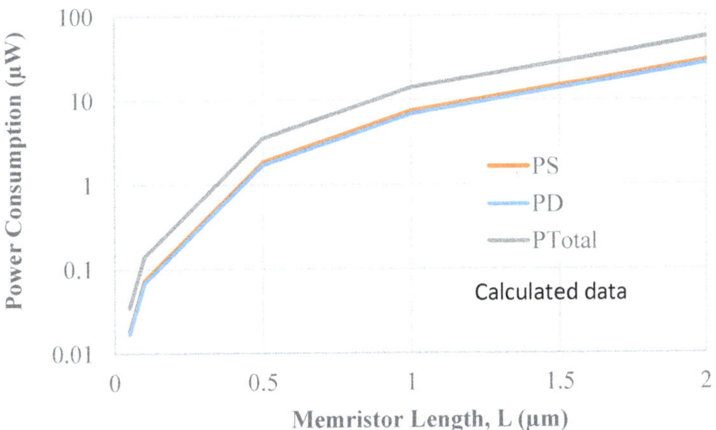

Fig. 5.   Calculated power dissipation of ZnO memristor-based implication logic as a function of device length.

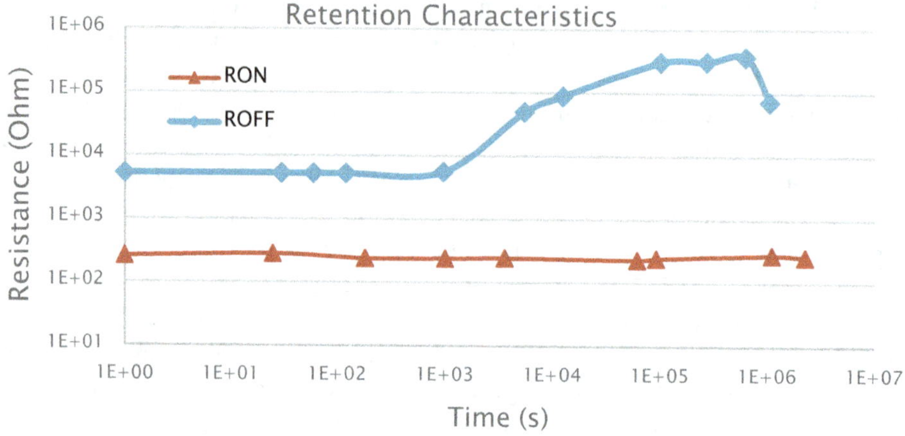

Fig. 6.   Measured retention characteristics of fabricated ZnO memristor.

Figure 6 shows the measured retention characteristics of ZnO memristors. R_{ON} and R_{OFF} are measured at 0.4 V by sweeping the applied bias from 0V to 0.4V using HP4145B Semiconductor Parameter Analyzer. The value of R_{ON} fluctuates between 229 Ω and 281 Ω for a duration up to 2.24×10^6 seconds or 26 days. ROFF maintained its high value for a duration of over 23 days. A minimum R_{OFF}/R_{ON} ratio of 23 is observed throughout the measurement period. Material implication coupled with FALSE logic provides the required functionality that, when coupled with the ability for memristors to remember or function as memory, paves the way to in-memory computing or non-von Neumann architectures.

4. Conclusion

ZnO NW memristors are fabricated using MOCVD followed by hydrothermal synthesis and EBL. Two memristors are used to implement IMP logic, where they serve as both the memory and logic elements. The IMP logic is successfully demonstrated with a R_{OFF}/R_{ON} ratio as low as 5, resulting in a smaller bit error rate of the logic function. Logic pulses of positive amplitudes are demonstrated demonstrating the viability to be integrated with existing binary logic systems. The capability to perform logic operations within the memory element shows the potential of designing a computer architecture different from the familiar paradigms of von Neumann systems. Power consumption of memristor-based logic gates is significantly lower than CMOS-based gates and even that can be reduced further by scaling down the memristors.

References

1. Di Ventra, M., Pershin, Y. V. & Chua, L. O. Circuit elements with memory: Memristors, memcapacitors, and meminductors. *Proc. IEEE* **97**, 1717–1724 (2009).
2. Borghetti, J. *et al.* A hybrid nanomemristor/transistor logic circuit capable of self-programming. *Proc. Natl. Acad. Sci. USA* **106**, 1699–1703 (2009).

3. Kvatinsky, S. *et al.* Memristor-based material implication (imply) logic: Design principles and methodologies (2013).

4. Shauly, E. N. CMOS leakage and power reduction in transistors and circuits: Process and layout considerations. *J. Low Power Electron. Appl.* **2**, 1–29 (2012).

5. Houssa, M. High k gate dielectrics (2003).

6. Yeap, G. C., Krishnan, S. & Lin, M. Fringing-induced barrier lowering (FIBL) in sub-100 nm MOSFETs with high-k gate dielectrics. *Electron. Lett.* **34**, 1150–1152 (1998).

7. Wu, C. *et al.* High performance 22/20nm FinFET CMOS devices with advanced high-K/metal gate scheme. In *2010 IEEE International Electron Devices Meeting (IEDM)* (IEEE, 2010).

8. Collaert, N. *et al.* Multi-gate devices for the 32nm technology node and beyond. *Solid-State Electron.* **52**, 1291–1296 (2008).

9. Manoj, C., Nagpal, M., Varghese, D. & Ramgopal Rao, V. Device design and optimization considerations for bulk FinFETs. *IEEE Trans. Electron Devices* **55**, 609–615 (2008).

10. Gaynor, B. D. & Hassoun, S. Fin shape impact on FinFET leakage with application to multithreshold and ultralow-leakage FinFET design. *IEEE Trans. Electron Devices* **61**, 2738–2744 (2014).

11. Maly, W. *et al.* Twin gate, vertical slit FET (VeSFET) for highly periodic layout and 3D integration. In *2011 Proceedings of the 18th International Conference on Mixed Design of Integrated Circuits and Systems (MIXDES)* (IEEE, 2011).

12. Kavalieros, J. *et al.* Tri-gate transistor architecture with high-k gate dielectrics, metal gates and strain engineering. In *2006 Symposium on VLSI Technology*, Digest of Technical Papers (IEEE, 2006).

13. Yeh, C. *et al.* A low operating power FinFET transistor module featuring scaled gate stack and strain engineering for 32/28nm SoC technology. In *2010 IEEE International Electron Devices Meeting (IEDM)* (IEEE, 2010).

14. Shahidi, G. Device architecture: Ultimate planar CMOS limit and sub-32nm device options. In *IEEE IEDM Short Course* (2009).

15. Strukov D. B., Snider G. S., Stewart D. R. & Williams R. S. The missing memristor found. *Nature* **453**, 80–83 (2008).

16. Mazady, A. & Anwar, M. Switching transients in memristors. *Int. J. Hi. Spe. Ele. Syst.* **28**(3–4), 1940023 (2019)

17. Borghetti, J. *et al.* 'Memristive' switches enable 'stateful' logic operations via material implication. *Nature* **464**, 873–876 (2010).

18. Whitehead, A. & Russel, B. In *Principia Mathematica* (Cambridge University Press, 1910).

19. Shannon, C. E. A symbolic analysis of relay and switching circuits. *Trans. Amer. Inst. Electric. Eng.* **57**, 713–723 (1938).

20. Kvatinsky, S., Kolodny, A., Weiser, U. C. & Friedman, E. G. Memristor-based IMPLY logic design procedure. In *2011 IEEE 29th International Conference on Computer Design (ICCD)* (IEEE, 2011).

21. Klimo, M. & Such, O. Memristors can implement fuzzy logic. arXiv:1110.2074.

22. Kvatinsky, S. *et al.* MRL—memristor ratioed logic. In *2012 13th International Workshop on Cellular Nanoscale Networks and Their Applications (CNNA)* (IEEE, 2012).

23. Chang, M., Chiu, P., Wu, W., Chuang, C. & Sheu, S. Challenges and trends in low-power 3D die-stacked IC designs using RAM, memristor logic, and resistive memory (ReRAM). In *2011 IEEE 9th International Conference on ASIC* (IEEE, 2011).

24. Rajendran, J., Manem, H., Karri, R. & Rose, G. S. Memristor based programmable threshold logic array. In *Proceedings of the 2010 IEEE/ACM International Symposium on Nanoscale Architectures* (IEEE Press, 2010).

25. Wang, J., Meng, H. & Wang, J. Programmable spintronics logic device based on a magnetic tunnel junction element. *J. Appl. Phys.* **97**, 10D509 (2005).

26. Kimura, H. *et al.* Complementary ferroelectric-capacitor logic for low-power logic-in-memory VLSI. *IEEE J. Solid-State Circuits* **39**, 919–926 (2004).

27. Zidan, M. A., Fahmy, H. A. H., Hussain, M. M. & Salama, K. N. Memristor-based memory: The sneak paths problem and solutions. *Microelectron. J.* **44**, 176–183 (2013).

28. Rivera, A., Mazady, A. & Anwar, M. Low temperature growth of horizontal ZnO nanorods. *MRS Proc.* **1707**, uu03-33 (2014).

29. Chang, W., Lai, Y., Wu, T., Wang, S., Chen, F. & Tsai, M. Unipolar resistive switching characteristics of ZnO thin films for nonvolatile memory applications. *Appl. Phys. Lett.* **92**, 022110 (2008).

30. Amsinck, C. J., Di Spigna, N. H., Nackashi, D. P. & Franzon, P. D. Scaling constraints in nanoelectronic random-access memories. *Nanotechnology* **16**, 2251 (2005).

31. Mazady, A. & Anwar, M. Memristor: Part I — The underlying physics and conduction mechanism. *IEEE Trans. Electron Devices* **61**, 1054–1061 (2014).

32. Mazady, A. & Anwar, M. Memristor: Part II — DC, transient and RF analysis. *IEEE Trans. Electron Devices* **61**, 1062–1070 (2014).

Magnetostrictive Fiber Sensors as Total Field Magnetometers

R. Dougenik[*,†,‡], R. Lacomb[*,†] and F. Jain[†]

[*]*Electromagnetic Systems Department, Naval Undersea Warfare Center Division Newport, RI, USA*
[†]*ECE, University of Connecticut, Storrs, CT, USA*
[‡]*robert.dougenik@uconn.edu*

A novel magnetostrictive thin film fiber sensor is presented which can be utilized in an interferometric architecture. The magnetostrictive fiber sensor utilizes novel fiber and thin film technology to achieve high sensitivity in the sub-nanotesla regime. The base sensor architecture utilizes optical fiber wound in a novel low stress flat coil resembling a record, coated with a thin magnetostrictive film. Two prototype variants were fabricated, incorporating FeCo sputtered films of different thicknesses on 4 μm polyimide jacket single mode fiber. The sensors were incorporated into a fiber optic interferometric measurement apparatus to characterize time-varying magnetic field sensitivity. Device results are presented which demonstrate sensitivity is a function of film thickness. The experimental data exhibited a two-order magnitude sensitivity improvement as the thin film thickness was doubled. Sensitivity projections are made based on film thickness.

Keywords: Magnetostrictive fiber sensor; sub-nanotesla magnetometer; sensitivity improvement.

1. Background

Magnetostriction is a property of ferromagnetic materials allowing a measurable change in size or shape in the presence of a magnetic field. FeCo is one example of a magnetostrictive film which exhibits large magnetostriction [1, 2]. It has been demonstrated that by tuning the presence of structural heterogeneity in textured Co_xFe_{1-x} thin films, effective magnetostriction as large as 1000 p.p.m. is feasible, which makes possible detection of fields in the nT-pT range [3]. The maximum magnetostrictive composition found near the (fcc+bcc)/bcc phase boundary is compositionally and process-wise dependent upon the synthesis technique and subsequent annealing [4]. Through co-sputtering, this thin metallic film can be deposited directly onto a fiber jacket.

The basis of the expected sensitivity of this sensor lies in the inherent sensitivity of interferometry, the large effective magnetostriction of FeCo, and the well-known modulation/demodulation scheme developed by Dandrich *et al.* [5]. The mechanism by which a magnetic field induces a phase shift in the fiber is known as the photoelastic effect. Stress induced on the fiber by the magnetostriction of the FeCo in the presence of a magnetic field

[‡]Corresponding author.

results in a change in optical path length. It is therefore expected that the sensitivity of the device will scale proportionally with magnetostrictive (MS) coated fiber length.

1.1. *Magnetostrictive Fiber Sensor Architecture*

A schematic of a fiber optic interferometric measurement apparatus incorporating the magnetostrictive (MS) fiber sensor is shown in Fig. 1.

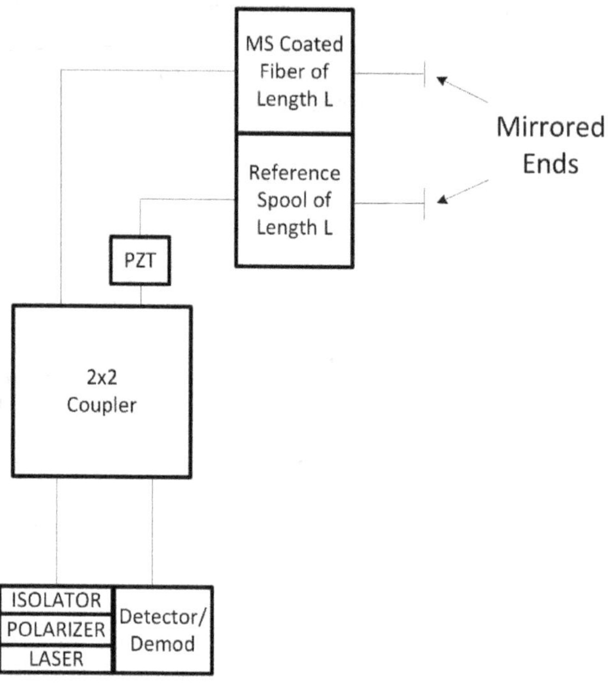

Fig. 1. Michelson interferometric architecture incorporating the FeCo fiber sensor.

Utilizing a mirrored end on both the magnetostrictive coated fiber and reference spool allows a single 2×2 coupler to act as an interferometer. The sensor and reference spool are of equal length such that the interferometer will be approximately matched when no magnetic field is present. The piezoelectric transducer (PZT) in the reference leg allows the signal of interest (i.e. a magnetic field incident on the sensor) to be modulated/demodulated as described in [6]. Coupled light in the sensor arm experiences a net phase change delay with respect to coupled light traveling through the reference arm which provides a detectable signal. A phase-generated carrier demodulation scheme is utilized to characterize the signal-to-noise ratio, which has been used as a sensitivity metric. The device sensitivity can be tailored through the magnetostrictive film, fiber–film interface and the number of fiber turns N of the sensor (similar to that of a fiber-optic gyroscope).

2. Sensor Fabrication

The fabrication strategy was to deposit magnetostrictive thin films directly on or as close to the fiber cladding as possible. Additionally, the sensor architecture is configured to minimize geometric stresses. This strategy is intended to maximize the magnetostrictive stress induced on the glass to improve sensitivity and utilize a sensor geometry compatible with standard semiconductor wafer processing technology.

A typical optical fiber has two jackets; an inner "hard jacket" which rests on the glass cladding of the fiber, and a softer external jacket. Typically, the inner jacket thickness is on the order of 100–200 µm. The custom fiber utilized in the fabrication of this sensor had only a polyimide inner jacket with a thickness of about 4 µm. The thin jacket allowed the magnetostrictive-induced stress to be better transferred to the glass resulting in a larger change in optical path length. In addition, the thin jacket enables a smaller form factor for a given length of fiber.

Prototype sensors were fabricated by winding 50 meter lengths of custom fiber into low stress single layer coils resembling a record. The fiber coil was subsequently glued together. The custom fiber coil was then coated with a thin FeCo film in a six-pocket sputtering machine. The sputtering machine used is designed for semiconductor processing and can only accommodate a wafer less than four inches in diameter and of little height. This fabrication method limited both the configuration of the sensor as well as the maximum length. Figure 2 portrays a completed FeCo-coated fiber sensor.

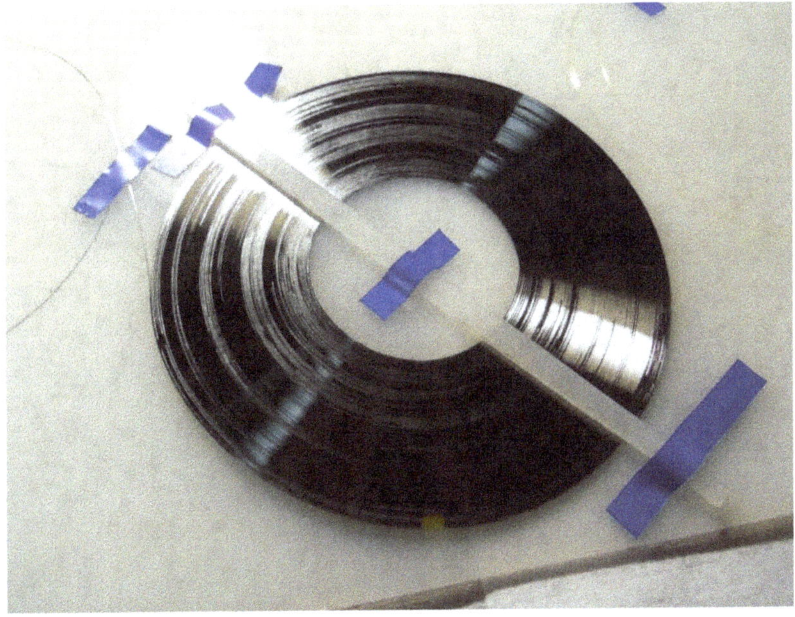

Fig. 2. Prototype FeCo-coated fiber sensor.

Two thin jacket fiber sensors were fabricated using the method described above. Critical prototype parameters are documented in Table 1.

Table 1. Sensor designs.

Parameter	Prototype #1	Prototype #2
Fiber jacket thickness (μm)	4	4
FeCo coated	Yes	Yes
FeCo thickness (μm)	0.5	1.5
Length (m)	44.2	48.3
Insertion loss (dB)	0.450	0.712

3. Test Results

The sensor characterized in the test setup is illustrated in Fig. 3. The sensors were placed in a Helmholtz coil, which produced a known uniform axial magnetic field proportional to drive current. The magnetic field incident on the sensor was verified with an off-the-shelf time-varying Gauss meter. The optical phase change due to the sensor was translated to an electrical signal by the demodulator and measured as signal-to-noise ratio (SNR) in a 3 Hz bandwidth at a spectrum analyzer.

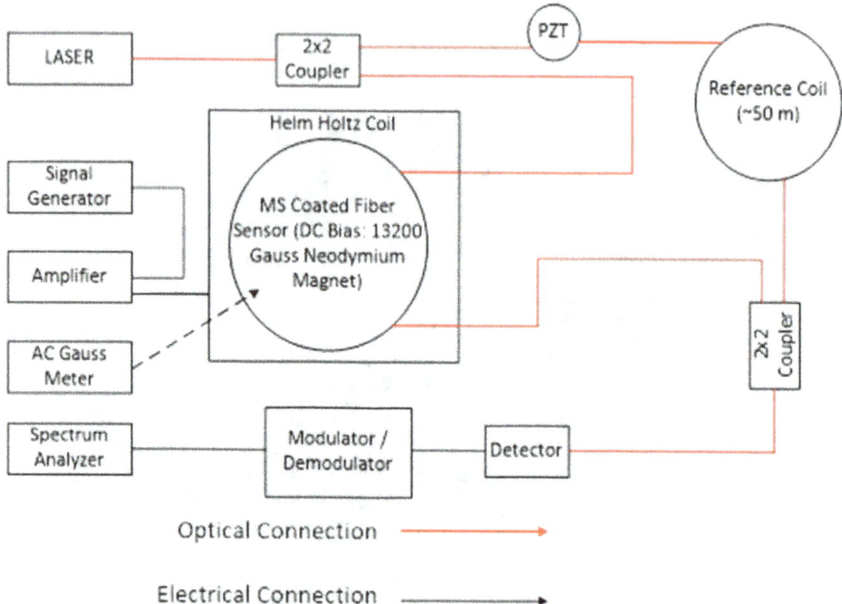

Fig. 3. FeCo-coated fiber sensor test setup.

All SNR measurements taken at the spectrum analyzer were measured in a 3 Hz bandwidth at 1.3 kHz. The lower limit of the magnetic field strengths tested was set by off-the-shelf time-varying Gauss meter used to verify the field strength. Figure 4 shows the output SNR versus magnetic field strength.

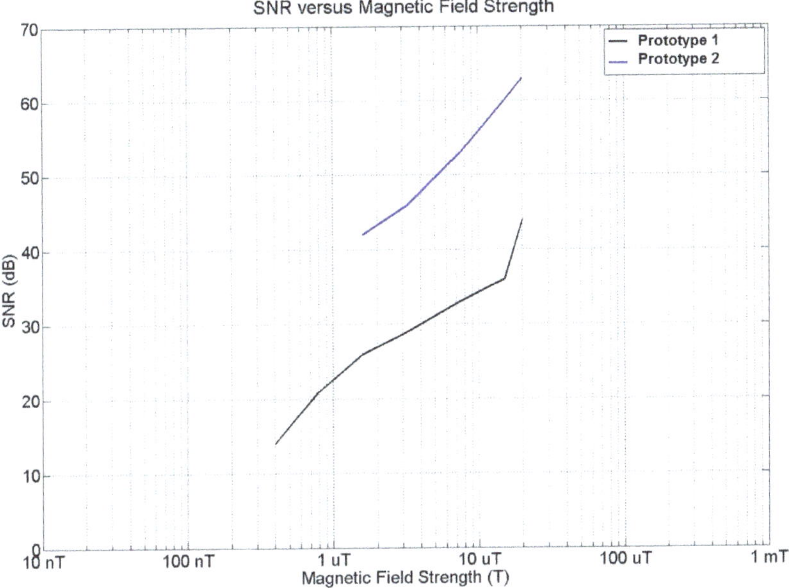

Fig. 4. Experimental data for two prototype sensors of different FeCo thickness (Prototype #1 film thickness: 0.5 μm, Prototype #2 film thickness: 1.5 μm). Dotted line represents extrapolation for minimum detectable signals assuming uncorrelated noise.

Prototype #1 produced a 14 dB SNR with a 400 nT incident magnetic field. By extrapolation, it is estimated that this sensor is capable of detecting magnetic fields on the order of 40 nT.

Prototype #2 produced a 42 dB SNR with approximately 1.6 μT incident magnetic field. By extrapolation, it is estimated that this sensor is capable of detecting magnetic fields on the order of 200–300 pT.

4. Discussion

The initial FeCo coated fiber sensor (Prototype #1) showed promising results, yielding an approximate sensitivity on the order of 40 nT. Prototype #2 showed a significant improvement in sensitivity due to increased FeCo film thickness yielding a sensitivity of about 200–300 pT. This result was expected due to the volumetric increase in aligned domains of the magnetic film. We expect that the sensor sensitivity can be increased by increasing the overall fiber length for the total change in optical path length is cumulative. It is expected that doubling the fiber length will result in a 3 dB sensitivity improvement while the data demonstrates a more significant improvement through changing the film thickness.

Fiber optic-based magnetic sensors are capable of detecting fields on the order of 10 pT with only superconducting quantum interference devices (SQUIDs) capable of detecting magnetic fields less than 1 pT. Figure 5 shows a performance comparison of several types of magnetic field sensors with Prototype #2 performance superimposed.

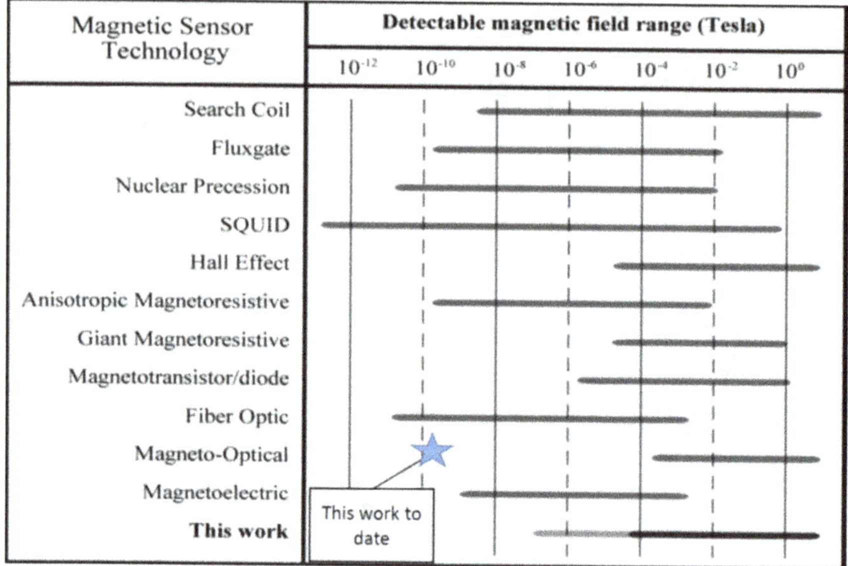

Fig. 5. Comparison of FeCo fiber sensor data to other magnetic field sensor device technologies [7].

References

1. N. Tiercelin, V. Preobrazhensky, P. Permod and A. Ostaschenko, "Enhanced magnetoelectric effect in nanostructured magnetostrictive thin film resonant actuator with field induced spin reorientation transition," *Appl. Phys. Lett.* 96(6), 062904 (2008).

2. Y. Fu, Z. Cheng and Z. Yang, "Magnetically soft and high-saturation-magnetization FeCo films fabricated by co-sputtering," *IPSS* 202(6), 1150–1154 (2005).

3. A. L. Herrera-May *et al.*, "Resonant magnetic field sensors based on MEMS technology," *Sensors* 9(10), 7785–7813 (2009).

4. D. Hunter *et al.*, "Giant magnetostriction in annealed Co1-xFex thin-films," *Nat. Commun.* 2, 518 (2011).

5. A. Dandridge, A. B. Tveten and T. G. Giallorenzi, "Homodyne demodulation scheme for fiber optic sensors using phase generated carrier," *IEEE J. Quantum Electron.* QE-18(10) (1982).

6. T. G. Giallorenzi, J. A. Bucaro, A. Dandridge, G. H. Sigel, Jr., J. H. Cole, S. C. Rashleigh and R. G. Priest, "Optical fiber sensor technology," *IEEE J. Quantum Electron.* QE-18(4) (1982).

7. B. Bahreyni, "A resonant micromachined magnetic field sensor," *IEEE Sensor J.* 7(9), 1326–1334 (2007).

Propagation Delay and Power Dissipation Analysis for a 2-Bit SRAM Using Multi-State SWS Inverter

A. Husawi[1,2], R. H. Gudlavalleti[1], B. Saman[3], A. Almalki[1,3], J. Chandy[1], E. Heller[4] and F. C. Jain[1,*]

[1]*ECE Department, University of Connecticut, CT, USA*
[2]*Department of Electrical and Electronics Engineering Technology, Yanbu Industrial College,*
Yanbu Industrial City, Al-Medina 46452, KSA
[3]*Electrical Engineering Department, Taif University, KSA*
[4]*Synopsys Inc., Ossining, NY, USA*
[*]fcj@engr.uconn.edu

This paper presents a study on the propagation delay and power dissipation of a 2-bit static random-access memory (SRAM) with two cross-coupled multi-state spatial wave function switching (SWS) CMOS inverters. The proposed SRAM design utilizes the advantages of the CMOS-SWS inverter, such as its small area, low power consumption, and high speed. The 2-bit SRAM circuit simulations were carried out in Cadence to analyze the power dissipation and propagation delay. An Analog Behavioral Model (ABM) and the Berkeley Short-channel IGFET Model (BSIM4.6) in 0.18-μm technology were combined to create this model. The analysis of the propagation delay shows that the multi-state CMOS-SWS SRAM significantly reduces the delay compared to other multi-state 6T SRAM memories. Additionally, the analysis of the power dissipation shows that the multi-state SWS-SRAM is comparable to conventional SRAMs. These results demonstrate the potential of multi-state SWS-SRAM for improving the performance of memory circuits and provide valuable insights for future design optimization.

Keywords: SWS FETs; power dissipation; multi-state circuit; 2-bit SWS-SRAM circuit; ABM simulations.

1. Introduction

The SWS-FET transistor incorporates vertically stacked quantum well/quantum dot channels to allow current to flow across multiple channels in a single transistor by varying the gate voltage [1, 2]. In addition, the SWS-FET was shown to perform multivalued logic computation, and storage in memories [3]. The multiple-quantum-well/quantum-dot structure encodes logic states (00), (01), (10), and (11) utilizing the spatial location of carriers [4–7]. Alternatively, these states could be written as (OFF, OFF), (OFF, ON), (ON, OFF), and (ON, ON) states.

Multi-state operation of SWS-FETs makes it possible to process two bits in the design of logic circuits. Primary logic cells, full-adders, latches, and static random-access memory

---

[*]Corresponding author.

(SRAM) have been shown to have 50% area savings employing 4-state/2-bit SWS-FETs [1, 6, 7]. These circuits made use of several n-channel SWS-FET source/drain topologies to process two or more bits at once. A 2-bit SRAM is implemented using 4-state inverters having additional threshold voltages.

This helps reduce the number of transistors and interconnects required in the SRAM circuit, resulting in low power consumption and smaller propagation delay. The propagation delay is calculated by measuring the time taken for the output of the circuit to reach a steady state after a change in the input. Power consumption is calculated by measuring the average energy consumed by the circuit.

This paper aims to evaluate the propagation delay and power consumption for a 2-bit SRAM using a 4-state CMOS-SWS inverter in a 180 nm technology node. Simulations model the circuit and perform transient and power analysis. An Analog Behavioral Model (ABM) and the Berkeley Short-channel IGFET Model (BSIM4.6) were combined to create the SWS-FET model [8]. Overall, the analysis of propagation delay and power consumption for the 2-bit SRAM cell using 4-state CMOS-SWS inverters provide insights into the efficiency and performance of this novel memory cell design in the 180 nm technology node.

2. Four-state SWS Inverter

The SWS-FET device is comprised of multiple Quantum Well/Quantum Dot channels that allow the electron wave functions to move from one channel to another in response to the gate voltage (VG). Jain *et al.* reported the fabrication process for SWS-FETs [1]. Figure 1 shows the cross-sectional schematic of n-channel and p-channel SWS-FET in CMOS-X configuration. Figure 2 illustrates the energy band diagram, and Fig. 3 shows the circuit schematic.

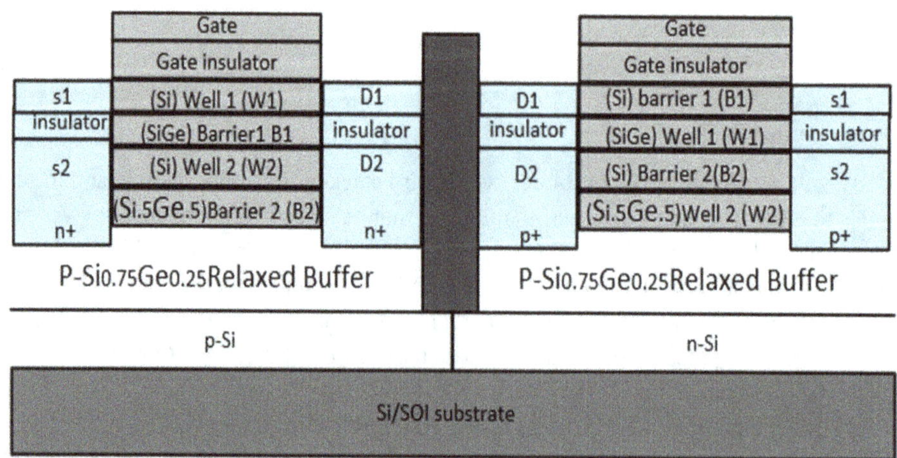

Fig. 1. Cross-sectional schematic of n- and p-channel SWS-FETs [8].

The left side shows n-SWS-FET with two Si Quantum Well (2-QW) channels. The device has an upper Si well (W1) and a lower Si well (W2) sandwiched between SiGe barriers. Carriers (electrons) flow in lower well W2 or upper well W1 in an n-SWS-FET depending on the gate voltage (VGS). The lower well W2 has a higher carrier density than the upper well at lower gate voltages. Carrier density in W1 increases as gate voltage increases. The p-SWS-FET (right side) is comprised of two SiGe quantum well sandwiched between Si barriers. The hole current flows in lower W2 or upper W1 well/channel depending on the applied gate voltage (VGS). When the gate voltage is low, the top well W1 has more holes than the lower well W2 [6].

Figure 2A shows the type I heterostructure energy band diagram, whereas the type II heterostructure energy band diagram is shown in Fig. 2B.

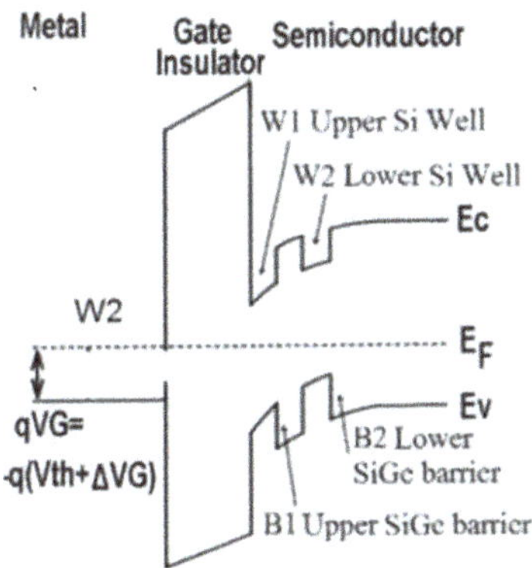

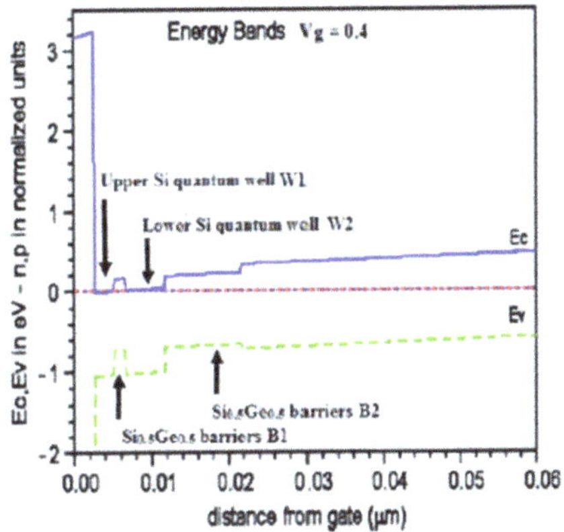

Fig. 2(A). Type I heterostructure of n-SWS-FET energy band diagram [4].

Fig. 2(B). Type II heterostructure of n-SWS-FET energy band diagram [4].

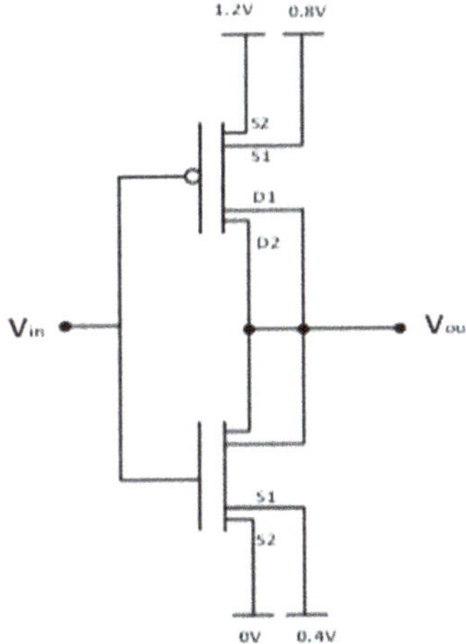

Fig. 3. The 4-state SWS-FET based inverter in CMOS-X configuration.

The proposed 4-state SWS-FET circuit operates with four voltage levels, corresponding to 0 V, 0.4 V, 0.8 V, and 1.2 V. One p-SWS-FET and n-SWS-FET make up the complementary 4-state inverter. Source, S2 of p-SWS-FET is maintained at 1.2 V and Source, S1 is kept at 0.8 V. In the case of n-SWS-FET Source, S2 is grounded and Source S1 is kept at 0.4 V. The (V_{th}) values corresponding to the transistor source-bulk are shown in Table 1. The logic is shown in the truth Table 2. Model parameters of the SWS-FET based inverter are listed in Table 3.

Table 1. Threshold voltage values

	V_{th_1}	V_{th_2}
p-SWS	−0.8 V	−0.2 V
n-SWS	0.8 V	0.2 V

Table 2. Four-state SWS truth table

IN	OUT
0	1.2
0.4	0.8
0.8	0.4
1.2	0

Table 3. Model parameters of the n/p SWS-FET-based inverter

	N-Channel SWS-FET	**P-Channel SWS-FET**
V_{DD}	1.2	1.2
Width (W1)	0.45 μm	2.14 μm
Width (W2)	0.9 μm	4.28 μm
Length (L)	0.18 μm	0.18 μm
V_{th1}	0.8 V	−0.8 V
V_{th2}	0.2 V	−0.8 V
VUL	0.28 V	−0.28 V

3. SWSFET based 6T 2-bit SRAM

Figure 4 shows the circuit schematic of 6T SRAM with two cross-coupled (SWS) inverters and the two access transistors Nmos1 and Nmos2. The cross-coupled inverters consist of two pairs of nSWS and pSWS transistors, while the access transistors are two single nMOS transistors that control the read and write operations of the cell. The gate of these access transistors is controlled by word line inputs (Word_Line). Whenever the WL is made high, the Data and Data_Bar are connected to the cell; hence cell can be written in and read out the data through the bit lines.

3.1. *Read and Write operation of 6T SWSFET based 2-bit SRAM*

Figure 5 shows the simulation result for the write operation of 2-bit SRAM based on 6T SWSFET unit cell shown in Fig. 4. The top panel in Fig. 5 shows the word_line (blue line), the input data (green line) and data stored and datastored_bar in the memory cell (red line) when wordline input goes high are shown in bottom two of the plot. The change in each state for a given input state depicts the storage of each state in this multistate SRAM cell.

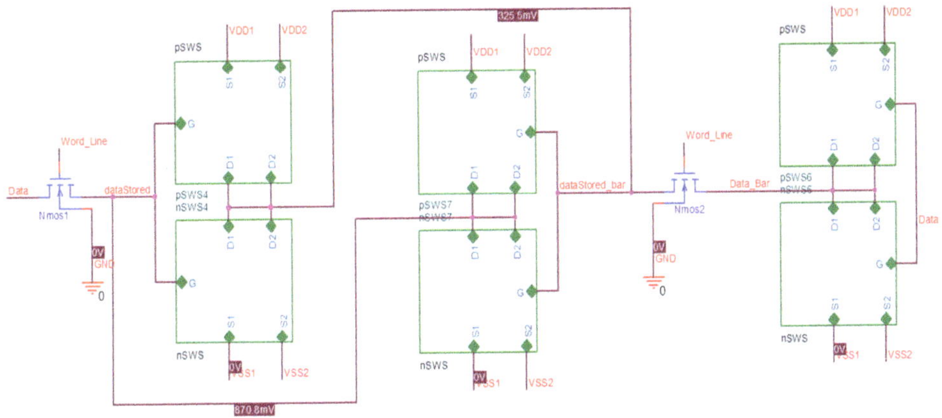

Fig. 4. Circuit schematic of 2-bit SRAM with cross-coupled SWS-inverters.

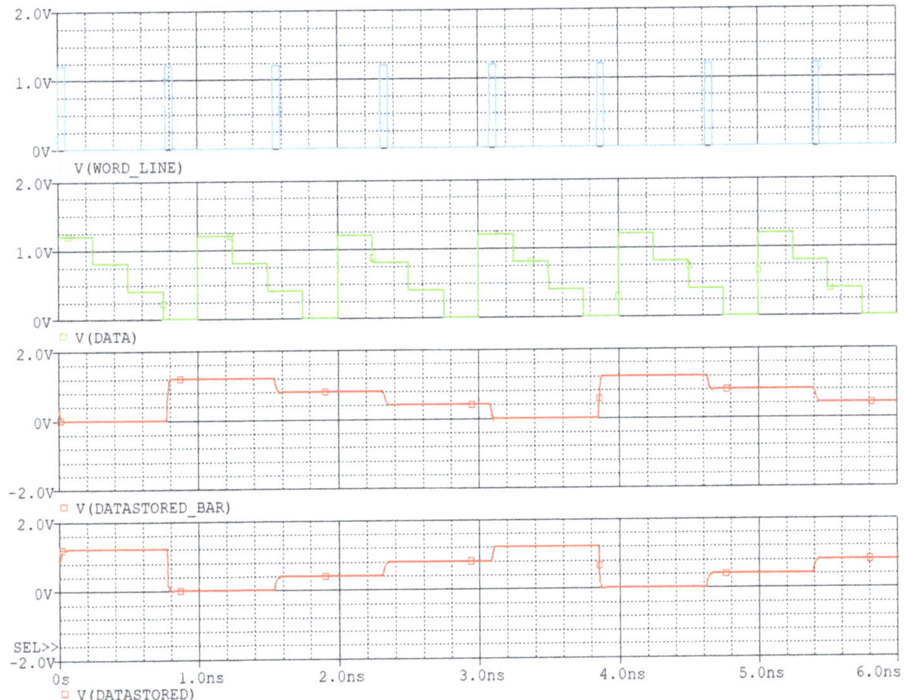

Fig. 5. Simulation result of Word line (blue line), Data (green line), Data stored (red line), and Data_stored_bar (bottom red line).

During read operation, the data and databar lines are initially pre-charged. When wordline (WL) goes high, the stored data from the SRAM cell is retrieved from memory cell to the data lines. Here, the sensing circuitry (not shown in the Fig. 4) converts the signal to appropriate rail voltage so that further processing the memory data is performed.

3.2. *Propagation delay*

Propagation delay is the time required for the output response for the given input signal. The delay is measured between 50% of the output transition to 50% of the applied input transition signal. In 2-bit/4-state SWSFET, the propagation delay for the four voltage states is measured when the output state changes from 'state 0' to 'state 3' for a given input state change from 'state 0' to 'state 3' Fig. 6A. Similarly, propagation delays in other output state transitions when state changes from 'state 1' to 'state 0', 'state 2' to 'state 1' and 'state 3' to 'state 2' are shown in Figs. 6B–6D, respectively. The propagation delay was calculated between input V (Word line) and output V (Data stored_bar) can be clearly seen in Figs. 8A–8D respectively. Results are tabulated in Table 4.

Table 4. Propagation delay in 4-state/2-bit SWS-SRAM

Logic States	State 0 (0)	State 1 (0.4)	State 2 (0.8)	State 3 (1.2)
Propagation Delay (ps)	5.7	18.5	13.5	5.8

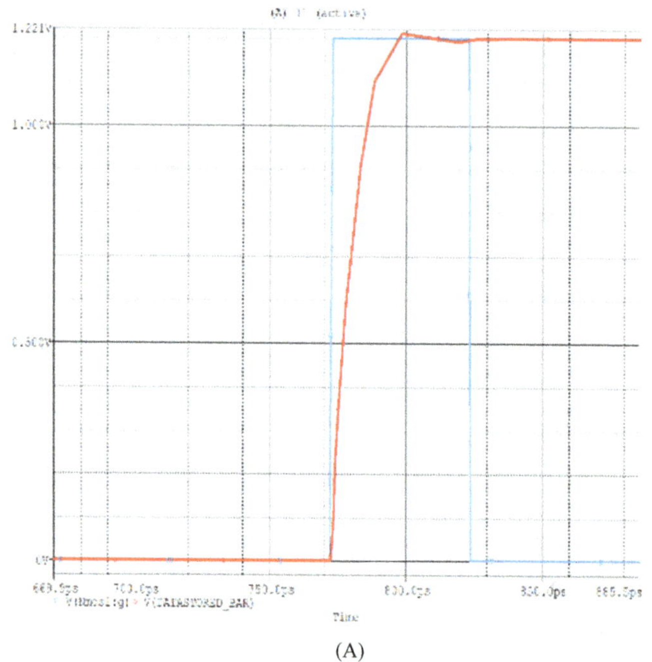

(A)

Fig. 6. The output response for a given input signal transition when output changes from (A) state '0' to '3' (B) state '1' to state'0' (C) state '2' to state '1' and (D) state '3' to state '2'.

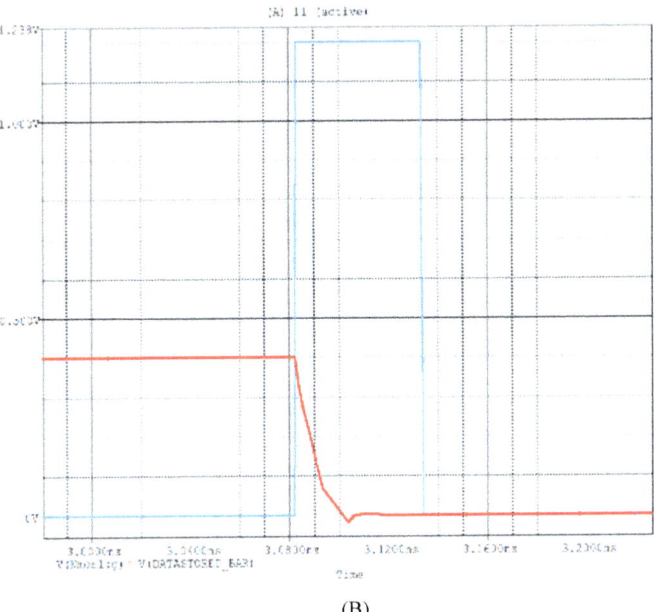

(B)

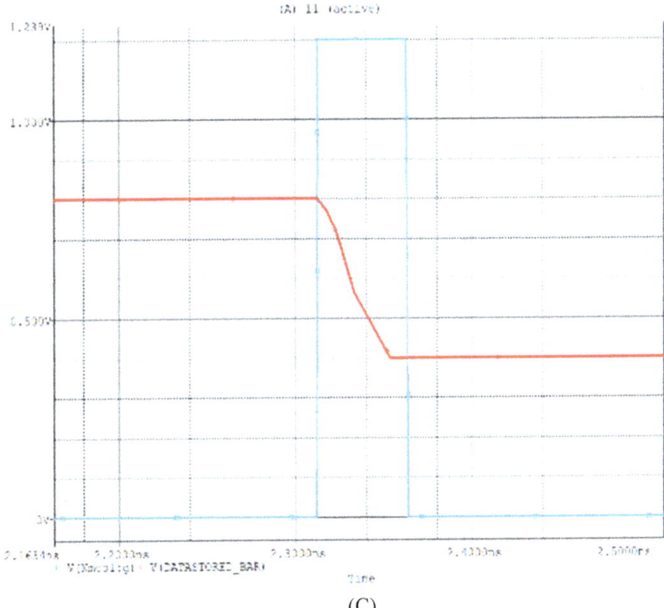

(C)

Fig. 6. (*Continued*)

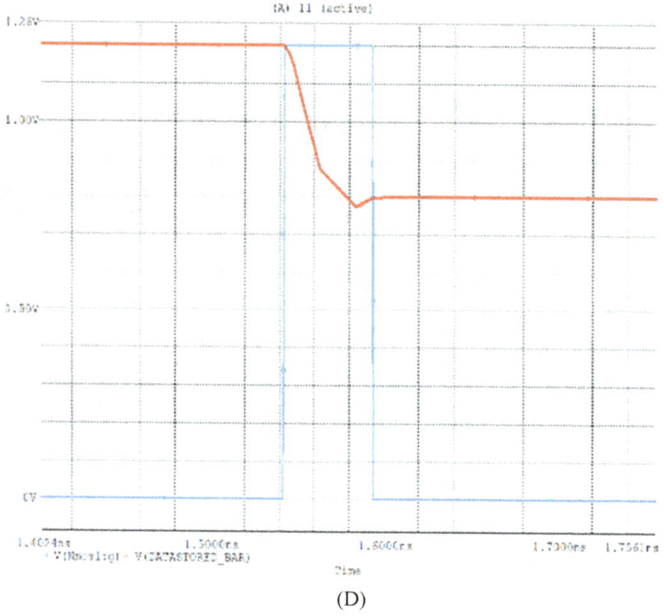

(D)

Fig. 6. (*Continued*)

3.3. *Power dissipation*

The power dissipation of the 6T 2-bit/4-state SRAM consists of static power and dynamic power. The average power dissipation for every state is calculated by adding the average dynamic power and the average static power for every state. However, for state 0, only average static power is considered which is equal to 26 nW. Figure 7 shows static and dynamic power dissipation as a function of time in SWS-SRAM cell. Dynamic power dissipation is the main factor that increases total power consumption [9]. The average power dissipation reported on Table 5 is obtained by calculating the average power on the four different states and was analyzed using the tools available in Cadence. The average power dissipation results in each state are given in Table 5.

Table 5. Average power dissipation of 2-bit/4-state SWS-SRAM

Input Volt	Average Power (nw)
0	26.3
0.4	0.0073
0.8	26.2
1.2	1290

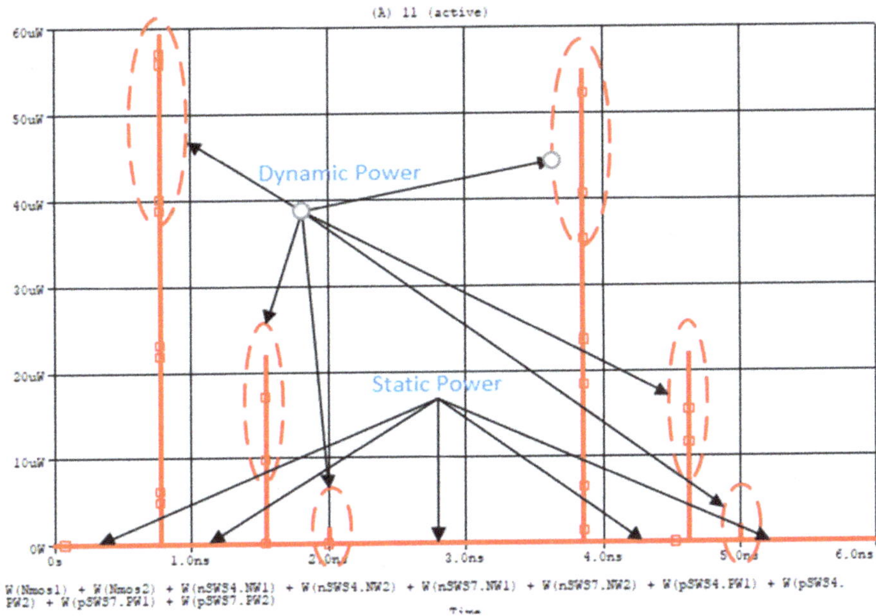

Fig. 7. Static and dynamic power dissipation as a function of time.

3.4. *Comparison of propagation delay and power dissipation*

The propagation delay and average power dissipation of this work on 2-bit/4-state SWS based SRAM unit is compared with the reported work conventional 6T SRAM cells in 180 nm technology, as shown in Table 6.

Table 6. Comparison of power dissipation and Propagation delay with reported work

Parameters	[10]	[11]	This work
Technology	180 nm	180 nm	180 nm
SRAM Configuration	6T CMOS	6T CMOS	6T SWSFET
Logic states	2-states	2-states	4-states/2-bit
Propagation delay	~42.5 ps	70 ps	18.5 ps
Average power dissipation	~20 µW	1.18 µW	1.29 µW
Supply voltage	1.8 V	1.2 V	1.2 V

The 6T SRAM reported in [10] is based on 2-states CMOS using 180 nm technology. SRAMs reported in [11] also use 2-state CMOS transistors using 180 nm at 1.2 V. In SWSFET SRAM design, we are using 6T with 4-state SWSFETs in the core cell. SWSFET SRAM design has 4-state or 2-bit memory storage capability which is twice as compared to 6T CMOS SRAM [11] which has 2-states and 6T CMOS SRAM [10]. To compare

SWSFET SRAMs with work reported in [10, 11], propagation delay and power dissipation should be *multiplied by a factor of 2*. Finally, SWSFET SRAMs are two-times faster and consume half the power which is not shown in Table 6.

4. Conclusion

The SWS-SRAM simulation combines an Analog Behavioral Model (ABM) and the Berkeley Short-channel IGFET Model (BSIM4.6). The simulations were carried out in Cadence.

The propagation delay and power dissipation of a 2-bit SWS-SRAM have been calculated. The 4-state logic is achieved using two threshold voltage levels for p-SWS-FET and n-SWS-FET. In conclusion, as compared to the conventional CMOS SRAMs reported in [10, 11], SWSFET SRAM is two-times faster and consumes half the power.

References

1. F. C. Jain, B. Miller, E. Suarez, P.-Y. Chan, S. Karmakar, F. Al-Amoody, M. Gogna, J. Chandy, and E. Heller. "Spatial wavefunction-switched (SWS) InGaAs FETs with II–VI gate insulators." *Journal of Electronic Materials* 40, no. 8 (2011): 1717-1726.

2. F. Jain, B. Saman, R. Gudlavalleti, R. Mays, J. Chandy, and E. Heller. "Low-threshold II–VI lattice-matched SWS-FETs for multivalued low-power logic." *Journal of Electronic Materials* 50, no. 5 (2021): 2618-2629.

3. R. H. Gudlavalleti, B. Saman, R. Mays, E. Heller, J. Chandy, and F. Jain. "A novel peripheral circuit for SWSFET based multivalued static random-access memory." *International Journal of High Speed Electronics and Systems* 29, no. 01n04 (2020): 2040010.

4. B. Saman, P. Gogna, E.-S. Hasaneen, J. Chandy, E. Heller, and F. C. Jain. "Spatial wavefunction switched (SWS) FET SRAM circuits and simulation." In *Microelectronics and Optoelectronics: The 25th Annual Symposium of Connecticut Microelectronics and Optoelectronics Consortium (CMOC 2016)*, 2017, pp. 37-48.

5. A. Almalki, B. Saman, J. Chandy, E. Heller and F. C. Jain. "Propagation delay evaluation for spatial wavefunction switched (SWS) FET-based inverter." *International Journal of High Speed Electronics and Systems* 31, no. 01n04 (2022): 2240008.

6. F. Jain, M. Lingalugari, B. Saman, P. Y. Chan, P. Gogna, E. S. Hasaneen, J. Chandy, and E. Heller. "Multi-state sub-9 nm QDC-SWS FETs for compact memory circuits." In *46th IEEE Semiconductor Interface Specialists Conference (SISC)*, 2015, pp. 2-5.

7. F. Jain, B. Saman, R. Gudlavalleti, R. Mays, J. Chandy, and E. Heller. "Low-threshold II-VI lattice-matched SWS-FETs for multi-valued low-power logic." *Journal of Electronic Material* 50 (2021): 2618-2629.

8. A. Husawi, B. Saman, A. Almalki, R. Gudlavalleti, and F. C. Jain. "Power dissipation and cell area: Quaternary logic CMOS inverter vs. four-state SWS-FET inverter." *International Journal of High Speed Electronics and Systems* 31, no. 01n04 (2022): 2240009.

9. S. Kumar *et al.* "Impact of different technology nodes on the delay and power dissipation of 6T SRAM cell." In *2021 7th International Conference on Signal Processing and Communication (ICSC)*. IEEE, 2021.

10. S. Mutum and K. Mondal. "Comparative performance analysis of 6T SRAM cell in 180nm and 90nm technologies." In *2021 Asian Conference on Innovation in Technology (ASIANCON)*, Pune, India, 2021.

11. D. Chaudhuri, K. Roy, A. Nag, Comparison of Different SRAM Cell Topologies Using 180 nm Technology. In: Biswas, U., Banerjee, A., Pal, S., Biswas, A., Sarkar, D., Haldar, S. (eds) Advances in Computer, Communication and Control. Lecture Notes in Networks and Systems, vol 41. Springer, Singapore.

Ultra-Short Pulse-Train Generation of 30-GHz Repetition Rate Using Rational Harmonic Mode Locking and Nonlinear Polarization Rotation

A. Rahman*, S. Fan and N. K. Dutta

Department of Physics, University of Connecticut, Storrs, Connecticut 06269, USA
ashiq.rahman@uconn.edu

A 30-GHz pulse-train is generated using the rational harmonic mode-locking technique, experimentally, using a Mach–Zehnder Lithium Niobate modulator. The width of the pulses is then reduced from 5.8-ps to 1.9-ps by incorporating nonlinear polarization rotation. This phenomenon arises due to the very high nonlinear behavior of the photonic crystal fiber (PCF) added to the ring laser cavity. Numerically solving the Generalized Nonlinear Schrödinger Equation provided insights into the pulse evolution behavior. The relative polarization angle and length of the PCF were varied to study their effects on the pulse-width.

Keywords: Fiber optics; fiber lasers; mode locking; nonlinear polarization rotation; RHML; PCF.

1. Introduction

Generating a stable pulse train with short pulse width and high speed is critical for high-bit-rate fiber-optic telecommunication systems. Active harmonic mode locking has been extensively studied as a way to generate high-repetition-rate pulses [1–4] by using electro-optical intensity modulators like $LiNbO_3$ modulators or electric absorption modulators inside the laser cavity [1, 5]. Rational harmonic mode locking is a technique that detunes the cavity-loss modulation frequency away from the exact harmonics of the cavity's fundamental frequency to further increase the repetition rate [6]. However, this technique has limitations in terms of temporal pulse duration and pulse train amplitude fluctuations. Several experiments have been conducted to improve laser stability and reduce fluctuations, by means of pulse width compression and pulse amplitude equalization [6, 7].

On the other hand, passively mode-locked fiber lasers can generate pulse trains with ultrashort pulse width, and recently, compact passively mode-locked fiber lasers have been successfully constructed using saturable absorbers like semiconductor saturable absorber mirrors, graphene, and nonlinear fiber loop mirrors [8–12]. The nonlinear polarization rotation (NPR) technique combined with a polarizer has been used to achieve ultrashort pulses in fiber lasers by inducing an intensity-dependent loss in the cavity [13–16]. However, NPR-based passively mode-locked fiber lasers suffer from low repetition rates,

---

*Corresponding author.

limiting their application in high-speed fiber-optic communications. To overcome this limitation, a hybrid mode-locking scheme that combines active and passive mode-locking methods has been proposed.

This paper provides both numerical and experimental evidence of the decrease in pulse-width of a 30-GHz pulse train from 5.8-ps to 1.9-ps. This reduction was achieved through Rational Harmonic Mode-Locking, which involved using a Mach–Zehnder Lithium Niobate modulator and adjusting the cavity frequency. The pulses were significantly shaped by a highly nonlinear PCF within the cavity. The paper also includes simulations of how changes in the polarization angles and length of the PCF would affect the output pulse-width.

2. Fiber Ring Laser Experimental Setup and Mathematical Model

2.1. *Experiment Setup*

The rational harmonic mode locked laser (RHML) setup that was used to generate the rational harmonic mode-locked pulses is shown in Fig. 1. The input seed pulse is provided by the 980-nm diode laser pump. A 980/1550-nm Wavelength Division Multiplexer (WDM) coupler takes the 980-nm input and outputs 1550-nm CW laser so that the experiment yields output in the wavelength that is most useful for standard telecommunication practices. The input is amplified by a 23-m long Erbium Doped Fiber Amplifier (EDFA). The modulation of the CW laser to a pulsed laser of frequency 10-GHz is achieved by Lithium Niobate ($LiNbO_3$) Mach Zehnder Modulator (MZM). The polarization controller (PC-1) just before the MZM input controls the polarization of the laser

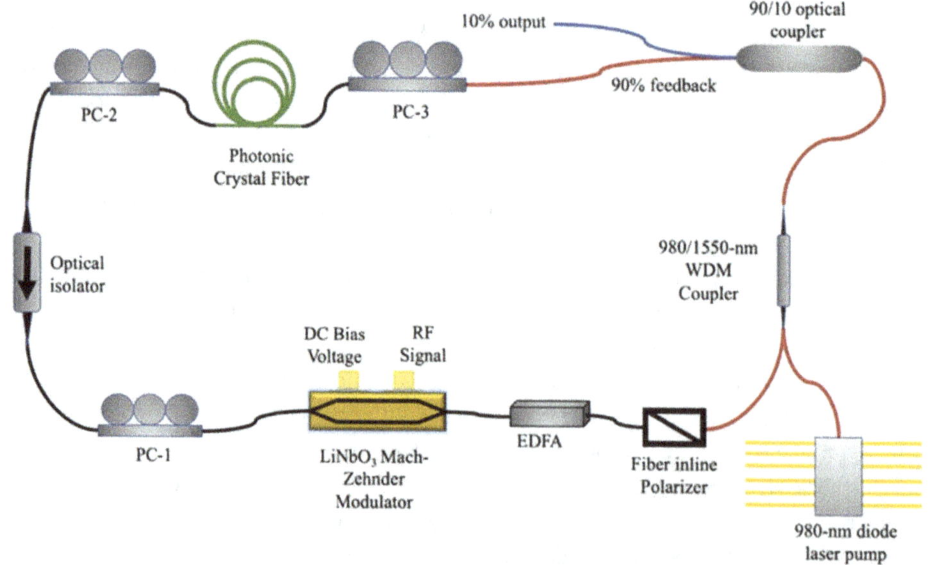

Fig. 1. Fiber ring laser setup incorporating RHML and NPR, powered by a 980-nm pump laser.

that goes into the MZM since the it is polarization sensitive. There are two more polarization controllers, (PC-2 and PC-3) and a polarizer in-line with the highly non-linear photonic crystal fiber (PCF). This combination induces the nonlinear polarization rotation (NPR) effect in the laser cavity. The optical isolator ensures unidirectional flow, which is necessary for the pulse formation and output from the ring cavity. After stable pulses are formed in the cavity, 10% of them are 'read-out' using the 90:10 optical coupler, whereas 90% of them are fed back into the cavity to ensure continuous operation. All the components in the optical circuit are connected using the standard commercial single mode fibers (SMF).

2.2. *Mathematical Model of Pulse Propagation*

The electric field in an optical pulse can be written as

$$E(x, y, z, t) = \frac{1}{2}\left[A(z,t)\psi(x,y)e^{i(\omega_0 t - \beta_0 z)} + cc \right]$$

where A(z,t) represents the slowly varying complex envelope of the pulse, ψ(x,y) represents the transverse electric field distribution of the mode, ω_0 represents center frequency, and β_0 represents the center propagation constant at ω_0. In the presence of optical loss or gain, second-order dispersion, and third-order nonlinearity, (by using Maxwell's electromagnetic wave equations) the complex envelope A(z,t) can be shown to satisfy the following equation [17, 18]:

$$\frac{\partial A(z,t)}{\partial z} + \frac{\alpha}{2}A(z,\tau) + \sum_{k \geq 2}\frac{i^{k-1}}{k!}\beta_k\frac{\partial^k A(z,\tau)}{\partial \tau^k}$$

$$= \frac{g}{2}A(z,\tau) + \frac{g}{2\Omega_g^2}\frac{\partial^2 A(z,\tau)}{\partial \tau^2} + i\gamma|A(z,\tau)|^2 A(z,\tau). \tag{1}$$

The fiber ring laser setup outlined above is simulated using MATLAB program by solving the Generalized Nonlinear Schrödinger Equation (GNLSE), eq. (1), using a pseudo-spectral numerical procedure known as Split Step Fourier Method (SSFM) [17]. This equation is used to model the pulse propagation in various fiber optic components in the fiber laser. Here, $A(z, t)$ is the envelope of the propagating pulse, that is varying slowly. It is a function of the propagating distance z and the time-delay parameter τ. The fiber loss is given by α, and the nonlinear parameter γ represents the strength of the nonlinearity of the optical component. β_k are the various dispersion coefficients. The gain bandwidth of the EDFA is the spectrum over which the laser is amplified, which is written as Ω_g. Equation (2) is a simplification of eq. (1), assuming that all the dispersion parameters higher than $k = 3$, and the fiber loss, are negligible.

$$\frac{\partial A(z,t)}{\partial z} + \frac{i}{2}\beta_2\frac{\partial^2 A(z,\tau)}{\partial \tau^2} - \frac{1}{6}\beta_3\frac{\partial^3 A(z,\tau)}{\partial \tau^3}$$

$$= \frac{g}{2}A(z,\tau) + \frac{g}{2\Omega_g^2}\frac{\partial^2 A(z,\tau)}{\partial \tau^2} + i\gamma|A(z,\tau)|^2 A(z,\tau). \tag{2}$$

It is also important to consider that the gain g of the EDFA is saturated when the input pulse energy E is equal to the saturation energy E_{sat}. When E is small, g is approximately equal to g_0, which is the small signal gain. This is mathematically written as in eq. (3). Single mode fibers provide no gain to the system, so $g = 0$ for that.

$$g = g_0 \left(1 + \frac{E}{E_{sat}}\right)^{-1}.$$

(3)

When stable pulses are formed in the cavity, 10% of the pulses are coupled out as mentioned in the previous section, which is simulated by including eq. (4) in the code.

$$A(z,\tau)_{out} = A_{in}(z,\tau) \cdot R.$$

(4)

The mathematical models for the Mach Zehnder Modulator and the Nonlinear Polarization Rotation are explained in detail in sections 3 and 4, respectively. $R = 0.1$ (10 percent). The values for all the various optical parameters used in the MATLAB code are summarized in Table 1 for quick reference.

3. Rational Harmonic Mode-Locking

Rational Harmonic Mode-Locking (RHML) is a form of active mode-locking scheme to generate very high frequency laser pulse trains. This experiment involves a Lithium Niobate (LiNbO$_3$) Mach-Zehnder modulator (MZM) to achieve mode-locking of the CW laser, and a deliberate detuning of the cavity frequency results in rational harmonic mode locking. The basic working principle is as follows [18]. Continuous wave laser is split and passed through the two arms of the device. The two arms are driven by two different electrical signals that cause the laser beams to change their optical path lengths, hence the phases, differently. They are then recombined, resulting in an intensity modulated output. Mathematically, the transmission through the MZM is modeled as in eq. (5) [19].

$$T = \cos^2\left(\pi \frac{V_b + V_m \sin(2\pi f_m \tau)}{V_\pi}\right),$$

(5)

where the DC voltage bias, voltage required for π phase shift, and the RF signal amplitude are given by V_b, V_π, V_m, respectively. f_m is the frequency of the input RF signal from the synthesizer. Now, to achieve RHML, the cavity mode separation f_c is deliberately detuned to a fraction of the modulation frequency f_m. Mathematically, when $f_m = (n + 1/p)f_c$ is satisfied, where n and p are integers, the output frequency is p times that of the modulation frequency. In this experiment $p = 3$, hence the output pulse train frequency was 30-GHz. The cavity mode separation is defined as follows: $f_c = c/n_i L$, where c is the speed of light in vacuum, n_i is the average index of the fiber, and L is the length of fiber loop.

4. Nonlinear Polarization Rotation

The nonlinear polarization rotation (NPR) effect is induced in the laser cavity using the setup shown in Fig. 2(a), where polarization controllers, PC2 and PC3, are connected in-line with the photonic crystal fiber (PCF) and a polarizer. The PCs are designed by

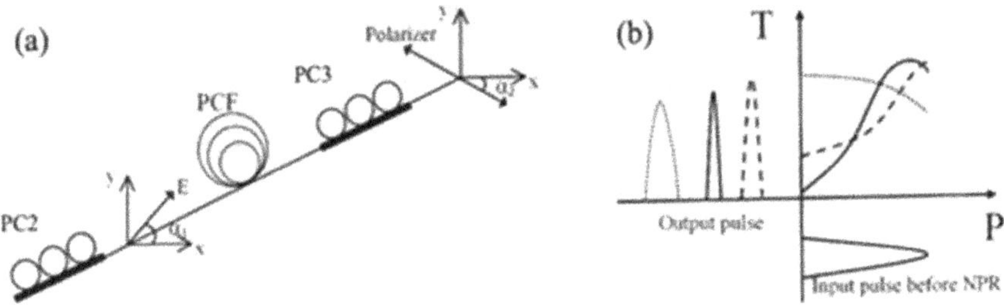

Fig. 2. (a) PCF placed in-line with polarization controllers to generate NPR (b) Output pulse behavior for three different transmission curves. Transmission vs. power is plotted.

having short length of single-mode fibers going around spools. When the orientation is adjusted stress is applied on the fibers, resulting in birefringence property of the controllers, causing the transmitted light to have fractional waveplates. Controlling the polarization of the transmitted light using the PCs and the polarizer, which affects the transmitted light intensity, induces Kerr nonlinearity in the PCF. This causes the polarization states to rotate, hence, nonlinear polarization rotation in the laser cavity. Equations (6)–(7) relate the transmission to the polarizer angles and parameters of the PCF.

$$T = \cos^2 \alpha_1 \cos^2 \alpha_2 + \sin^2 \alpha_1 \sin^2 \alpha_2 \frac{1}{2} \sin 2\alpha_1 \sin 2\alpha_2 \cos \left(\Delta\phi_L + \Delta\phi_{NL} \right) \tag{6}$$

$$\Delta\phi_L = \frac{(n_x - n_y)}{\beta L}, \quad \Delta\phi_{NL} = -\frac{1}{3}\gamma PL \cos \alpha_2 \tag{7}$$

Here, E is the electric field vector of the input light, α_1 is the angle between the input polarization and the fast axis of the PCF, and α_2 is angle between the fast axis of the PCF and polarization of the inline polarizer, $\Delta\phi_L$ and $\Delta\phi_{NL}$ are the linear and nonlinear changes in phase, respectively, L is the PCF length, $\beta = 2\pi/L$ is the propagation constant, n_x, n_y, are the linear birefringence coefficients of the fast (x) and slow (y) axes of the PCF, $\Delta n = n_x - n_y$ used in the simulation, γ is the nonlinear coefficient of the PCF, and P is the instantaneous power of the input signal. Values for the different PCF parameters used in the simulation are summarized in Table 1.

Varying the angles α_1 and α_2 in eq. (6), done by carefully tuning the polarization controllers PC2 and PC3, yields different transmission curves, three sets of which are plotted in Fig. 2(b). The sets of values affect the amplitude, offset and period of the cosine curves of the transmittance. When the laser pulses pass through the setup which has the 'dotted' transmission curve, it lets the low-power portion of a pulse (pulse wings) pass through, while throttling the pulse center, which has the highest power. This broadens the pulse-width and equalizes the pulse amplitude. The 'dashed' curve allows the high-power parts of the pulses to pass through with negligible loss, but the low-power pulse wings to undergo high loss, hence, shaping the pulse. This effect is similar to that obtained by using saturable absorbers in the cavity instead. Now, when the angles are further adjusted, the 'solid' line transmission property can be achieved, where both pulse-shaping as well as

Table 1. Modulator parameters used for generating the hybrid mode-locked pulse train.

Component	Parameter	Value
Erbium Doped Fiber	β_2	$-0.13 \times 10^{-3}\ ps^2 m^{-1}$
	β_3	$0.135 \times 10^{-3}\ ps^3 m^{-1}$
	γ	$3.69\ W^{-1} km^{-1}$
	g_0	$1.09\ dBm^{-1}$
	L	$23\ m$
Single Mode Fiber	β_2	$-22.1 \times 10^{-3}\ ps^2 m^{-1}$
	β_3	$0.171 \times 10^{-3}\ ps^3 m^{-1}$
	γ	$1.20\ W^{-1} km^{-1}$
	g_0	$0\ dBm^{-1}$
PCF	β_2	$-1.66\ ps^2 km^{-1}$
	β_3	$-0.03\ ps^3 km^{-1}$
	γ	$11\ W^{-1} km^{-1}$
	Δn	3.5×10^{-5}
	L	$0.6\ km$
Coupler	R	$90\ \%$
Modulator	V_π	$6\ V$
	V_m	$6\ V$
	V_b	$0\ V$
	f_m	$\sim 10\ GHz$
	f_c	$4.29\ MHz$
	n	2329
	p	3

peak equalization occurs. This is achieved by absorbing the pulse-edges and simultaneously increasing the transmission loss for the high pulse amplitudes that arise randomly. The effect of these cosine transmission curves on pulse-width is simulated in the next section.

5. Simulation Results

The fiber ring laser setup shown in Fig. 1 is simulated in MATLAB, by solving the GNLSE (eq. 1) using SSFM. The parameters mentioned in Table 1 were incorporated in the code. On the other hand, using only RHML in the laser cavity yields 5.3-ps pulse-width. The optimal length and angular settings were found by varying α_2 and L, as shown in Fig. 3. When the length of the PCF and polarization angle α_1 is kept constant and α_2 is varied, the maximum pulse compression is achieved when $\alpha_2 = 30°$. Figure 3(b) shows the range

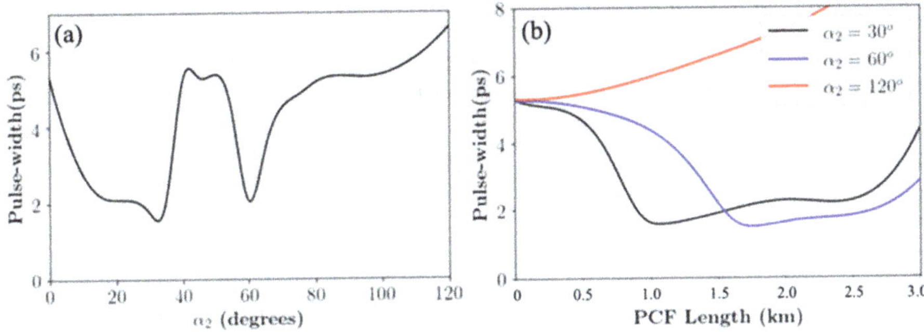

Fig. 3. Keeping $\alpha_1 = 45°$ constant, variation of pulse width with respect to (a) angle α_2 (PCF length fixed) and (b) PCF length (α_2 fixed).

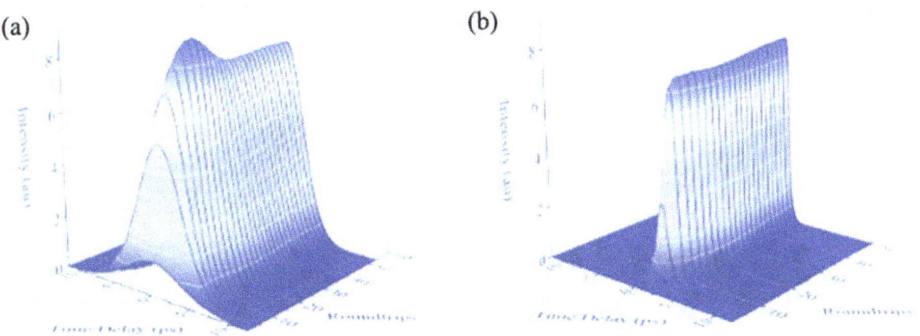

Fig. 4. Simulation of intensity vs pulse-width vs round-trips required (a) with RHML only (b) with both RHML and NPR in the cavity.

of PCF lengths that would result in the most amounts of pulse-shaping for three different α_2 angles. As seen in both Figs. 3(a) and 3(b), the pulse-widths at angle $\alpha_2 = 120°$ are actually broadened instead of being compressed. The NPR only equalizes the pulse-amplitudes in this scenario. For the given PCF length $L = 0.6$ km, setting the polarization angles $\alpha_1 = 45°$ and $\alpha_2 = 30°$, the output pulse-width obtained was 1.9-ps. The pulse evolution process for this set of parameters is plotted in Fig. 4(b). Figure 4(a) is generated when the laser pulses are simulated using RHML only. It takes more roundtrips around the ring laser setup for stable output pulses to form. On the other hand, Fig. 4(b) is a simulation of the same laser setup but incorporating NPR in the cavity as well. In this case, stable pulses are formed much quicker due to the pulse-shaping action, resulting in a compressed output pulse.

6. Experiment Results

The fiber laser experiment yields a compressed output pulse when both RHML and NPR is incorporated. Throughout this experiment, the PCF length was kept constant at 0.6 km and the two polarizer angles were $\alpha_1 = 45°$ and $\alpha_2 = 30°$. Figure 5 shows the RF-spectra of the output pulses when RHML and Hybrid-ML is incorporated. Both cases show a peak frequency at around 30-GHz, at which the experiment is operated. It is seen that there are supermode noise for the RHML only experiment. No amount of adjustment of the modulation frequency is able to get rid of the sidebands. This is because the RF-driving frequency and the fundamental cavity frequency are not equal, resulting in an overall amplitude with a periodic envelope function. However, when NPR is incorporated along with RHML supermode noise is suppressed and the output pulses are more stable, due to decreased pulse amplitude fluctuation, which is useful in RHML generation of higher orders [1]. This is desirable in optical telecommunication purposes. NPR in the cavity also increases the signal-to-noise ratio to over 25 dB. This strong pulse-shaping phenomenon brought about by the set transmission property of the NPR effect. Figure 6 shows the autocorrelator traces

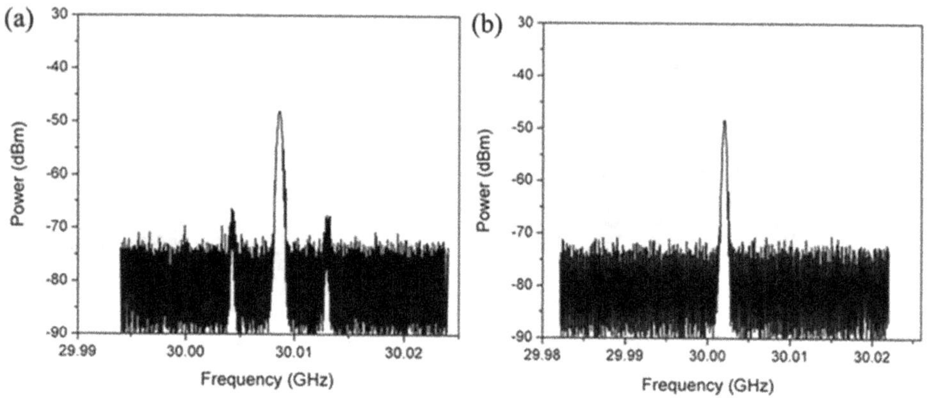

Fig. 5. RF spectra of the output pulses with (a) only RHML and (b) both RHML and NPR, showing a suppression of super-mode noise.

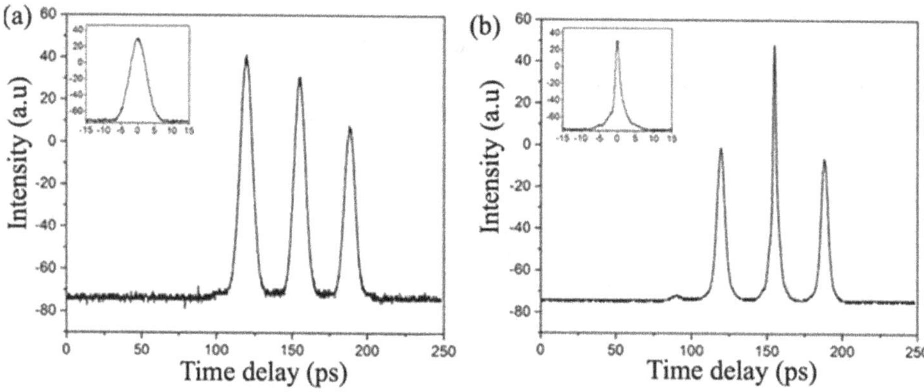

Fig. 6. Autocorrelator trace of 30-GHz output pulse trains for (a) RHML only and (b) both RHML and NPR. Insets showing a reduction in FWHM from 5.8-ps to 1.9-ps.

for the RHML and Hybrid-ML experiments. The pulse-delay between successive peaks is ~33.2-ps, indicating a pulse-train frequency of 30-GHz for both the experiments. The inset plots show a reduction of pulse-width from 5.8-ps to 1.9-ps when Hybrid-ML is incorporated, as compared to using only RHML in the laser cavity.

7. Conclusion and Discussion

This paper numerically and experimentally demonstrates the reduction of pulse-width of a 30-GHz pulse train from 5.8-ps to 1.9-ps highly nonlinear photonic crystal fiber (PCF) in the optical cavity. Using Rational Harmonic Mode-Locking, facilitated by a Mach-Zehnder Lithium Niobate modulator and careful detuning of the cavity frequency, the pulse train was generated. This short pulse is important for optical time division transmission at very high data rates.

Disclosures

The authors declare no conflicts of interest.

References

1. C. Wu and N. K. Dutta, "High-repetition-rate optical pulse generation using a rational harmonic mode-locked fiber laser," IEEE J. Quantum Electron. **36**, 721–727 (2000).
2. A. O. Wiberg, C.-S. Brès, B. P. Kuo, J. X. Zhao, N. Alic, and S. Radic, "Pedestal-free pulse source for high data rate optical time-division multiplexing based on fiber-optical parametric processes," IEEE J. Quantum Electron. **45**, 1325–1330 (2009).
3. S. Ma, W. Li, H. Hu, and N. K. Dutta, "High speed ultra short pulse fiber ring laser using photonic crystal fiber nonlinear optical loop mirror," Opt. Commun. **285**, 2832–2835 (2012).
4. S. Nolte, S. Döring, A. Ancona, J. Limpert, and A. Tünnermann, "High repetition rate ultrashort pulse micromachining with fiber lasers," in Fiber Laser Applications (Optical Society of America, 2011), paper FThC1.
5. Y. Fukuchi and J. Maeda, "Characteristics of rational harmonic mode-locked short-cavity fiber ring laser using a bismuth-oxide-based erbium-doped fiber and a bismuth-oxide-based highly nonlinear fiber," Opt. Express **19**, 22502–22509 (2011).
6. G. R. Lin, J. J. Kang, and C.-K. Lee, "High-order rational harmonic mode-locking and pulse-amplitude equalization of SOAFL via reshaped gain-switching FPLD pulse injection," Opt. Express **18**, 9570–9579 (2010).
7. Z. Li, C. Lou, K. T. Chan, Y. Li, and Y. Gao, "Theoretical and experimental study of pulse-amplitude-equalization in a rational harmonic mode-locked fiber ring laser," IEEE J. Quantum Electron. **37**, 33–37 (2001).
8. W. Li, "Different methods to achieve hybrid mode locking," Cogent Phys. **6**, 1707624 (2019). https://doi.org/10.1080/23311940.2019.1707624.
9. S. Thapa, A. Rahman, and N. K. Dutta, "Mode-locked fiber ring laser using graphene nano-particles as saturable absorbers," Int. J. High Speed Electron. Syst. **31** (01n04), 2240002 (2022). https://doi.org/10.1142/s012915642240002x.
10. W. Li, H. Hu, X. Zhang, S. Zhao, K. Fu, and N. K. Dutta, "High-speed ultrashort pulse fiber ring laser using charcoal nanoparticles," Appl. Opt. **55**, 2149–2154 (2016).

11. X. He, Z.-B. Liu, and D. Wang, "Wavelength-tunable, passively mode-locked fiber laser based on graphene and chirped fiber Bragg grating," Opt. Lett. **37**, 2394–2396 (2012).

12. A. F. Runge, C. Aguergaray, R. Provo, M. Erkintalo, and N. G. Broderick, "All-normal dispersion fiber lasers mode-locked with a nonlinear amplifying loop mirror," Opt. Fiber Technol. **20**, 657–665 (2014).

13. J. L. Luo, L. Li, Y. Q. Ge, X. X. Jin, D. Y. Tang, S. M. Zhang, and L. M. Zhao, "L-band femtosecond fiber laser mode locked by nonlinear polarization rotation," IEEE Photon. Technol. Lett. **26**, 2438–2441 (2014).

14. X. Liu, T. Wang, C. Shu, L. Wang, A. Lin, K. Lu, T. Zhang, and W. Zhao, "Passively harmonic mode-locked erbium-doped fiber soliton laser with a nonlinear polarization rotation," Laser Phys. **18**, 1357–1361 (2008).

15. X. Zhang, H. Hu, W. Li, and N. K. Dutta, "High-repetition-rate ultrashort pulsed fiber ring laser using hybrid mode locking," Appl. Opt. **55**, 7885–7891 (2016).

16. X. Fang, P. Wai, C. Lu, and J. Chen, "Flattop pulse generation based on the combined action of active mode locking and nonlinear polarization rotation," Appl. Opt. **53**, 902–906 (2014).

17. G. P. Agrawal, *Nonlinear Fiber Optics* (Academic, 2007).

18. N. K. Dutta, X. Zhang, *Optoelectronic Devices* (World Scientific Publishing Co. Pte. Ltd., 2018), Appendix 3.

19. H. Dong, H. Sun, G. Zhu, Q. Wang, and N. Dutta, "Clock recovery using cascaded $LiNbO_3$ modulator," Opt. Express **12**, 4751–4757 (2004).

https://doi.org/10.1142/9789811283765_0022

Threshold Inverter Quantizer (TIQ)-Based 2-Bit Comparator Using Spatial Wavefunction Switched (SWS) FET Inverters

W. Alamoudi[1], B. Saman[2], R. H. Gudlavalleti[1], A. Almalki[1], J. Chandy[1], E. Heller[3] and F. Jain[1,*]

[1]*Department of Electrical and Computer Engineering, University of Connecticut, CT, USA*
[2]*Department of Electrical Engineering, College of Engineering, Taif University, Saudi Arabia*
[3]*Synopsys Corporation, Ossining, NY, USA*
faquir.jain@uconn.edu

A Threshold Inverter Quantizer (TIQ)-based voltage comparator is used to quantize analog input signal in flash ADC designs. This quantizer is based on the systematic sizing of CMOS inverter thus eliminating resistor array which is used for conventional comparator array. Such an implementation removes static power during quantization of analog input signal. This paper presents a simulation of TIQ 2-bit-based comparator using spatial wavefunction switched (SWS) field effect transistor (FET)-based CMOS inverters. The inverters use 4-state SWSFETs. Unlike conventional FETs, SWSFETs consist of two or more vertical coupled arrays of either quantum dot or quantum well channels, where the spatial location of carriers within these channels is used to encode the logic states (00), (01), (10), and (11). The TIQ-based comparator circuit presented here is based on the 2-bit SWS-CMOS inverter. The schematic of the ADC comparator circuit is demonstrated as well as the 2-bit ADC configuration cascading two 2-bit SWSFET-based inverters in CMOS-X. The circuit simulation was done in Cadence and SWSFET was modeled by integrating Berkeley Short-Channel IGFET Model (BSIM) and the Analog Behavioral Model (ABM). The 2-bit comparator circuit provides a four-state logic output voltage for any given analog input signal.

Keywords: Threshold Inverter Quantizer (TIQ); 2-bit SWS-CMOS inverter; SWSFETs.

1. Introduction

The demand for high performance integrated circuits (ICs) has been met over the years by continuous scaling-down of metal-oxide semiconductor (MOS) devices. The digital logic circuits developed using MOSFET exhibit two logic states i.e., logic '0' and '1'. Another approach towards improving the performance of the ICs is by utilizing multistate logic circuits. A spatial wavefunction switched field effect transistor (SWSFET) is one such device which can be utilized to develop multistate logic circuits. SWSFET was introduced first by Jain, *et al.* [1], which works based on the electron wavefunction switching between two channels depending on the voltage applied on its gate. A SWSFET has two or more vertical coupled arrays of quantum dot or quantum well channels, where the spatial location

---

*Corresponding author.

of carriers within these channels is used to encode the logic states (00), (01), (10), and (11) [1]. The application of two strained layers of $Si/Si_{0.5}Ge_{0.5}$ in SWSFET promotes improved performance and increased device integration [1, 2].

A flash analog-to-digital comparator (ADC), mainly known for its high speed conversion as compared to other ADC architectures such as pipelined, successive-approximation register (SAR) ADCs. However, flash ADC consumes more power compared to other ADCs. A Threshold Inverter Quantizer (TIQ) flash ADC is one approach towards reducing power consumption in flash ADC while maintaining the conversion speed of the flash ADC. In this work, a Threshold Inverter Quantizer (TIQ)-based voltage comparator is used to quantize analog input signal in flash ADC designs. This paper presents a simulation of TIQ-based 2-bit/4-state comparator using SWSFET-based inverters.

Section 2 describes SWSFET device structure and experimental characteristics reported in the literature [1]. Section 3 introduces flash ADC and section 4 describes the implementation of TIQ-based flash ADC along with Cadence simulation results. The circuit simulation was done in Cadence and SWSFET was modeled by integrating Berkeley Short-Channel IGFET Model (BSIM) and the Analog Behavioral Model (ABM).

2. SWSFET Device Structure

A SWSFET device consists of four layers of SiO_x-cladded Si quantum dots (QD) which serve as transporter channel [1–3]. The schematic cross-section is illustrated in Fig. 1(a). SWSFET has two drains, in which D_1 represents shallow drain whereas D_2 represents deep drain. The upper two QD layers are interfaced to the shallow drain D_1 while the lower two QD layers are connected to the deep drain D_2. The current-voltage (I-V) characteristics of the device were demonstrated in Fig. 1(b) [10]. The charge transport in QD layers occurs based on the level of gate voltage. When the gate voltage is in the range 1.8–2.0 V, the lower QD layers are conducting. When the gate voltage is in the range 2.2–2.4 V, the upper QD layers are conducting [3].

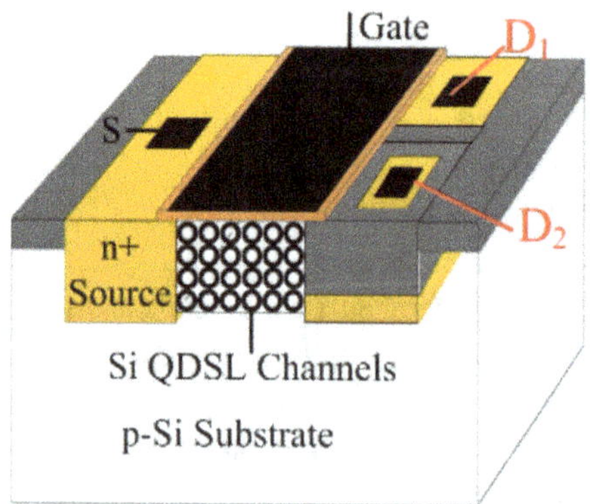

Fig. 1(a). Schematic cross-section of n-SWSFET structure [10].

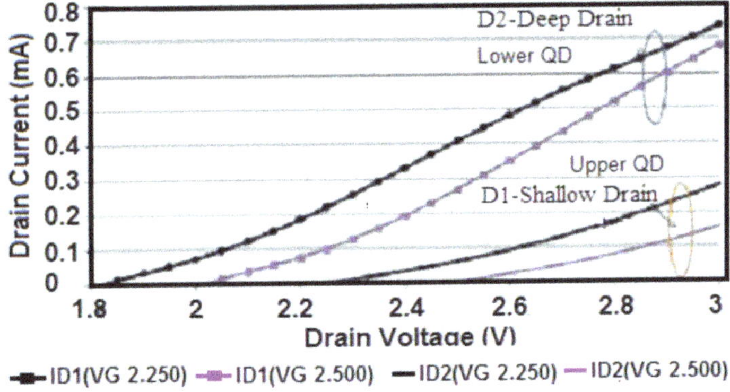

Fig. 1(b). ID-VD plot of SWSFET demonstrating the conduction in lower and upper channels.

The complementary structure of SWSFET is illustrated in Fig. 2. The n-MOS and p-MOS devices are separated by an oxide/insulator layer. In the n-MOS side, the two Si quantum wells are inserted between SiGe barriers, while in the p-MOS side, Si layers serve as barriers [1-3]. Quantum simulations display the switching of wavefunctions from the upper SiGe quantum well to the lower well in the p-channel and from lower Si quantum well to the upper well in the n-channel as the gate voltage increased Fig. 3(a) and 3(b). As the gate voltage decreased, inversion charge moved toward accumulation.

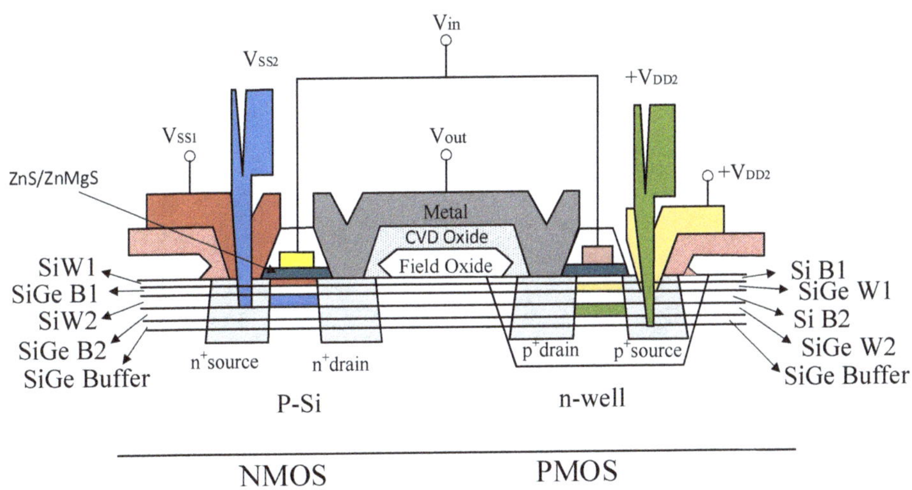

Fig. 2. Schematic cross-section of a 4-State/2-bit SWS-CMOS Inverter.

3. Flash ADC

A conventional 2-bit flash ADC comprises a resistive ladder that contains $2^N = 4$ resistors and $2^N\text{-}1 = 3$ comparators Fig. 4.

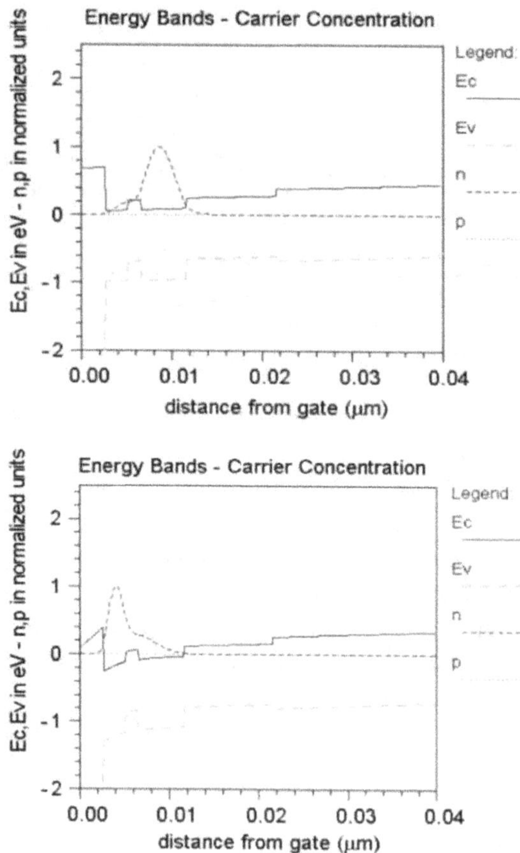

Fig. 3. Different states of electron wavefunctions on W1 and W2. (a) Electron confinement to lower quantum well W2 at Vg=0.2V. (b) Electron transferred to upper well W1 at Vg=0.8V [10].

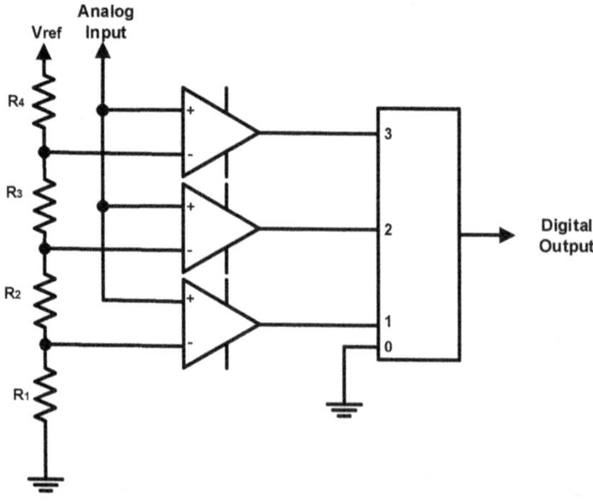

Fig. 4. Conventional 2-bit flash ADC comparators.

The flash ADC architecture uses a resistor ladder circuit where each comparator has a reference voltage that is 1 least significant bit (LSB) higher than the one below it in the ladder. As a result, comparators below a certain point output a logic 1 since their input voltage is greater than the reference voltage. On the other hand, comparators above that point output a logic 0 since their input voltage is smaller than the reference voltage. The digital output is produced through a priority encoder that encodes the output of 2N-1 comparators in thermometer encoding. This architecture is very fast because the analog input is applied to all the comparators at once. However, it requires many resistors, resulting in large area and high-power consumption.

4. Threshold Inverter Quantizer using 4-state SWSFETs

TIQ comparators compare the input voltage with internal threshold voltage (V_T) of the inverter, which is determined by the transistor sizes in the inverters. Hence, we do not need the resistor ladder circuit used in a conventional flash ADC (Fig. 5) [9]. The comparator outputs a binary code in two steps through an encoder. Comparator's role is to convert an input voltage (Vin) into a logic '1' or '0' by comparing a internal reference voltage (Vref) with the Vin. If Vin is greater than Vref, the output of the comparator is '1', otherwise '0'. The TIQ comparator uses two cascading CMOS inverters as a comparator for high speed and low power consumption.

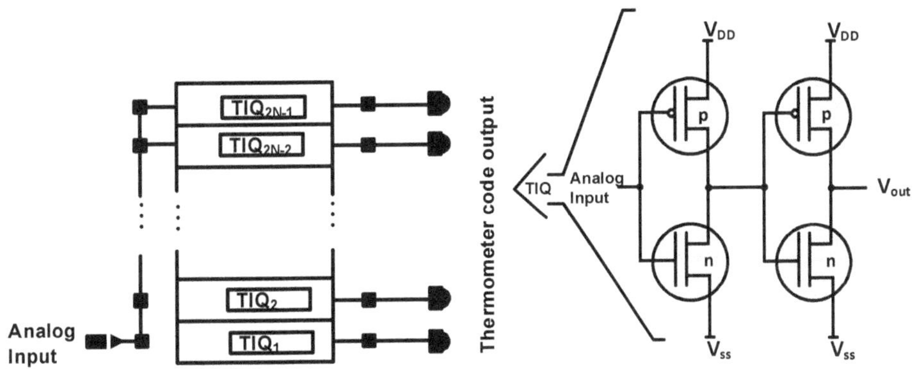

Fig. 5. 2-bit threshold inverter quantizer (TIQ)-based CMOS.

4.1. *Device structure of 2-bit TIQ SWS CMOS-X-based inverter*

This work presents simulation of a 2-bit TIQ-based comparator using SWSFET-based CMOS-X inverters. The inverters use 4-state spatial wavefunction switched SWSFETs as shown in Fig. 6. For higher resolution ADC, an array of SWS-CMOS-based TIQ comparators can be used with a fixed length of the P-SWSFET and N-SWSFET devices, and we can get desired threshold values by increasing only the width of the transistors. More research is required to encode the 2-bit output of each comparator in the ADC design.

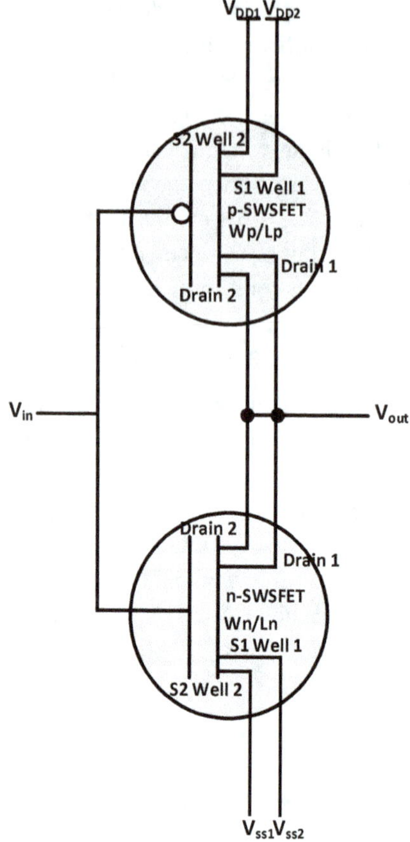

Fig. 6. 4-State/2-bit SWS-(CMOS) inverter.

The schematic of the 2-bit TIQ comparator is illustrated in Figs. 7 and 8. The SWS-CMOS-X inverter-based 2-bit TIQ has two internal reference voltages Vref1 and Verf2 corresponding to Vth1 and Vth2 (Fig. 9). The 2-bit TIQ comparator uses two cascading SWS-CMOS inverters as a comparator for high speed and low power consumption.

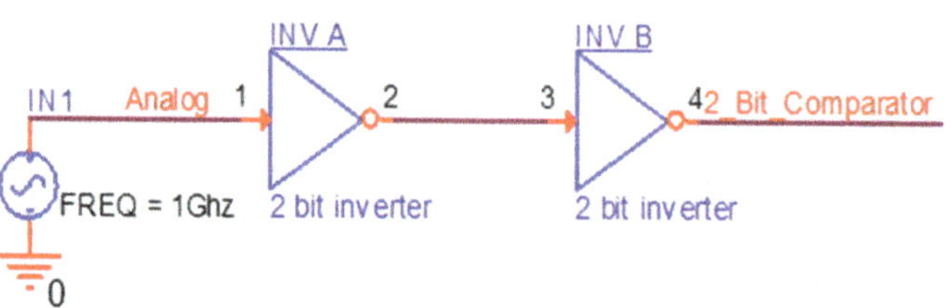

Fig. 7. Schematic of a 2-bit TIQ-based ADC comparator.

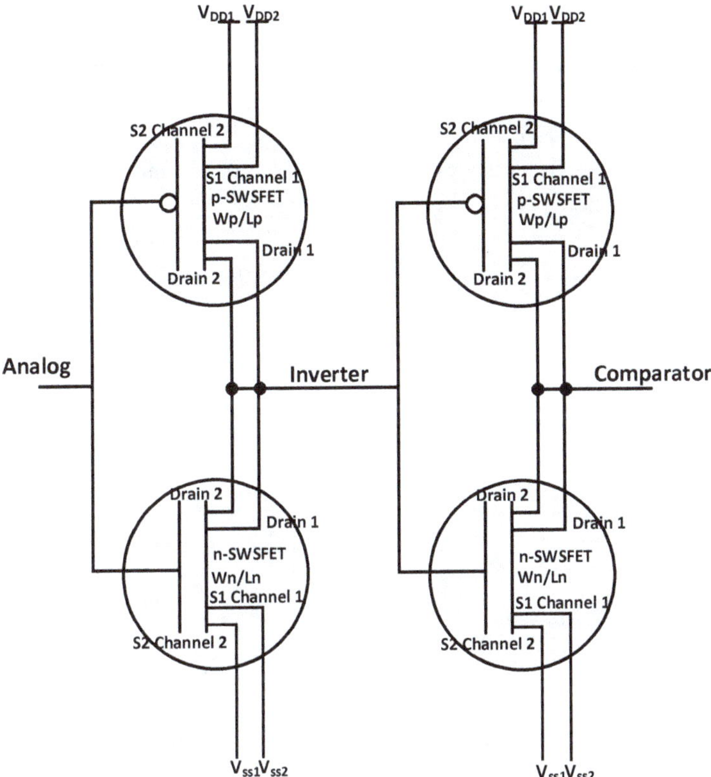

Fig. 8. Schematic of 2-bit TIQ-based ADC comparator using SWS-FET-based inverters in CMOS-X.

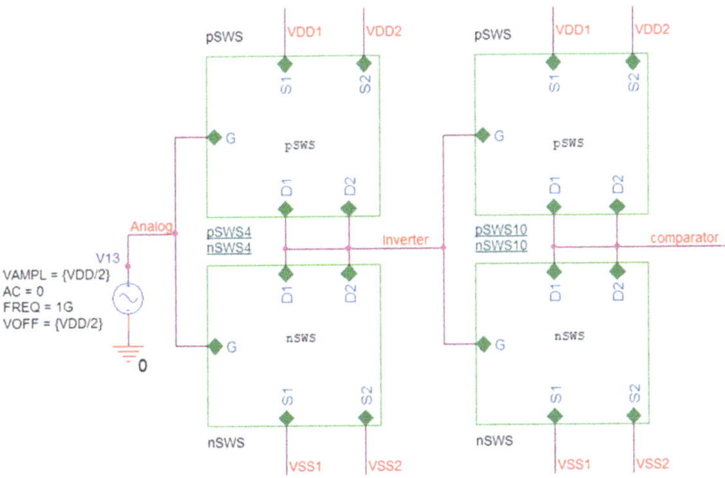

Fig. 9. Design 2-bit TIQ-based ADC comparator using SWS-FET-based inverters in CMOS-X.

4.2. *SWSFET Model and Parameters*

The SWSFET model has been developed by integrating two conventional MOS transistors and an analog behavioral model (ABM). The two transistors use 0.18 µm technology Berkeley Short-Channel IGFET Model (BSIM) models. The ABM model uses equations which depict the spatial wavefunction switched behavior of the SWSFET. The model parameters of the SWSFET are presented in Table 1.

Table 1. Parameter used for the SWSFET-based inverter.

Parameter	N Channel SWSFET	P Channel SWSFET
V_{TH1}	0.80 V	−0.80 V
V_{TH2}	0.2 V	−0.2 V
Width (W1)	0.450 µm	2.14 µm
Width W2)	0.9 µm	4.28 µm
Lengh (L)	0.18 µm	0.18 µm

4.3. *Simulation of TIQ-based Comparator*

Simulation results of the TIQ-based 2-bit comparator are shown in Fig. 10 which depict 4 levels of quantization of an analog input using only one SWS-CMOS-X-based TIQ comparator. A 2-bit ADC requires four comparators to perform ADC conversion and the same can be performed using single SWSFET-based TIQ comparator. Future work involves simulation of the design and simulation of flash ADC for higher bits including multistate encoder and analyze various performance parameters such as resolution, conversion speed and area.

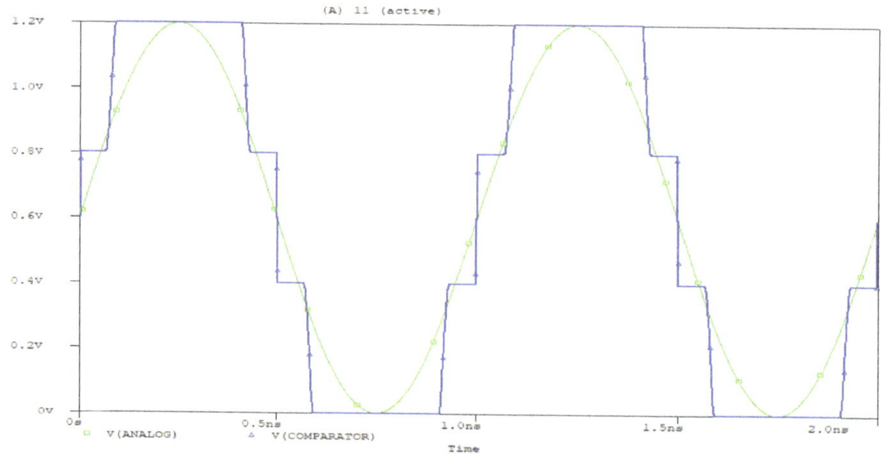

Fig. 10. Simulation results of the TIQ-based 2-bit ADC comparator showing the analog input signal (green), and the 2-bit 4-state comparator output (blue).

5. Conclusion

Spatial Wavefunction Switched FETs (SWSFETs) are unique as they generate multistate logic, which means they can produce four output states rather than the two output states (i.e., 0 or 1) produced by traditional CMOS transistors. This work has presented a 2-bit TIQ comparator design using SWS-CMOS-X inverters. This design takes advantage of the two internal reference voltages corresponding to Vth1 and Vth2 of the SWS-FETs to achieve higher resolution. Cadence simulation of the proposed 2-bit ADC comparator was carried out to show the 4 levels of quantization using only one SWS-CMOS-X-based TIQ comparator. This design can be expanded by using an array of SWS-CMOS-based TIQ comparators with different SWSFET aspect ratios leading to high resolution, low power, and high-speed ADCs.

References

1. Jain, F.C., Chandy, J., Miller, B., Hasaneen, E.S. and Heller, E., 2011. Spatial wavefunction-switched (SWS)-FET: A novel device to process multiple bits simultaneously with sub-picosecond delays. International Journal of High Speed Electronics and Systems, 20(03), pp. 641-652.
2. Saman, B., Gogna, P., Hasaneen, E.S., Chandy, J., Heller, E. and Jain, F.C., 2017. Spatial wavefunction switched (SWS) FET SRAM circuits and simulation. International Journal of High Speed Electronics and Systems, 26(03), p. 1740009.
3. Almalki, A., Saman, B., Chandy, J., Heller, E. and Jain, F., 2022. Propagation delay evaluation for spatial wavefunction switched (SWS) FET-based inverter. International Journal of High Speed Electronics and Systems, 31(01n04), p. 2240008.
4. Jain, F., Lingalugari, M., Saman, B., Chan, P.Y., Gogna, P., Hasaneen, E.S., Chandy, J. and Heller, E., 2015. Multi-state sub-9 nm QDC-SWS FETs for compact memory circuits. In 46th IEEE Semiconductor Interface Specialists Conference (SISC) (pp. 2-5).
5. Gudlavalleti, R.H., Saman, B., Mays, R., Heller, E., Chandy, J. and Jain, F., 2020. A novel peripheral circuit for SWSFET based multivalued static random-access memory. International Journal of High Speed Electronics and Systems, 29(01n04), p. 2040010.
6. Agrawal, N. and Paily, R., 2010. A threshold inverter quantization based folding and interpolation ADC in 0.18 μm CMOS. Analog Integrated Circuits and Signal Processing, 63, pp. 273-281.
7. Talukder, A.A. and Sarker, M.S., 2017, September. A three-bit threshold inverter quantization based CMOS flash ADC. In 2017 4th International Conference on Advances in Electrical Engineering (ICAEE) (pp. 352-356). IEEE.
8. Husawi, A., Saman, B., Almalki, A., Gudlavalleti, R. and Jain, F.C., 2022. Power dissipation and cell area: Quaternary logic CMOS inverter vs. four-state SWS-FET inverter. International Journal of High Speed Electronics and Systems, 31(01n04), p. 2240009.
9. Aytar, O., Tangel, A. and Dundar, G., 2008, June. A 9-bit 1GS/S CMOS folding ADC implementation using TIQ based flash ADC cores. In 2008 15th International Conference on Mixed Design of Integrated Circuits and Systems (pp. 159-164). IEEE.
10. Jain, F., Saman, B., Gudlavalleti, R. *et al.*, 2021. Low-threshold II–VI lattice-matched SWS-FETs for multivalued low-power logic. Journal of Electronic Materials, 50, pp. 2618-2629.

Novel Multi-State QDC-QDG FETs and Gate All Around (GAA) FETs for Integrated Logic and QD-NVRAMs

F. Jain[*,‡], R. H. Gudlavalleti[*], J. Chandy[*] and E. Heller[†]

[*]*University of Connecticut, CT, USA*
[†]*Synopsis Inc., Ossining, NY, USA*
[‡]*faquir.jain@uconn.edu*

This paper presents experimental I-V characteristics of a QDC-QDG FET that exhibited 5-states and has the potential to introduce additional states (e.g. 8) by utilizing Ge QDSL mini-energy sub-bands. Mini-energy bands are formed in an asymmetric Si quantum dot channel (QDC) comprising of two silicon oxide cladded Si quantum dots (QDs), where the upper layer has a smaller core diameter and thicker upper oxide cladding serving as tunnel oxide. Quantum simulations are presented to show more states when additional two germanium oxide cladded Ge dots are added on top of Si QD layers in the gate region. This paper also proposes Gate all around (GAA) FETs, when integrated with nonvolatile random access memories (NVRAMs) that have the potential for wafer scale integration, similar to vertical NANDs. Novel Si and Ge Quantum-dot-based device configurations discussed in this paper open the pathway forward to implement hardware platform for emerging applications using low power consumption and smaller footprint.

Keywords: Quantum dot FETs; multi-state FETs; Si QDC-Ge QDG; wafer scale integration.

1. Introduction

Cladded Si Quantum dot channel (QDC) FETs incorporating Si Quantum dot gate layers have been reported to experimentally exhibit distinct 4-state ID-VD and ID-VG characteristics [1]. In addition to ON and OFF, two intermediate states are attributed to carrier transfer in the inversion layer mini-energy sub-bands manifested in QDC due to the formation of quantum dot superlattice (QDSL) shown in Figs. 1(a) and 1(b).

2. Experimental Si QDC and Ge QDG Exhibiting 5-6 States in I-V Characteristics

Figure 2(a) shows the schematic of a FET replacing Si QDs in the gate region with Ge QDs. The experimental I-V characteristics are shown in Fig. 2(b). Unlike SiO_x-cladded Si quantum dots in the gate region, the GeO_x-cladded Ge dot forms additional mini-energy bands that are separated by smaller energies and hence have the potential of manifesting additional logic states.

[‡]Corresponding author.

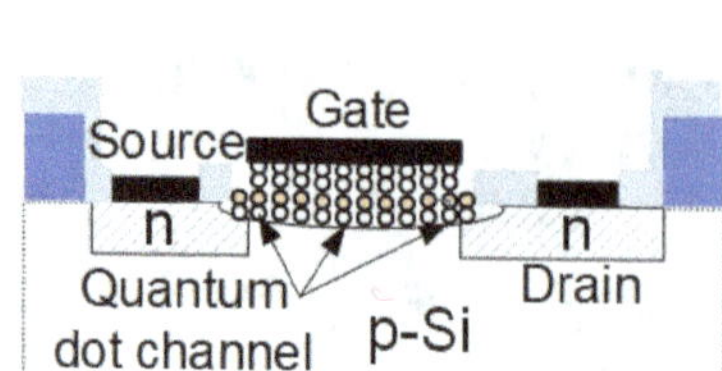

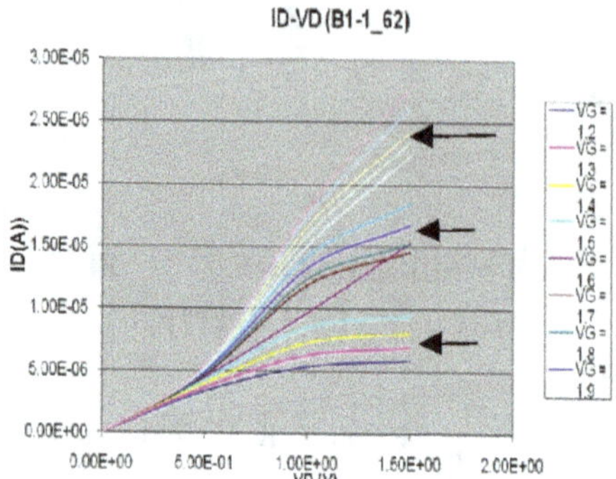

Fig. 1(a). Schematic of QDC-QDG FET.   Fig. 1(b). ID-VD characteristics showing distinct characteristics as a function of gate voltage.

The experimental 5-state ID-VG characteristics are shown in Fig. 2(b). The variation is in the use of HfO_2 layer tunnel oxide over upper SiO_x cladding on Si quantum dot layer, which improves sub-threshold slope. The schematic is shown in Fig. 2(c). We believe that the upper SiO_x cladding has an effective dielectric constant ($\kappa\sim8$). This contrasts with I-V shown in Fig. 1(b).

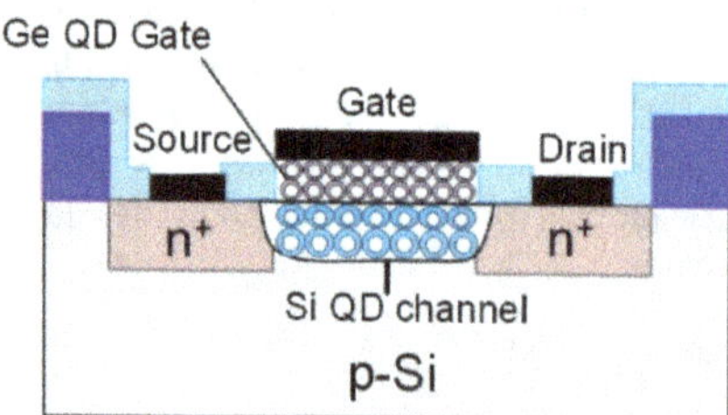

Fig. 2(a). QDC-QDG FET with asymmetric Si QD channel and Ge QDG.

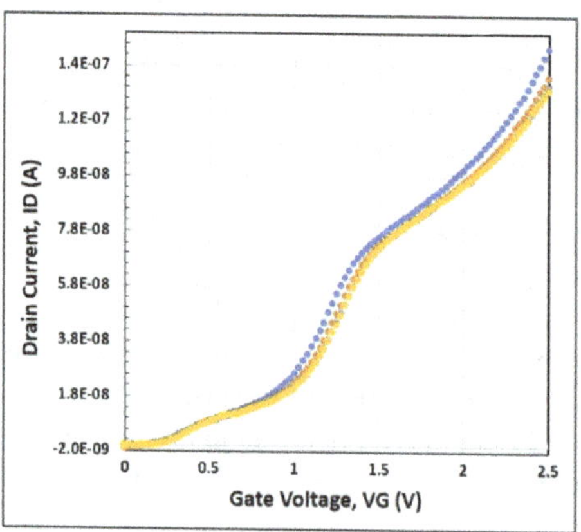

Fig. 2(b). 5-state experimental ID-VG as a function of VD.

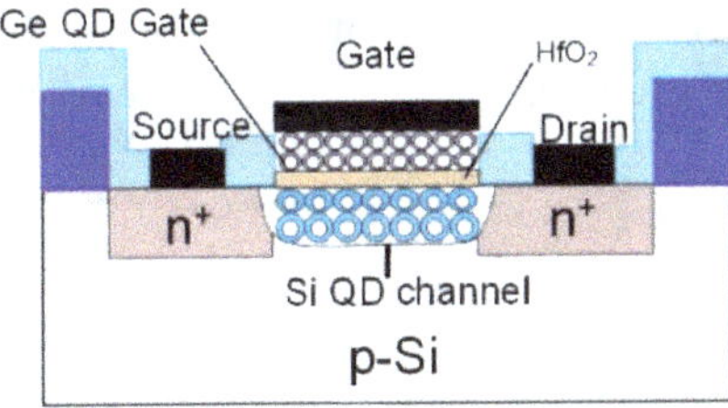

Fig. 2(c). Cross-sectional schematic of QDC-QDG FET exhibiting 5-state experimental ID-VG.

2.1. *Quantum simulations: QDC-QDG FETs*

Simulation of asymmetric Si QD channel with Ge QD gate layers, shown in Fig. 2(a), is presented in Fig. 3(a). Table 1 shows the parameters used.

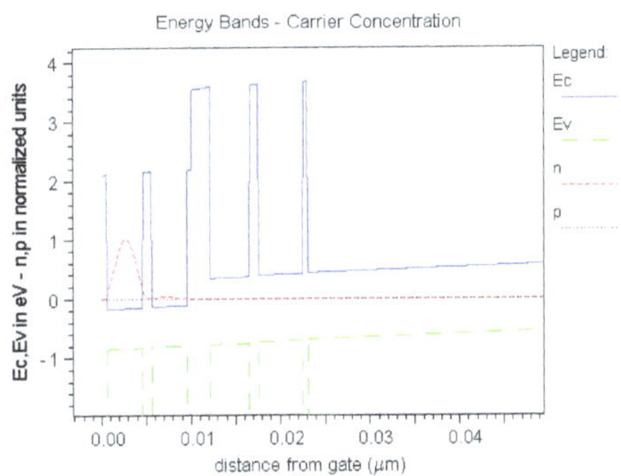

Fig. 3(a). Quantum simulations showing transfer of wavefunction in upper Ge QD.

Table 1. List of parameters used in Fig. 3(a).

Layer	Thick (μm)	χ(eV)	E_g (eV)	m_e	m_h	ε_r	N_d (cm^{-3})	N_a (cm^{-3})
GeO$_x$ clad	0.0005	2.25	5.70	0.16	0.16	4.4	0.0e00	0.0e00
GeQD core	0.0040	4.55	0.67	0.08	0.28	16.0	0.0e00	0.0e00
GeO$_x$ clad	0.001	2.25	5.70	0.16	0.16	4.4	0.0e00	0.0e00
GeQD core	0.0040	4.55	0.67	0.08	0.28	16.0	0.0e00	0.0e00
GeO$_x$ clad	0.0005	2.25	5.70	0.16	0.16	4.4	0.0e00	0.0e00
SiO$_x$ clad	0.002	0.9	9.0	0.5	0.5	3.9	0.0e00	0.0e00
Si QD	0.0045	4.15						
SiO$_x$ clad	0.0010	0.9	9.0	0.5	0.5	3.9	0.0e00	0.0e00
Si QD	0.0050	4.15						
SiO$_x$ clad	0.0005	0.9	9.0	0.5	0.5	3.9	0.0e00	0.0e00
Si	0.5000	4.15	1.12	0.19	0.49	11.9	0.0e00	1.0e16

Figures 3(b1) and 3(b2) show the effect of changing the diameter of Ge QD layers to 3 nm in the gate region and changing the oxide cladding over upper Si QD channel at two different gate voltages.

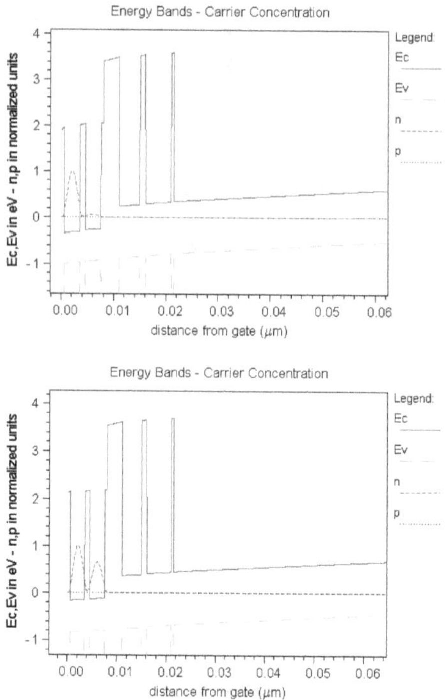

Fig. 3(b1-top) and (b2). Wavefunction transfer in Ge QDs due to the effect of changing the diameter of Ge QDs to 3nm.

Table 2. Parameters used in quantum simulations for Figs. 3(b1) and 3(b2).

Layer	Thick (μm)	χ (eV)	E_g (eV)	m_e	m_h	er	N_d (cm^{-3})	N_a (cm^{-3})
GeO$_x$ clad	0.0005	2.25	5.70	0.16	0.16	4.4	0.0e00	0.0e00
GeQD core*	0.0030	4.55	0.67	0.08	0.28	16.0	0.0e00	0.0e00
GeO$_x$ clad	0.0010	2.25	5.70	0.16	0.16	4.4	0.0e00	0.0e00
GeQD core*	0.0030	4.55	0.67	0.08	0.28	16.0	0.0e00	0.0e00
GeO$_x$ clad	0.0005	2.25	5.70	0.16	0.16	4.4	0.0e00	0.0e00
SiO$_x$ clad	0.0030	0.9	9.0	0.5	0.5	3.9	0.0e00	0.0e00
Si QD core*	0.0040	4.15	1.12	0.19	0.49	11.9	0.0e00	0.0e00
SiO$_x$ clad	0.0010	0.9	9.0	0.5	0.5	3.9	0.0e00	0.0e00
Si QD core*	0.0050	4.15	1.12	0.19	0.49	11.9	0.0e00	0.0e00
.SiO$_x$ clad	0.0005	0.9	9.0	0.5	0.5	3.9	0.0e00	0.0e00
Si	0.1000	4.15	1.12	0.19	0.49	11.9	0.0e00	1.0e16

* simulated as QWs

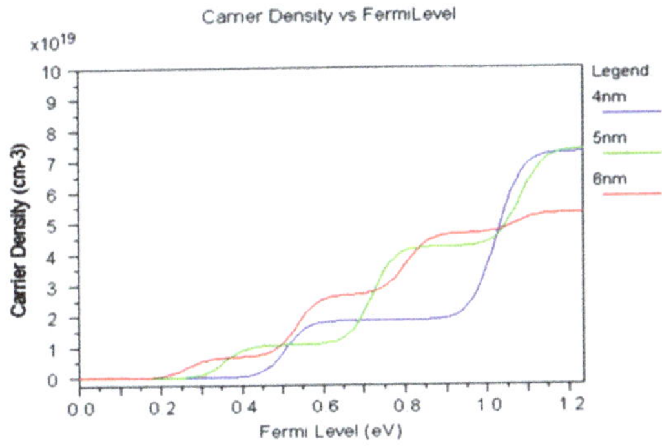

Fig. 3(c). Simulated carrier density in direct mini-energy sub-bands.

Figure 3(c) shows charge density vs Fermi-level location when Ge QDs are incorporated in the gate region [13]. The effect of Ge core size is shown for direct energy gap density of states (DOS) due to QDSL in the gate region.

3. Gate All Around (GAA) QDC-QDG FETs and NVRAMs

FETs using p-poly Si have been demonstrated [13] using Si QDC. The inversion channel exhibited electron mobility exceeding 80 cm^2/V-s.

A variation of Fig. 2(a) in GAA configuration using poly-Si core is shown in Fig. 4. The use of p-type poly-Si core has the potential for 3D stacking. The methodology is adapted from vertical nonvolatile memories (V-NAND) [14].

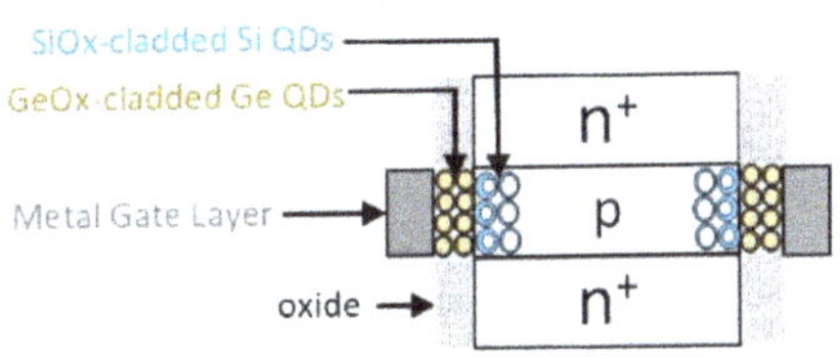

Fig. 4. QDC-QDG- FET in GAA configuration.

This structure lends itself to NVRAMs with Ge quantum dot access channel (QDAC) [7]. This is shown in Fig. 5.

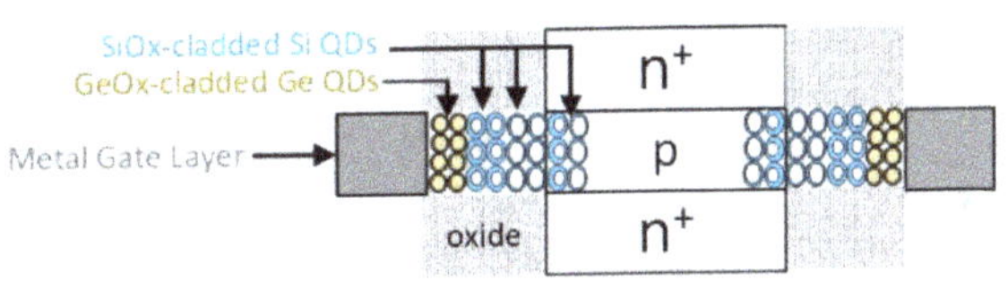

Fig. 5. Recessed QDC channel with Si QD floating gate and Ge QDAC.

4. Multi-state QDC-SWS FET Configurations with Single Drain

Figure 6(a) shows the schematic of a spatial wavefunction switched (SWS) FET and Fig. 6(b) shows the conduction in lower and upper channels for a fabricated device exhibiting 4-states [2].

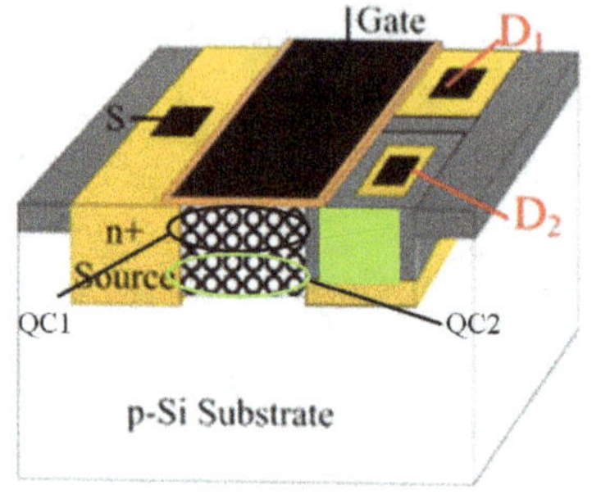

Fig. 6(a). Schematic of 2-Si quantum dot channel SWS-NMOS. Deep drain D2 is connected to QC2 comprising of lower most two QD layers, and shallow drain D1 is connected to QC1, upper QD layers.

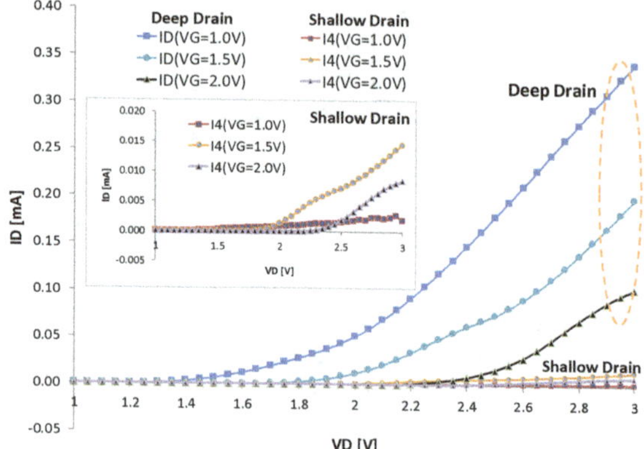

Fig. 6(b). I_D-V_D characteristics of 2-QD channel SWS-NMOS.

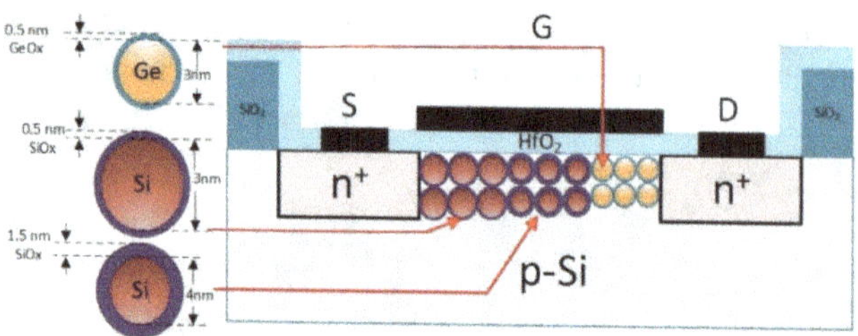

Fig. 7(a). QDC integrating Si QDs (4nm), Si QDs (3nm) and Ge QD (4nm) to implement 8-states.

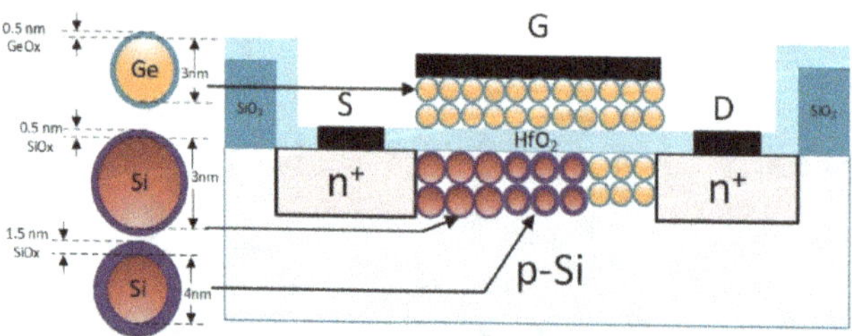

Fig. 7(b). Quantum dot channel integrating Si QDs (4nm), Si QDs (3nm) and Ge QD (4nm) along with using HfO2 as tunnel oxide to implement two layers of Ge QDs in the gate region.

Figure 7(a) shows the schematic cross-section of a QDC structure having SiO_x-Si QDs near the source end with larger core diameter (d1), and GeO_x-Ge QDs near the drain end. In between the Si QDs and Ge QDs, there is a set of Si QD layers with diameter (d2<d1).

Figure 7(b) shows the quantum dot channel integrating Si QDs (4nm), Si QDs (3nm) and Ge QD (4nm) along with using HfO_2 as tunnel oxide to implement two layers of Ge QDs in the gate region.

5. Conclusion

Novel QD structures, experimental data, and quantum simulations show the potential of achieving 5-8 states in NanoFETs. Circuit simulation of 4-state and 8-states (3-bit) inverters and SRAMs have been reported [8, 10]. This work opens the pathway for 3-bit/ 8-state QD-based logic and memory circuits in QDC-QDG configuration. Multi-state SWS-FET structures are also presented with one drain.

Increasing the states is obtained in QDC, SWS and QDG configuration by methods including (1) changing the thickness of upper cladding of upper QD of QD channel, (ii) changing the SiO_x cladded thickness in Si QD layers, (iii) making two QD layers asymmetrical by controlled oxidation of upper SiO2 layer (see Fig. 1(a)), (iv) changing the Ge core diameter of two uppermost Ge QD layers.

Wafer scale integration could be achieved using the methodology employed in vertical nonvolatile memories (V-NAND) [14]. Figure 5 schematically shows a QD-NVRAM implemented in GAA structure using poly-Si core and SiO_x-Si cladded quantum dot channel. Si Quantum dot channel is self-assembled on p-poly Si core where the inversion layer is formed with higher mobility than in poly-Si. In addition, cladded Si or Ge QD layers serve as floating gate in gate all around (GAA) configuration.

The QD structures discussed in this paper could provide advancement in multi-state FETs and NVRAMs and be transformative for future computing systems including the implementation of edge computing, artificial intelligence and machine learning platforms.

References

1. F. Jain, S. Karmakar, P.-Y. Chan, E. Suarez, M. Gogna, J. Chandy and E. Heller, "Quantum Dot Channel (QDC) Field-Effect Transistors (FETs) using II-VI Barrier Layers," Journal of Electronic Materials, 41, 2775, 2012.
2. F. Jain, M. Lingalugari, B. Saman, P.-Y. Chan, P. Gogna, E.-S. Hasaneen, J. Chandy and E. Heller, "Multi-State Sub-9 nm QDC-SWS FETs for Compact Memory Circuits," 46th IEEE Semiconductor Interface Specialists Conference (SISC), December 2–5, 2015.
3. F. Jain, R. Gudlavalleti, R. Mays, B. Saman, P-Y. Chan, J. Chandy, M. Lingalugari, and E. Heller, Multi-State Quantum Dot Channel (QDC) FETs for Multi-Bit Computing, *52nd IEEE Semiconductor Interface Specialists Conference (SISC)*, December 8–11, 2021.
4. F. Jain, M. Lingalugari, J. Kondo, P. Mirdha, E. Suarez, J. Chandy and E. Heller, "Quantum dot channel (QDC) FETs with Wraparound II-VI Gate Insulators: Numerical Simulations," Journal of Electronic Materials, 45, 5663, 2016.
5. E. K. Heller, S. K. Islam, G. Zhao and F. C. Jain, Solid-State Electronics, 42, 901–914, 1999.

6. C-W. Jiang and M. A. Green, Journal of Applied Physics, 99, 114902, 2006.

7. M. Lingalugari, P.-Y. Chan, E.K. Heller, J. Chandy and F.C. Jain, "Quantum Dot Floating Gate Nonvolatile Random Access Memory Using Quantum Dot Channel for Faster Erasing," Electronic Letters, 54, 36, 2018.

8. F. Jain, R. Gudlavalleti, R. Mays, B. Saman, P-Y. Chan, J. Chandy, M. Lingalugari and E. Heller, "Two-dimensional SiOx-cladded Si and GeOx-cladded Ge Quantum Dot Arrays vertically stacked for CMOS-X Logic," SRAMs, NVRAMs and IR imaging and multi-bit Computing, SISC, December 2022.

9. R. Meyer, Y. Fukuzumi and Y. Dong, "3D NAND Scaling in Next Decade," Proceedings of International Electron Devices Meeting, 26.1.1–26.1.4, 599–602, December 2022.

10. R. H. Gudlavalleti, E. Heller, J. Chandy and F. Jain, "Computing-In-Memory SRAM Cell using Multistate Spatial Wavefunction Switching (SWS)-Quantum Dot Channel (QDC) FET," *Int. J. of High Speed Electron. Syst.*, **32**, 2350012, 2023..

11. D. Lelmini and H.-S. P. Wong, "In-memory computing with resistive switching devices," Nature Electronics, 1, 333–337, 2018.

12. A. J. Sigillito, J. C. Loy, D. M. Zajac, M. J. Gullans, L. F. Edge and J. R. Petta, "Site-selective Quantum Control in an Isotopically Enriched 28Si/Si0.7Ge0.3 Quadruple Quantum Dot," Physics Review Applied, 11, 061006, 2019.

13. F. Jain, R. H. Gudlavalleti, A. Almalki, B. Saman, P-Y. Chan, J. Chandy, F. Papadimitrakopoulos and E. Heller, *Int. J. of High Speed Electron. Syst.*, **32**, 2350018, 2023.

14. Chen *et al.*, IEEE Transactions on Very Large Scale Integrated Systems, 2018.

Author Index